Implications of Mobility as a Service (MaaS) in Urban and Rural Environments:

Emerging Research and Opportunities

António Manuel Amaral
University of Minho, Portugal

Luís Barreto
Applied Research Center for Digital Transformation (ARC4DigiT), Escola Superior de Ciências Empresariais, Instituto Politécnico Viana do Castelo, Portugal & Instituto de Telecomunicações (IT), Universidade de Aveiro, Portugal

Sara Baltazar
Escola Superior de Ciências Empresariais (ESCE), Instituto Politécnico Viana do Castelo (IPVC), Portugal & Research Center for Systems and Technologies (SYSTEC), University of Porto, Portugal & Institute of Systems and Robotics (ISR), University of Coimbra, Portugal

João Pedro Silva
Escola Superior de Tecnologia e Gestão, Politécnico de Leiria, Portugal

Luísa Gonçalves
Câmara Municipal de Leiria, Portugal & Escola Superior de Tecnologia e Gestão, Instituto Politécnico de Leiria, Portugal & INESCC, NOVA IMS, Universidade NOVA de Lisboa, Portugal

A volume in the Practice, Progress, and Proficiency in Sustainability (PPPS) Book Series

IGI Global
DISSEMINATOR OF KNOWLEDGE

Published in the United States of America by
IGI Global
Engineering Science Reference (an imprint of IGI Global)
701 E. Chocolate Avenue
Hershey PA, USA 17033
Tel: 717-533-8845
Fax: 717-533-8661
E-mail: cust@igi-global.com
Web site: http://www.igi-global.com

Library of Congress Cataloging-in-Publication Data

Names: Amaral, António Manuel, 1981- editor.
Title: Implications of mobility as a service (MaaS) in urban and rural environments : emerging research and opportunities / António Manuel Amaral, Luís Barreto, Sara Baltazar, João Pedro Silva, Luísa Gonçalves, editors.
Description: Hershey, PA : Engineering Science Reference, 2020. | Includes bibliographical references and index. | Summary: "This book explores the recent advances in the urban and rural mobility contexts which will have a major impact in the future citizen's quality of life"-- Provided by publisher.
Identifiers: LCCN 2019031682 (print) | LCCN 2019031683 (ebook) | ISBN 9781799816140 (hardcover) | ISBN 9781799816157 (paperback) | ISBN 9781799816164 (ebook)
Subjects: LCSH: Mobile computing. | Cloud computing. | Cell phones. | Portable computers.
Classification: LCC QA76.59 .I5884 2020 (print) | LCC QA76.59 (ebook) | DDC 004--dc23
LC record available at https://lccn.loc.gov/2019031682
LC ebook record available at https://lccn.loc.gov/2019031683

This book is published in the IGI Global book series Practice, Progress, and Proficiency in Sustainability (PPPS) (ISSN: 2330-3271; eISSN: 2330-328X)

British Cataloguing in Publication Data
A Cataloguing in Publication record for this book is available from the British Library.

For electronic access to this publication, please contact: eresources@igi-global.com.

Practice, Progress, and Proficiency in Sustainability (PPPS) Book Series

ISSN:2330-3271
EISSN:2330-328X

MISSION

In a world where traditional business practices are reconsidered and economic activity is performed in a global context, new areas of economic developments are recognized as the key enablers of wealth and income production. This knowledge of information technologies provides infrastructures, systems, and services towards sustainable development.

The **Practices, Progress, and Proficiency in Sustainability (PPPS) Book Series** focuses on the local and global challenges, business opportunities, and societal needs surrounding international collaboration and sustainable development of technology. This series brings together academics, researchers, entrepreneurs, policy makers and government officers aiming to contribute to the progress and proficiency in sustainability.

COVERAGE

- Socio-Economic
- Sustainable Development
- Outsourcing
- Global Content and Knowledge Repositories
- Knowledge clusters
- Eco-Innovation
- Technological learning
- ICT and knowledge for development
- Innovation Networks
- E-Development

The Practice, Progress, and Proficiency in Sustainability (PPPS) Book Series (ISSN 2330-3271) is published by IGI Global, 701 E. Chocolate Avenue, Hershey, PA 17033-1240, USA, www.igi-global.com. This series is composed of titles available for purchase individually; each title is edited to be contextually exclusive from any other title within the series. For pricing and ordering information please visit http://www.igi-global.com/book-series/practice-progress-proficiency-sustainability/73810. Postmaster: Send all address changes to above address.

Titles in this Series

For a list of additional titles in this series, please visit:
https://www.igi-global.com/book-series/practice-progress-proficiency-sustainability/73810

African Perspectives on Reshaping Rural Development
Mavhungu Abel Mafukata (University of South Africa, South Africa) and Khathutshelo Alfred Tshikolomo (Limpopo Department, South Arica)
Information Science Reference • © 2020 • 300pp • H/C (ISBN: 9781799823063) • US $195.00

Social, Economic, and Environmental Impacts Between Sustainable Financial Systems and Financial Markets
Magdalena Ziolo (University of Szczecin, Poland)
Business Science Reference • © 2020 • 383pp • H/C (ISBN: 9781799810339) • US $245.00

Developing and Designing Circular Cities Emerging Research and Opportunities
Elżbieta Ryńska (Warsaw University of Technology, Poland)
Engineering Science Reference • © 2020 • 194pp • H/C (ISBN: 9781799818861) • US $135.00

Technological Innovations for Sustainability and Business Growth
Geetika Jain (UP Technical University, Lucknow, India) Harjit Singh (Amity University, Noida, India) Shahriar Akter (University of Wollongong, Australia) Alka Munjal (Amity University, Noida, India) and Harpal S. Grewal (Claflin University, USA)
Business Science Reference • © 2020 • 338pp • H/C (ISBN: 9781522599401) • US $225.00

Handbook of Research on the Conservation and Restoration of Tropical Dry Forests
Rahul Bhadouria (University of Delhi, India) Sachchidanand Tripathi (University of Delhi, India) Pratap Srivastava (University of Allahabad, India) and Pardeep Singh (University of Delhi, India)
Engineering Science Reference • © 2020 • 465pp • H/C (ISBN: 9781799800149) • US $295.00

Innovative Waste Management Technologies for Sustainable Development

For an entire list of titles in this series, please visit:
https://www.igi-global.com/book-series/practice-progress-proficiency-sustainability/73810

IGI Global
DISSEMINATOR OF KNOWLEDGE

701 East Chocolate Avenue, Hershey, PA 17033, USA
Tel: 717-533-8845 x100 • Fax: 717-533-8661
E-Mail: cust@igi-global.com • www.igi-global.com

Table of Contents

Preface.. viii

Acknowledgment.. xiv

Introduction.. xvi

Section 1
Mobility as a Service (MaaS) System: An Overview

Chapter 1
Challenges and Implications of Mobility as a Service (MaaS)..............................1
Luísa Gonçalves, Câmara Municipal de Leiria, Portugal & Escola Superior de Tecnologia e Gestão, Instituto Politécnico de Leiria, Portugal & INESCC, NOVA IMS, Universidade NOVA de Lisboa, Portugal
João Pedro Silva, Escola Superior de Tecnologia e Gestão, Politécnico de Leiria, Portugal
Sara Baltazar, Escola Superior de Ciências Empresariais (ESCE), Instituto Politécnico Viana do Castelo (IPVC), Portugal & Research Center for Systems and Technologies (SYSTEC), University of Porto, Portugal & Institute of Systems and Robotics (ISR), University of Coimbra, Portugal
Luís Barreto, Applied Research Center for Digital Transformation (ARC4DigiT), Escola Superior de Ciências Empresariais, Instituto Politécnico Viana do Castelo, Portugal & Instituto de Telecomunicações (IT), Universidade de Aveiro, Portugal
António Amaral, Escola Superior de Tecnologia e Gestão, Polytechnic Institute of Porto, Portugal

Chapter 2
Public-Private-People Partnership Networks and Stakeholder Roles Within MaaS Ecosystems 21
Jenni Eckhardt, VTT Technical Research Centre of Finland Ltd., Finland
Aki Aapaoja, VTT Technical Research Centre of Finland Ltd., Finland
Harri Haapasalo, University of Oulu, Finland

Chapter 3
Potential User Groups of Mobility as a Service in Finland 51
Timo Liljamo, Tampere University, Finland
Heikki Liimatainen, Tampere University, Finland
Markus Pöllänen, Tampere University, Finland
Roni Utriainen, Tampere University, Finland
Riku Viri, Tampere University, Finland

Section 2
MaaS as a Differentiating Mobility System in Rural and Urban Contexts

Chapter 4
Challenge, Coordination, and Collaboration for Effective Rural Mobility Solutions 83
Kate Pangbourne, Institute for Transport Studies, University of Leeds, UK

Chapter 5
Public Transport as a Provider of MaaS in Rural Areas: Mission (Im) Possible? 109
Lars E. Olsson, Karlstad University, Sweden
Margareta Friman, Karlstad University, Sweden

Chapter 6
Mobility on Demand: Evolving and Growing Shared Mobility in the Suburbs of Northern Virginia 125
Susan A. Shaheen, University of California, Berkeley, USA
Adam Cohen, University of California, Berkeley, USA
Emily Farrar, University of California, Berkeley, USA

Chapter 7

Innovative Mobility Solutions in Rural Areas With Case Studies....................156
Marian Gogola, University of Žilina, Slovakia
Dana Sitányiová, University of Žilina, Slovakia

Section 3
Emerging Trends and Innovations in Mobility Systems

Chapter 8
Sustainable Mobility in Smart Cities: The Key Role of Gamified Motivational Systems for Citizens' Engagement and Behavior Change............211
Annapaola Marconi, Fondazione Bruno Kessler, Italy
Enrica Loria, Fondazione Bruno Kessler, Italy

Chapter 9
The Future of Mobility as a Service (MaaS): Driving Through the Internet of Mobility (IoM)..247
Sara Baltazar, Escola Superior de Ciências Empresariais (ESCE), Instituto Politécnico Viana do Castelo (IPVC), Portugal & Research Center for Systems and Technologies (SYSTEC), University of Porto, Portugal & Institute of Systems and Robotics (ISR), University of Coimbra, Portugal
António Amaral, Escola Superior de Tecnologia e Gestão, Polytechnic Institute of Porto, Portugal
Luís Barreto, Applied Research Center for Digital Transformation (ARC4DigiT), Escola Superior de Ciências Empresariais, Instituto Politécnico Viana do Castelo, Portugal & Instituto de Telecomunicações (IT), Universidade de Aveiro, Portugal
João Pedro Silva, Escola Superior de Tecnologia e Gestão, Politécnico de Leiria, Portugal
Luísa Gonçalves, Instituto Politécnico de Leiria, Portugal

Related Readings..273

About the Contributors...286

Index..292

Preface

Mobility as a Service (MaaS) can be briefly described as a concept that aims to fulfil individual's mobility needs in a sustainable way by combining different transport services to seamless trips (Utriainen & Pöllänen 2018), offering an appealing alternative to owning and using a private car (Lyons et al. 2019). The core characteristics of MaaS include integration of multiple transport modes and actors, various payment options, use of various technologies through one user interface platform as registered user as well as personalization and customization to offer demand oriented user-centric mobility services (Jittapirom et al. 2017). The history of MaaS by name and related research is not long, but MaaS can be seen as an evolutionary continuation of integrating mobility services (Lyons et al. 2019). MaaS research is accumulating rapidly and the focus has been on describing the concept (Jittapirom et al. 2017, Sochor et al. 2018), identifying the roles of various stakeholders (Hensher 2017, Smith et al. 2018), exploring the user potential (Ho et al. 2018) and lately taking a critical view on the effects of MaaS and the needs for governing the development (Matyas & Kamargianni 2019, Pangbourne et al. 2019, Smith et al. 2019).

The visions of MaaS focus on the possibility to offer seamless and effortless, instant accessibility that is smarter, faster and greener than using a private car (Pangbourne et al. 2019). Indeed, if the aim of implementing MaaS is to change the modal split of transport to achieve sustainable mobility, as it should be, then the vision should be to offer a better freedom of mobility and with a lower cost than with a private car. The possibility to offer better freedom of mobility through MaaS is largely dependent on the cities offering high quality public transport services that are the backbone of MaaS (Matyas & Kamargianni 2019). Without the backbone of high quality public transport services, mobility services are based on ride-hailing services, even publicly subsidized ones, which decrease the use of public transport (Clewlow & Mishra 2017).

While the number of MaaS pilots is expanding, the commercial operation of a fully integrated MaaS offering seems to be difficult and operational offerings are scarce. Whim by MaaS Global remains the only offering available in several countries, but even it only operates in one city in each country (MaaS Global 2019). Hence, the expectations of operational MaaS before 2020 by the Delphi panel of experts seem to have been optimistic and the constraints of MaaS having limited value, difficulties in integrating public transport and lack of collaboration between crucial actors seem to have prevailed (Jittapirom et al. 2018). Based on the MaaS research literature it seems that MaaS is slipping into the 'trough of disillusionment' of the Gartner's Hype Cycle (Lyons et al. 2019). In addition to the difficulties in establishing the collaboration needed to enable MaaS operations, one of the biggest challenges is the need for a cultural shift from ownership to access (Mulley 2017). This challenge may be greater than expected and MaaS is in danger of becoming a hyped socio-technical phenomenon (Giesecke et al 2016) with lacking full-scale implementations reaping the alleged benefits (Smith et al. 2018).

The challenge of a cultural shift from ownership to access seems to decide the faith of MaaS operations. Lyons et al. (2019) propose a behavioral schema on the individual choice making related to the adoption of MaaS, which can be used to analyze the trends affecting the viability of MaaS in the long-term.

1. Choice of getting a driving license. There are trends in the UK and Finland that young people do not get driving licenses as often as they used to (Chatterjee et al. 2018, Traficom 2019).
2. Choice of owning a car. Trends in the UK and USA show that car ownership is declining, due to demographics, living situation and socio-economic situation, but also due to a cars having less symbolic value than before (Chatterjee et al. 2018).
3. Modal choices. Socio economic factors, life stage delays as well as re-urbanisation made possible by land-use and transport system policies seem to have decreased the share of trips and absolute number of miles travelled by car in the UK, USA, Australia and Austria (Buehler et al. 2017, Melia et al. 2018, Delbosc et al. 2019).
4. Choice between MaaS offering and using mobility services separately. Vast majority of the trips are made within a relatively small area around peoples' homes. Hence, the number of possible mobility services is likely to be rather limited and people are familiar with the available mobility services. This leads to situations where the cognitive user effort (Lyons et al. 2019) related to choosing a MaaS offering is likely to be higher than when choosing the existing mobility services and even adopting an additional service, such as an electric scooters, to the mix of existing mobility services.

5. Choice of best MaaS plan. MaaS offerings in a certain area typically has a limited number of plans, from which the customer can choose the most suitable one. However, People have very different needs and it is impossible to tailor a plan for every customer. This may lead to lost customers but also to extra discretionary trips in order to 'make the most' of the prepaid MaaS plan (Pangbourne et al. 2019).

To sum up, it seems that the hyped expectations and optimistic timeframes of fully integrated MaaS offerings bringing benefits for the society and customers while creating revenue for the MaaS operators are changing towards facing the difficulties in the trough of disillusionment. Current mobility market is vastly dominated by private car use, which in many areas offers superior level of service in terms of effort and journey times, although with high fixed costs, compared to alternative mobility services. The superior level of service is the result of decades of determined land use and transport system planning and it is very difficult to break the pattern. Indeed, what would an ideal transport system for MaaS look like? However, emerging trends of socio-economic factors, attitudes as well as in land use and transport system policy favor modal shift from car to mobility services. Ultimately the question remains, whether MaaS offerings bring more value to the customers than owning and using private car, or using the existing and possible new mobility services separately.

ORGANIZATION OF THE BOOK

The book includes nine chapters, of which the first three focus on describing the concept of MaaS, identifying the stakeholders and exploring the potential users. Chapters 4-7 focus on the opportunities and challenges of MaaS and shared mobility services in general in rural areas. The last two chapters open new research avenues through multidisciplinary approach. A brief description of each of the chapters follows:

Chapter 1 provides a synopsis of the underlying factors that enable the existence of the MaaS concept and its role towards the improvement of the performance of transportation systems. It aims to contextualize MaaS in modern societies and explains its main functions and advantages but also the challenges facing its widespread implementation.

Chapter 2 presents Public-Private-People Partnership (PPPP) network for MaaS and considers the integration of market-based mobility services and subsidized public transport. The roles and responsibilities of MaaS stakeholders in the PPPP network are also described.

Chapter 3 aims to identify the potential user groups of MaaS through the results of an extensive public survey carried out in Finland. Young adults and people living in urban areas and in households without a car are found to be more interested in MaaS than other groups. The findings highlight that the greatest potential for MaaS can be found in urban areas.

Chapter 4 turns the focus on rural areas and considers the collaboration of authorities to provide accessibility through mobility services. The authors emphasize that strong governance is needed to develop MaaS that do not exclude rural areas.

Chapter 5 takes a practical example of integrating public transport service with carpooling to solve the challenges of MaaS in rural areas. Management and user perspectives are analyzed and the chapter also identifies barriers concerning legal aspects.

Chapter 6 reviews the applications of shared mobility services in a suburban area. As opposed to previous chapters the focus is on shared services instead of public transport based services, reflecting the difference between American and European approach to the concept of MaaS.

Chapter 7 deals with mobility in rural areas by taking the perspective of demographic and social changes as starting points. Rural areas are found to have disadvantages compared to urban areas in terms of accessibility of jobs, education and services, but various innovative solutions for improving the accessibility in rural areas are also presented.

Chapter 8 highlights the importance of civic engagement in developing sustainable mobility. Gamification is used as a tool to improve civic engagement and two cases of using gamification to promote sustainable mobility are presented.

Chapter 9 considers the concepts of Internet of Mobility (IoM) related to the technological developments in mobility and MaaS related to the service developments in mobility and introduces a new concept of Internet of Mobility as a Service (IoMaaS).

Heikki Liimatainen
Transport Research Centre Verne, Tampere University, Finland

REFERENCES

Buehler, R., Pucher, J., & Altshuler, A. (2017). Vienna's path to sustainable transport. *International Journal of Sustainable Transportation*, *11*(4), 257–271. doi:10.1080/15568318.2016.1251997

Chatterjee, K., Goodwin, P., Schwanen, T., Clark, B., Jain, J., Melia, S., . . . Stokes, G. (2018). *Young People's Travel – What's Changed and Why? Review and Analysis. Report to Department for Transport*. UWE Bristol. Available at: http://eprints.uwe.ac.uk/34640/

Clewlow, R. R., & Mishra, G. S. (2017). *Disruptive Transportation: The Adoption, Utilization, and Impacts of Ride-Hailing in the United States*. Davis, CA: Institute of Transportation, UC Davis.

Delbosc, A., McDonald, N., Stokes, G., Lucas, K., Circella, G., & Lee, Y. (2019). Millenials in cities: Comparing travel behavior trends across six case study regions. *Cities (London, England)*, *90*, 1–14. doi:10.1016/j.cities.2019.01.023

Giesecke, R., Surakka, T., & Hakonen, M. (2016). Conceptualising mobility as a service – A user centric view on key issues of mobility services. *Proceedings of the Eleventh International Conference on Ecological Vehicles and Renewable Energies (EVER)*. 10.1109/EVER.2016.7476443

Hensher, D. A. (2017). Future bus transport contracts under a mobility as a service (MaaS) regime in the digital age: Are they likely to change? *Transportation Research Part A, Policy and Practice*, *98*, 86–96. doi:10.1016/j.tra.2017.02.006

Ho, C., Hensher, D. A., Mulley, C., & Wong, Y. Z. (2018). *Prospects for switching out of conventional transport services to mobility as a service subscription plans – a stated choice study*. UTSG.

Jittrapirom, P., Caiati, V., Feneri, A. M., Ebrahimigharehbaghi, S., González, M. J. A., & Narayan, J. (2017). Mobility as a service: A critical review of definitions, assessments of schemes, and key challenges. *Urban Planning*, *2*(2), 13–25. doi:10.17645/up.v2i2.931

Jittrapirom, P., Marchau, V., van der Heijden, R., & Meurs, H. (2018). *Future Implementation of Mobility as a Service (MaaS): Results of an International Delphi Study*. Radboud University.

Lyons, G., Hammond, P., & Mackay, K. (2019). The importance of user perspective in the evolution of MaaS. *Transportation Research Part A, Policy and Practice*, *121*, 22–36. doi:10.1016/j.tra.2018.12.010

Maa, S. Global. (2019). *Find your plan*. Available at: https://whimapp.com/plans/

Matyas, M., & Kamargianni, M. (2019). The potential of mobility as a service bundles as a mobility management tool. *Transportation*, *46*(5), 1951–1968. doi:10.100711116-018-9913-4

Melia, S., Chatterjee, K., & Stokes, G. (2018). Is the urbanization of young adults reducing their driving? *Transportation Research Part A, Policy and Practice*, *118*, 444–456. doi:10.1016/j.tra.2018.09.021

Mulley, C. (2017). Mobility as a services (MaaS)–does it have critical mass? *Transport Reviews*, *37*(3), 247–251. doi:10.1080/01441647.2017.1280932

Pangbourne, K., Mladenovic, M. N., Stead, D., & Milakis, D. (2019). Questioning mobility as a service: Unanticipated implications for society and governance. *Transportation Research Part A, Policy and Practice*. doi:10.1016/j.tra.2019.09.033

Smith, G., Sochor, J., & Karlsson, I. M. (2018). Mobility as a Service: Development scenarios and implications for public transport. *Research in Transportation Economics*, *69*, 592–599. doi:10.1016/j.retrec.2018.04.001

Smith, G., Sochor, J., & Karlsson, I. M. (2019). Public–private innovation: Barriers in the case of mobility as a service in West Sweden. *Public Management Review*, *21*(1), 116–137. doi:10.1080/14719037.2018.1462399

Sochor, J., Arby, H., Karlsson, I. C. M., & Sarasini, S. (2018). A topological approach to Mobility as a Service: A proposed tool for understanding requirements and effects, and for aiding the integration of societal goals. *Research in Transportation Business & Management*, *27*, 3–14. doi:10.1016/j.rtbm.2018.12.003

Traficom. (2019). *Uusien henkilöautonkuljettajien liikenneturvallisuus*. Available at: https://www.liikennefakta.fi/turvallisuus/tieliikenne/uusien_henkiloautonkuljettajien_liikenneturvallisuus

Utriainen, R., & Pöllänen, M. (2018). Review on mobility as a service in scientific publications. *Research in Transportation Business & Management*, *27*, 15–23. doi:10.1016/j.rtbm.2018.10.005

Acknowledgment

The editors would like to acknowledge the support and the help of all the involved in this project: chapter's authors, editorial advisor board elements that also took part in the review process, reviewers, and preface author. Therefore, the editors would like to express their sincere gratitude: a) to each one of the chapter's authors for their contributions, and time and expertise invested to this book project; b) to the reviewers for their valuable contributions regarding the improvement of quality, coherence, and content presentation of chapters. Most of the authors also served as referees and the editors also wish to acknowledge how highly appreciated their double tasks and efforts have been. Finally a special word to the preface author for his generosity towards this project namely by presenting it with his vision as an opening door to the book.

It should be also mentioned that this book project was developed in the scope of the Project "ALTO MINHO. SMOB - Mobilidade Sustentável para o Alto Minho", Ref. POCI-01-0145-FEDER-024043, supported by FEDER (Fundo Europeu de Desenvolvimento Regional) through the Programa Operacional Competitividade e Internacionalização.

António Manuel Amaral
University of Minho, Portugal

Luís Barreto
Applied Research Center for Digital Transformation (ARC4DigiT), Escola Superior de Ciências Empresariais, Instituto Politécnico Viana do Castelo, Portugal & Instituto de Telecomunicações (IT), Universidade de Aveiro, Portugal

Sara Baltazar
Escola Superior de Ciências Empresariais (ESCE), Instituto Politécnico Viana do Castelo (IPVC), Portugal & Research Center for Systems and Technologies (SYSTEC), University of Porto, Portugal & Institute of Systems and Robotics (ISR), University of Coimbra, Portugal

João Pedro Silva
Escola Superior de Tecnologia e Gestão, Politécnico de Leiria, Portugal

Luísa Gonçalves
Câmara Municipal de Leiria, Portugal & Escola Superior de Tecnologia e Gestão, Instituto Politécnico de Leiria, Portugal & INESCC, NOVA IMS, Universidade NOVA de Lisboa, Portugal

Introduction

This book focuses on the current mobility thematic: Mobility as a Service (MaaS), which appears at a critical junction in time where a paradigm shift is on course to change, in a quite profound manner, the way that we, as a collective society, view and embrace one of our most basic perceived needs: mobility. MaaS can be seen as an innovative approach and can provide a toolkit of tailored solutions that can, hopefully, with the aid of other measures, assist modern societies in shifting their approach on solving mobility problems, while creating a more inclusive and efficient society that respects environmental and sustainability values.

Based on a diversified set of MaaS analyses aimed at an interdisciplinary audience, this book presents new solutions for the development and innovation of MaaS and its features, while presenting implemented strategies in the mobility field. The covered topics include: sustainable mobility; MaaS users; networks and stakeholder roles within MaaS ecosystems; MaaS evolution and its connection to public transport; transport demand and shared schemes; mobility in urban and rural environments and its challenges; social inclusion; gamification; Internet of Mobility (IoM) and mobility future trends.

This book is mainly aimed at researchers and professionals in the areas of mobility, planning, and MaaS but can be useful for everyone with an interest in keeping up with the latest trends in mobility. Therefore, readers are invited to take part in this mobility paradigm shift, finding new perspectives and the latest improvements and solutions for inclusive mobility, transport efficiency and sustainability.

Section 1

Mobility as a Service (MaaS) System: An Overview

The first section of the book provides some introductory concepts and highlights the advantages of the MaaS concepts, solutions, and applications for its main users and providers. This is naturally relevant as it endows a sound foundation for the development of tailor-made applications and solutions, thus enhancing the potential success rate.

Chapter 1
Challenges and Implications of Mobility as a Service (MaaS)

Luísa Gonçalves
https://orcid.org/0000-0002-6265-8903
Câmara Municipal de Leiria, Portugal & Escola Superior de Tecnologia e Gestão, Instituto Politécnico de Leiria, Portugal & INESCC, NOVA IMS, Universidade NOVA de Lisboa, Portugal

Sara Baltazar
Escola Superior de Ciências Empresariais (ESCE), Instituto Politécnico Viana do Castelo (IPVC), Portugal & Research Center for Systems and Technologies (SYSTEC), University of Porto, Portugal & Institute of Systems and Robotics (ISR), University of Coimbra, Portugal

João Pedro Silva
https://orcid.org/0000-0003-4206-243X
Escola Superior de Tecnologia e Gestão, Politécnico de Leiria, Portugal

Luís Barreto
https://orcid.org/0000-0001-6173-433X
Applied Research Center for Digital Transformation (ARC4DigiT), Escola Superior de Ciências Empresariais, Instituto Politécnico Viana do Castelo, Portugal & Instituto de Telecomunicações (IT), Universidade de Aveiro, Portugal

António Amaral
https://orcid.org/0000-0002-7910-2418
Escola Superior de Tecnologia e Gestão, Polytechnic Institute of Porto, Portugal

ABSTRACT

This introductory chapter provides a synopsis of the underlying factors that enable the existence of the mobility as a service (MaaS) concept and its role regarding the improvement of the performance of transportation systems. Therefore, several considerations are made to contextualize MaaS in modern societies explaining not

DOI: 10.4018/978-1-7998-1614-0.ch001

only its main functions and advantages but also the challenges facing its widespread implementation.

INTRODUCTION

Driving is, for most of us, an essential part of our everyday lives (Hu *et al.*, 2015). People are in constant movement in order to work, for leisure and to socialize (Boero *et al.*, 2016) increasing the need to have access to their main interest points. In all regions, may they be urban, suburban or rural, mobility has a vital impact on accessibility and in ensuring social inclusion. Mobility choices have therefore significant implications not only in social and economic terms but also for the environment.

Mobility as a Service (MaaS) is a recent innovative transport concept with the goal of introducing changes in the current transport practices namely, by integrating public and private transport services in a single mobility service, which is accessible on demand (Meurs, Sharmeen, Marchau, & van der Heijden, 2019).

In this context the MaaS concept stems from people's necessity to access different points, planning their own trips and adopting the most suitable proper means of transport to use. Therefore, it can have an important role in the mobility paradigm shift towards more sustainable transport systems (Hesselgren, Sjöman, & Pernestål, 2019), encouraging the use of cleaner transport modes (Boero *et al.*, 2016), thus reducing car usage as well as releasing urban space and reducing congestion (Hesselgren, Sjöman, & Pernestål, 2019).

However, there is still a high degree of ambiguity surrounding the concept with multiple sources offering their own definitions of MaaS, dealing with different aspects and views of the concept. In addition several challenges and implications still exist and several authors consider that there is a lack of an assessment framework to classify MaaS characteristics in a systematic manner (Jittrapirom *et al.*, 2017).

The present chapter gives an overview of some Challenges and Implications of MaaS, which are later developed and discussed in detail in each chapter. In recent years, the degree of awareness of the need to integrate city planning practices and measures with the mobility infrastructures has gradually increased which in turn has improved the quality of life for citizens. This has also positively impacted the environment. These procedures play an important role in the definition and development of the next generation of mobility systems.

MOBILITY CONCEPTS AND FEATURES

The mobility and the accessibility concepts are common nowadays, representing simultaneously a challenge and a concern about the achievement of a sustainable, inclusive and happy society. The need to move and access main points is not a new concept and mobility is an essential part of modern society and a prerequisite for autonomous living (Stein, Meurer, Boden, & Wulf, 2017). The mobility evolution has been felt differently in urban and rural areas, since each one of the contexts has its features and geographical characteristics. The urban population aims for cities which are more livable, namely making them less vehicle-centric (Goodal, Fishman, Bornstein, & Bonthron, 2017). Therefore, a change in urban mobility namely by decreasing the usage of the private car and increasing the usage of public transport (PT) and active modes is crucial. Passenger transportation is experiencing a serious restructuring, with the surely technology as a vital driving force, allowing the sustainable and smart cities concepts to be connected. On the other hand, the rural population aims for social integration, comfort and a transport system which serves these areas. An adequate mobility is an important key factor to enhance not only social inclusion but also quality of life among the population of the rural region (Barreto, Amaral, & Baltazar, 2019a). his pressure on urban and rural mobility and the transport system has increased the demand for new and innovative solutions in order to increase its efficiency (Kamargianni, Li, Matyas, & Schäfer, 2016). Thus, a new mobility concept that integrates existing and new mobility services into one single digital platform is emerging, providing customized door-to- door transport and offering personalized trip planning, as well as payment options (Durand, Harms, Hoogendoorn-Lanser, & Zijlstra, 2018 A relevant issue regarding electronic, technological and telecommunications innovations is the integrated mobility offer, including extra services. Future mobility will increasingly run on algorithms, a near future example being the autonomous cars by using AI/ML, with algorithms processing the data generated by the IoT-powered transportation system (i.e., connected travelers, vehicles and infrastructure) (Barreto, Amaral, & Baltazar, 2019c).). In this era geared towards the creation of mobility "as-a-Service" and society's Digitalization, a mobility concept following this tendency is forthcoming and necessary. Heikkilä, in 2014, in her Master Thesis (Heikkilä, 2014) proposed visionary mobility scenarios for different types of users. This vision of the integrated mobility system together with the extra services was named by Mobility-as-a-Service (MaaS).

Next, some generic concepts regarding mobility are presented, which also has features and essential elements of the current mobility.

Table 1. Fundamental concept of mobility

Expression/Fundamental Concept	Explanation
Accessibility	The accessibility concept can be connected to the relationship between the land use (territory) and the transportation system that serves it (Nuzzolo, Coppola, & Papa, 2014), which nowadays is the contemporary notion of urbanism.
Mobility	The mobility scope contemplates the physical movements that can be measured by trips, distance and speed, among others (Litman, 2015).
Sustainable mobility	Sustainable Mobility is designed to promote equal opportunities for travelling, with facilities for accessing the multiple activities of a region, promoting the reduction of transport energy consumption (reduction of environmental pollution) and thus in quest of an improvement in efficiency of the resources invested in transport (Campos & Ramos, 2005). This mobility includes concepts as PT as well as non-motorized transport; consider the needs of poor and wealthy alike; and incorporate environmentally-friendly private vehicles and 'shared' use practices (BMZ, 2016).
Smart mobility	Applicable in a society with technological advances which delivered a benign mobility system that all users can access seamlessly and on-demand, avoiding much of the waste, pollution and environmental degradation leaved nowadays (Docherty, Marsden, & Anable, 2018)
Mobility as a Service (MaaS)	Mobility as a Service (MaaS) is a recent mobility concept were transportation converge and is tailored according to users' needs as monthly packages (Hietanen & Sahala, 2014), with a promising perspective to substitute private car, since demand responsive on-demand services will play an important role in MaaS (Jittrapirom *et al.*, 2017).

Fundamental Concepts

The fundamental concepts of mobility and its explanation are available in their content, disclosed in Table 1.

As suggested by Table 1, in order to create a smart mobility and a MaaS ecosystem, smart technologies and devices are a critical component, as well as for the success of urbanization and societal development.

Features and Mobility Elements

There exist different stimuli which impel the redefinition of current mobility, which can be present and be influenced by various features/elements with positive or negative characteristics. Thus, connecting all of them and focusing on future mobility, the MaaS concept imposes its existence.

Sustainability

The personalization of mobility, supported by the availability of several services, will profoundly change the way users live and interact (Barreto, Amaral, & Baltazar, 2019c), as well as their behaviors towards sustainable mobility habits.

Individual Car Dependency

In the latest years, the city/regions mobility planning and disposition was focused on the private car. The massive use of the private car, almost for single occupancy, creates urban congestion, contributes to the rising cost and inconvenience of private ownership of cars, resulting in an increasing indirect cost in terms of quality of life and environmental degradation (Webb, Wilson, & Kularatne, 2018).

Shared Mobility

The shared use of a vehicle, bicycle, or other mode is considered shared mobility and can be viewed as an innovative transportation strategy, enabling users to gain short-term access to transportation modes on an as-needed basis, and having a transformative impact on many cities by enhancing transportation accessibility, increasing multimodality, reducing vehicle ownership and vehicle miles travelled, providing new ways to access goods and services (Shaheen, Cohen, & Zohdy, 2016).

Integrated Mobility

The MaaS concept will only be valid as long as it integrates PT. The future mobility system must encourage its usage, by bringing together multimodal transportation and allowing users to choose and facilitate their intermodal trips, creating a service beyond the urban boundaries that embraces long-distance buses and trains, flights, and ferries (Jittrapirom *et al.*, 2017) - creating a truly integrated mobility system and service. The shift towards shared mobility services (car-sharing, bike-sharing and so on), together with PT can be seen as a substitute to private vehicles (Kamargianni, Li, Matyas, & Schäfer, 2016). Thus, the integration of all transport offers and means is vital to the promotion of a transportation change that can lead the population to adopt new mobility habits and behaviors (Barreto, Amaral, & Baltazar, 2019b). Nevertheless, the complexity of using diverse transport modes, with different payment methods and subscriptions, each one with its own mobile applications, lack of integrated information and so on, discourages the user from taking advantage of the situation (Kamargianni, Li, Matyas, & Schäfer, 2016) and enjoying the choice. Durand, Harms, Hoogendoorn-Lanser, and Zijlstra (2018), defined MaaS's levels of integration as: 1) information integration, 2) ticketing and payment integration, 3) service integration, and 4) integration of societal goals, knowing that an approach which combines several of these levels of integration has the potential to encourage users to opt for integrated modes.

Unified Fares, Ticketing and Payment

Achieving a unified fares system is a challenge that can benefit the users. The main aim of an integrated ticketing system is to improve the service quality for (potential) PT users and therefore, to encourage the use of tangible alternatives to the use of their own car (Puhe, 2014). Integrated tickets are not smart per se; combining the services of several operators on one single ticket can likewise be paper-based; and one of the main advantages of smart ticketing is that tickets can be sold and stored in an electronic device, such as smart cards or mobile phones (Barreto, Amaral, & Baltazar, 2019c). Nowadays, there are diverse payment methods that can facilitate this process, such as: contactless devices, mobile wallets, P2P payments, closed loop, QR codes, among others (Mörch-Pierre, 2019).

APP's/Platforms

Nowadays, MaaS is being created and accessed through the development of platforms – either a multimodal trip-planning app – or webpages and, instead of making a fierce competition among these platforms (Goodal, Fishman, Bornstein, & Bonthron, 2017), a common base such be promoted. An integrated platform must include registration procedure and travel packages, an inter-modal journey planning, booking options, smart ticketing and payment functions so that the entire chain of transport can be managed in a centralized platform (Ambrosino, Nelson, Boero, & Pettinelli, 2016), as the MaaS-London proposal. To better address the mobility needs of travelers, information from different transport providers (e.g. buses, metro, ambulances, taxis) needs to be handled and made available in a central system (Fontes *et al.*, 2017). A mobility platform assistance system should support users with urban trip-planning and several comparable approaches inasmuch as the inter-modal options (Masuch, Lützenberger, & Keiser, 2013) should take the dynamic interactions between platform owners, service providers, and consumers into consideration (Meurs & Timmermans, 2017). Only with cooperation and a unified vision can MaaS be properly expanded and cover different and complementary services offers in order to have a robust system. An important challenge related to the security of intelligent mobility is how to handle all the information about the users, guaranteeing the protection, confidentiality, and privacy of any personal information without conditioning the operational efficiency of the system. Another challenge that may also be considered is related to the integrity of these data, together with the assurance of the availability of the applications and systems, regarding the integration of mobility technologies platforms (Barreto, Amaral, & Baltazar, 2019c).

Total Cooperation, Collaboration and Combination Between the Actors

The success of MaaS depends on the cooperation and collaboration of all stakeholders, regarding the notion of a cooperative and interconnected transport system (including services, infrastructure, information, and payment), extended to public and private operators (Karlsson, Sochor, & Strömberg, 2016). Polydoropoulou, Pagoni, and Tsirimpa (2018), based on the analysis of workshops with stakeholders and focus groups with end-users, identified that stakeholders are highly motivated to join a MaaS partnership in order to receive better quality on-demand data and improve their market shares and incomes; nevertheless, the lack of data and Application Programming Interfaces (APIs) was recognized as an operational/technical barrier, together with the desire of people to use their own car. Therefore, the global adoption of the MaaS paradigm requires a closer strategy among a vast number of institutions from different countries in terms of enhancing the concept adoption on a larger scale, possibly at a European level. A systemic adoption of the MaaS concept requires an integrated effort among all the mobility' stakeholders, that should be supported and stimulated by the development of the right policies towards ensuring cooperation and collaboration between all of them.

WHAT DRIVES THE MaaS EXISTENCE?

During the nineteenth century and the beginning of the twentieth century, railways and fixed PT networks supported the concentration of population and employment in the cities that were growing due to the industrial revolution. During the twentieth century the mass production of private motor vehicles contributed to the rise of suburbanization and decentralization of activities outside the city urban area. Nowadays, there is an ongoing and urgent need for the transport sector to address local and global problems that are playing a significant role in creating the following issues: urban congestion, noise, air pollution, public health, transport safety, unequal access to services, and climate change emissions (Stead, 2016).

With the technological advancements, namely Information and Communications Technology (ICT), new opportunities have opened up like the introduction of an innovative MaaS concept world (Goodall, Fishman, Bornstein, & Bonthron, 2017). As a recent mobility concept, the definition of MaaS is not fully solidified, representing a significant opportunity for all participants in the mobility ecosystem (Streeting *et al.*, 2016), with a promising perspective to change the role of the private car (Mulley, Nelson, & Wright, 2018 and Jittrapirom *et al.*, 2017), especially for short trips (Ison & Sagaris, 2016). Several definitions of MaaS have been presented, for example, Atkins (2015) defines MaaS as a new way to provide transport, which assists users in getting from point A to B by combining available mobility options and presenting them in a completely integrated manner. Kamargianni *et al.* (2018a), defines "Mobility as a Service (MaaS) as a user-centric, intelligent mobility management and distribution system, in which an integrator brings together offerings of multiple mobility service providers, and provides end-users access to them through a digital interface, allowing them to seamlessly plan and pay for mobility.". Li and Voege (2017) consider that MaaS would only be operated in a city when the city offers different transport modes. So, to have a MaaS system it is necessary that a multitude of mobility services can be successfully integrated. The combination of diverse transport modes and services lead to a MaaS services that can succeed when a robust, transparently organized, and well managed cooperation between all the stakeholders is achieved (Meurs, Sharmeen, Marchau, & van der Heijden, 2019). MaaS is a general concept which can embrace diverse themes, but, in particular, one that involves mobility integration (Kamargianni, Li, Matyas, & Schäfer, 2016). Goodall, Fishman, Bornstein and Bonthron (2017) consider that at its core, MaaS relies on a digital platform that integrates end-to-end trip planning, booking, electronic ticketing, and payment services across all modes of transportation, either public or private. Another widely mentioned definition for MaaS comes from the European MaaS Alliance, which defines the concept as "the integration of various forms of transport services into a single mobility service accessible on demand." Thus, MaaS can include many different features, such as a unified mobile app, multimodal journey planning, service bundles, a fixed monthly subscription, or pay-as-you-go billing (Dalton, 2018). Therefore it can combine different transport modes to offer a tailored mobility package, similar to a monthly mobile phone contract and includes other complementary services, such as trip planning, reservation, and payments, through a single interface (Hietanen, 2014). Moreover, the MaaS concept could expand to include urban logistics and other services (e.g. gym, cinema or restaurant bookings), integrating these with the transport service needed to access them (Pangbourne, Stead, Mladenovic, & Milakis, 2018). The expected result is provision of door-to-door mobility services, with the promise of greater efficiency and the opportunity to break, or at least diminish, car dependency (Pangbourne, Stead, Mladenovic, & Milakis, 2018). Thus, MaaS is also an opportunity to decarbonize the transport sector by reducing the use of private cars especially within the urban areas.

In recent years, in dense urban areas, ride hailing, ride sharing, car share, bike share, and microtransit, have all seen massive growth. In addition, planning apps, which help users identify and compare different modal options for getting to their destinations, have become commonplace, with local and global offerings available in every city. However, without a meaningful multi-modal integration with fixed-route public transit and other services, these "New Mobility Services" (NMS) are not themselves MaaS. Dalton (2018) considers that "MaaS is the framework by which we can enable, multi-modal journeys utilizing current and future technologies and services, which include but are not limited to NMS and Mobility on Demand (MOD) services as a way of addressing Transportation Demand Management (TDM) goals". The benefits of bundling various transport services together through one digital interface and a unique selling point are presented as a very positive feature for both the customer and the operator (Pangbourne, Stead, Mladenovic, & Milakis, 2018). However, the choice of the business model and the detailed design of the value offer is not trivial, raising important questions about costumers' freedom choice, inclusiveness, sustainability, potentially threatening the common good (Pangbourne, Stead, Mladenovic, & Milakis, 2018). This can turn out to be potentially threatening if an entire city becomes unable to move if the MaaS system is compromised, for example through a power failure, ICT failure or a deliberate denial of service such as a cyberattack. In addition, some of the business models of early Smart Mobility entrants are aggressive (e.g. Uber) and it is not clear that all the providers or stakeholders are looking to be part of an integrated model.

AN OVERVIEW OF THE CHALLENGES AND IMPLICATIONS OF MaaS

Throughout mankind´s history almost every main change in society has been characterized, in its initial phase, by generating opposing opinions, criticism, controversy and also some healthy debate, which normally leads to the understanding and acceptance of a new way of seeing things. Open and constructive discussion of a problem usually leads to a better final solution. Therefore viewing and accepting this process in a positive way is undoubtedly always the best choice. Fortunately, most aspects of MaaS are generally not very prone to fierce arguments since they usually present a new solution for a mobility problem without immediately excluding old and engrained habits. They just enrich the number of available options and sometimes offer solutions for previously excluded groups. Flexibility and convenience are the main selling arguments, which are greatly enhanced by the widespread use of smartphones and the ubiquitous presence of high-speed internet coverage. Nevertheless, their acceptance commonly implies some challenges for all of those involved, therefore requiring sensible and efficient planning, implementation and monitorization (Li & Voege, 2017).). Change, challenge and implications are, most of the time, intensely intertwined and mobility habits are generally very prone to this phenomenon.

MaaS is based on three main elements: 1) Ticket & Payment integration; 2) Mobility Package; and 3) ICT integration (Kamargianni, Li, Matyas, & Schäfer, 2016). The main challenges facing MaaS derive essentially from its core nature and concept of being (among other important things) an aggregator and disseminator of information from multiple sources and with the ultimate goal of aiding people to plan their trips in a manner that is better for them, society and the environment (Sochor, Arby, Karlsson, & Sarasini, 2018). Consequently, the earliest challenges usually are related to data gathering and organization, which have to be done in a structured manner that can easily be algorithmically processed and scalable to be able to include new mobility solutions added in the future. To ensure the reliability, effectiveness and continuous use of the system, keeping it up to date is clearly non-negotiable. Therefore, this data collection phase is, in reality, a never-ending process. To increase the effectiveness of MaaS, on all its potential levels, it is obviously advisable to conduct a critical evaluation of the supply and demand sides of the transportation system (Kamargianni *et al.*, 2018b). If necessary, adjustments should be made to, for instance, timetables, routes of PT, ticketing uniformization and so on. As this procedure normally involves several stakeholders, which can be public or private, negotiations are essential to ensure the alignment of interests in the pursuit of a better mobility system. Financial issues can play an important role here and, not infrequently, compromise the results. On the demand side it is crucial to identify the population groups that are having difficulties with their daily mobility needs. The transportation system, as a whole, must always thrive to provide the maximum inclusiveness for all members of the society and therefore the development of solutions for these cases should have a high priority. MaaS applications can certainly improve the use of a transportation system but cannot hide its underlying deficiencies. Therefore, finding and especially solving weak aspects of a local transport system is an important step towards a successful MaaS implementation. There are common points supporting the delivery of MaaS in the different regions. The disequilibrium between the urban and rural population, benefiting the urban population, started in 2010 (Corwin, Dinamani, & Pankratz 2019). This shift highlights the challenges and opportunities that arise, also give rural populations the need to adjust and deal with their own depopulation. In these areas – with specific demographic characteristics and economic features – the transport system needs to consider the population's mobility habits and the perspective of the authorities is imperative to promote an

Figure 1. MaaS key challenges' enhancers for rural environments
Source: (Barreto, Amaral & Baltazar, 2019a)

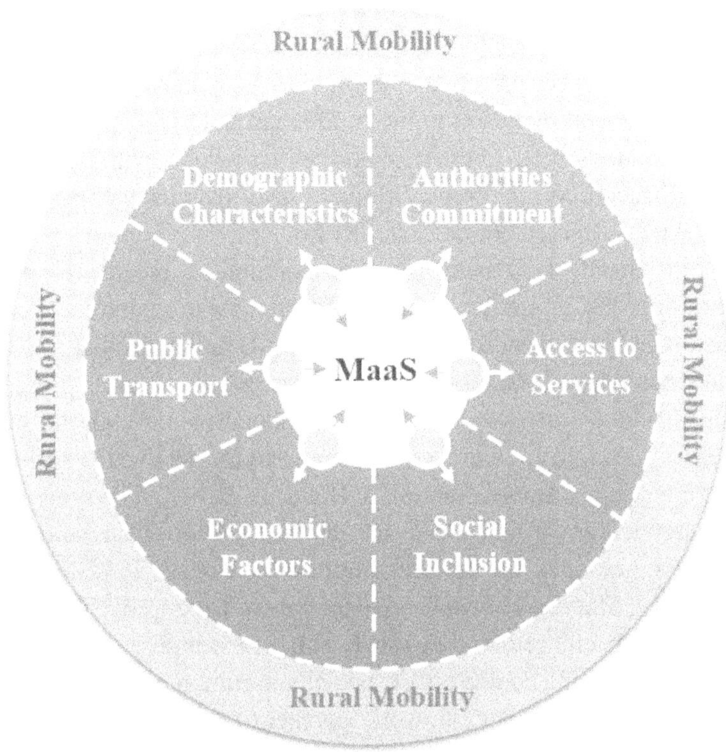

integrated and sustainable mobility, together with the inaccessibility of MaaS for illiterate people and the lack of trust of rural populations in such systems (Barreto, Amaral, & Baltazar, 2019a). In Figure 1 the key challenges for MaaS to enhance rural environments are presented.

The success of a rural MaaS depends on preventing social exclusion, promoting the use of PT, establishing a committed relationship with all the authorities (Barreto, Amaral, & Baltazar, 2019b) and collaborating with stakeholders as well as all other users of the system (Berselli, 2017), which can be achieved by showing the benefits of a MaaS service in rural areas to all stakeholders (Litman, 2017). in this context, the integration between the different actors and how they communicate with the population plays a vital role since it is the only way to reach the individuals and to achieve a fully responsive mobility ecosystem. However, the use of the Information and Communications Technologies' (ICT) to also support the faster development of a Rural MaaS, must be taken into consideration, using its potentialities for promoting inclusion and the benefits of those systems. The MaaS transport solution is being tested in several regions and specific places in order to create a viable solution so as to properly respond to the mobility needs of rural regions. Based on the overview performed, MaaS is thought to be of an important use to dynamize the rural mobility, as well as to create a sustainable solution that aims to integrate all forms and types of mobility solutions. This practice is pointed to as a very important approach to the promotion of social inclusion in the rural areas and to its residents. Especially for those with limited access to the transport, either by not owning a car, by having difficulties with locomotion or even by illiteracy in terms of using these technological solutions (Barreto, Amaral, & Baltazar, 2019a).

On the other hand, the growth of urban populations creates the necessity of having cities built faster and more effectively in order to guarantee a smooth and well-defined strategy to meet the above-mentioned challenges that arise. This fact justifies the emergency of changing the urban mobility, which is presently based on the usage of the private car, and which is imperative, through the support and development of technology, which could progressively involve the sustainable and smart cities concepts (Barreto, Amaral, & Baltazar, 2019c). The growing pressure of urban passengers regarding the usage of the transport systems has increased the demand for new and innovative solutions to increase its overall efficiency (Kamargianni, Li, Matyas, & Schäfer, 2016). One of the approaches used to tackle this challenge has been the slow but steady shift towards shared mobility services (car-sharing, bike-sharing, and so on.), especially in combination with traditional PT so that various transport modes can work together to serve as substitutes to private vehicles. Nevertheless, the difficulty in using a huge variety of transport modes without being integrated (i.e. different payment methods, subscriptions, different mobile applications for each operator, lack of integrated information, and so on.) discourages many people from using them. As such, integrating different transport modes and providing seamless door-to-door mobility is one of the priorities of decision makers and transport authorities, as well as of MaaS which is one of the novel mobility concepts that could assist in achieving seamless mobility (Kamargianni, Li, Matyas, & Schäfer, 2016).

The MaaS paradigm has been a disruptive approach to the current models of transport provision, especially in urban areas, with the potential to encourage the adoption of transportation options from more sustainable modes (Polis Traffic Efficiency & Mobility Working Group, 2017), to solve public accessibility and urban congestion problems, reduce car ownership and car use, as well as to increase public accessibility to major urban spots (Jittrapirom, Marchau, van der Heijden, & Meurs, 2018). The prospect of MaaS being a potential solution for urban mobility has raised the level of interest from various transport sectors. Nevertheless, the concept is still in its early stage of development (conceptualization) with limited evidence from real cases to support its anticipated benefits (Jittrapirom, Marchau, van der Heijden, & Meurs, 2018).

The next phase is focused on developing the MaaS application. Here the types of challenges encountered have numerous similarities with the development of other applications. In MaaS application and integration process, there are barriers and challenges that need to be properly identified and addressed, such as: 1) the integration with external transport systems, with corporate policy, culture and norms that could be in conflict with using the services; 2) society's laws and regulations that, in some cases, can limit system development (Hesselgren, Sjöman, & Pernestål, 2019); 3) security and privacy issues in data transition (Utriainen & Pöllänen, 2018); 4) the fares and ticketing integration (Chowdhury, Hadas, Gonzalez, & Schot, 2018); 5) the competition with cars; 6) and the collaboration between key actors, stakeholders, including PT agencies and authorities, local governments and town planning offices (Sochor & Sarasini, 2017), users, public administration and operators, including platform and transportation operators. As a result, a broad and extensive work has to be carried out to enable the development and implementation of an efficient and user friendly MaaS system. In broad terms the MaaS app must be easy to use, run on multiple platforms, be reliable, up to date (real time data is a must), effective and guarantee an unquestionable personal data privacy. Once again, this leads to another set of challenges that are essentially technical. To finalize the implementation and monitorization phase also requires some attention and commitment. A good marketing approach is necessary to achieve a generalized use and its intended improvement of the quality of the service provided by the transportation system of an area and the consequential increase in the quality of life of its citizens.

Around the world, several MaaS schemes have been implemented. However, researchers (e.g. Jittrapirom *et al.*, 2017) consider that there still exists a lack of an assessment framework to classify their unique characteristics in a systematic manner. As far as MaaS is concerned, mobility APPs have been developed, tested and upgraded in order to achieve an integrated and sustainable mobility system (Boero *et al.*, 2016). They allow users to plan, pay and have access to real time data about the trip along with the development of projects that consolidate ideas and approaches for future MaaS implementations. The first large-scale implementations, from which the current effects of MaaS can be studied, started in 2017. Therefore, it is an ongoing challenge for the scientific community (Utriainen & Pöllänen, 2018). Jittrapirom *et al.* (2017) have summarized the core characteristics of MaaS schemes that should be apparent when implementing such concept in practice, based on literature review that includes: Integration of transport modes, Tariff option, Unified platform, Multiple actors, Use of technologies, Demand orientation, Registration requirement, Personalization, Customization.

MaaS also appears at a critical junction in time where a paradigm shift is on its course to change, in a quite profound manner, the way that we, as a collective society, view and embrace one of our most basic perceived needs: mobility. Since ancient times, greater mobility has been eagerly desired by people and civilizations as a means to achieve better living standards or to validate their power and advancement. In the past decades, technological developments have, at a breathtaking pace, made these advancements, once a privilege of the higher classes, accessible to a very high percentage of the society. While this has to be viewed as a very positive byproduct of our technical prowess it is, nevertheless, reaching a point where its multiple advantages and widespread use are leading modern societies to address some very serious issues such as: pollution, climate changes, poor urban spaces and a general decrease in the quality of life to name just a few. MaaS is certainly no "white knight" which can, by itself, save us from the problems we have created. However, it is a very useful approach and can provide a toolkit of tailored solutions that can, hopefully with the aid of other measures, assist modern societies in shifting their approach in terms of solving our mobility problems while creating a more inclusive and efficient society that respects environmental and sustainability values.

Due to recent developments of technology, new opportunities are emerging like gamification, (Kazhamiakin *et al.*, 2015, Yen *et al.*, 2018). Thus, research studies (Sherly and Somasundareswari, 2015) have focused on the development of digital technology applications aiming at raising the awareness and changing the behavior of citizens to embrace a more sustainable and active mobility.

The potential is certainly there to change the way we live in a positive and sustainable way. In the end, it will be up to us, as an increasingly unified collective community, to develop, implement and embrace a new way of viewing mobility making use of our technological advancements for the benefit of all and the sustainability of our resources.

CONCLUSION

The future aim is to create a fully sustainable, universal and simultaneously personalized mobility service. Thus, besides creating opportunities and contributing to the evolution of a territory, mobility services should also provide social inclusion creating the opportunity to connect and interact with different means of transportation as well as with various additional service offers. Along with the more sustainable and healthier lifestyle and personalized mobility habits according to the user's patterns, MaaS can simultaneously promote an individual trip in a useful manner for society as a whole.

The technological trends drive the pathway of MaaS. However, the success of MaaS will depend on the integration of all actors and stakeholders through the user's accessibility and mobility satisfaction. Another crucial point is to establish users' preferences and behaviors to set a base for the development and improvement of MaaS. The dissemination and communication of mobility actions, which will lead to an increased political support in terms of promoting a change of transport policies and of the strategy's adoption beyond the partner areas. Establishing the cost for the user guarantees the viability of the economic support by the government, along with the economic success of the transport companies. The MaaS roadmap development needs to firstly establish some common ground of requirements and common understanding among all the involved mobility actors before attempting large-scale MaaS applications. This step is vital to be used in the definition of which developments and practices should be included in the MaaS design in each phase. After this initial consensus, a larger political and regulatory context must also be created and suitably handled regarding the stimulation of the adoption of MaaS, as well as discouraging the traditional forms of mobility that are not sustainable.

ACKNOWLEDGMENT

This research was supported by FEDER (Fundo Europeu de Desenvolvimento Regional) through Programa Operacional Competitividade e Internacionalização, in the scope of the Project "ALTO MINHO. SMOB - Mobilidade Sustentável para o Alto Minho", Ref. POCI-01-0145-FEDER-024043.

REFERENCES

Ambrosino, G., Nelson, J. D., Boero, M., & Pettinelli, I. (2016). Enabling intermodal urban transport through complementary services: From Flexible Mobility Services to the Shared Use Mobility Agency Workshop 4. Developing inter-modal transport systems. *Research in Transportation Economics*, *59*, 179–184. doi:10.1016/j.retrec.2016.07.015

Atkins. (2015). *Journeys of the future. Introducing Mobility as a Service.* Retrieved from https://www.atkinsglobal.com/~/media/Files/A/Atkins-Corporate/uk-and-europe/uk-thought-leadership/reports/Journeys%20of%20the%20future_300315.pdf

Barreto, L., Amaral, A., & Baltazar, S. (2019a). Mobility as a Service (MaaS) in rural regions: An overview. In *9th International Conference on Intelligent Systems 2018: Theory, Research and Innovation in Applications, IS 2018 - Proceedings* (pp. 856–860). Academic Press.

Barreto, L., Amaral, A., & Baltazar, S. (2019b). Flexible on Demand Transport Services (FDTS): The future of mobility systems. *25th ICE/IEEE International Technology Management Conference (ITMC)*. 10.1109/ICE.2019.8792649

Barreto, L., Amaral, A., & Baltazar, S. (2019c). Urban Mobility Digitalization: Towards Mobility as a Service (MaaS). *9th International Conference on Intelligent Systems 2018: Theory, Research and Innovation in Applications, IS 2018 - Proceedings*, 850–855.

Berselli, D. (2017). *D.T2.4.4 Pilot launch report - Castelfranco Emilia*. Modena: Interreg Central Europe – RUMOBIL.

Boero, M., Garré, M., Fernandez, J., Persi, S., Quesada, D., & Jakob, M. (2016). MyWay Personal Mobility: From Journey Planners to Mobility Resource Management, in 6th Transport Research Arena, April 18-21, 2016. *Transportation Research Procedia*, *14*, 3265–3273.

Campos, V. B. G., & Ramos, R. A. R. (2005). Proposta de Indicadores de Mobilidade Urbana Sustentável Relacionando Transporte e uso do Solo. 1o Congresso Luzo-Brasileiro para o Planejamento Urbano,Regional, Integrado e Sustentável (PLURIS).

Chowdhury, S., Hadas, Y., Gonzalez, V. A., & Schot, B. (2018). Public transport users' and policy makers' perceptions of integrated public transport systems. *Transport Policy*, *61*, 75–83. doi:10.1016/j.tranpol.2017.10.001

Corwin, S., Dinamani, A., & Pankratz, D. (2019). *Toward a mobility operating system: Establishing a lingua franca for urban transportation*. Deloitte Center for Integrated Research.

Dalton, J. (2018). *What is "New Mobility" Anyway? Making sense of MaaS, MOD, IoM, and how we got here*. Retrieved from https://method.city/what-is-new-mobility-anyway-581cbabb55a4

Docherty, I., Marsden, G., & Anable, J. (2018). The governance of smart mobility. *Transportation Research Part A, Policy and Practice*, *115*, 114–125. doi:10.1016/j.tra.2017.09.012

Durand, A., Harms, L., Hoogendoorn-Lanser, S., & Zijlstra, T. (2018). *Mobility-as-a-Service and changes in travel preferences and travel behaviour: a literature review*. Ministry of Infrastructure and Water Management. KiM Netherlands Institute for Transport Policy Analysis.

Federal Ministry for Economic Cooperation and Development (BMZ). (2016). *Urban Mobility: Strategies for Liveable Cities*. doi:10.1080/15568318.2013.821008

Fontes, T., Correia, J., de Sousa, J. P., de Sousa, J. F., & Galvão, T. (2017). A Multi-User Integrated Platform for Supporting the Design and Management of Urban Mobility Systems. In Transportation Research Procedia (Ed.), *20th EURO Working Group on Transportation Meeting, EWGT 2017* (Vol. 27, pp. 35–42). Elsevier B.V. 10.1016/j.trpro.2017.12.158

Goodall, W., Fishman, D. T., Bornstein, J., & Bonthron, B. (2017). The rise of mobility as a service-Reshaping how urbanites get around. Deloitte Review, 20.

Heikkilä, S. (2014). *Mobility as a Service - A Proposal for Action for the Public Administration Case Helsinki*. Aalto University.

Hesselgren, M., Sjöman, M., & Pernestål, A. (in press). Understanding user practices in mobility service systems: Results from studying large scale corporate MaaS in practice. *Travel Behaviour and Society*.

Hietanen, S. (2014). "Mobility as a Service"—The new transport model? *Eurotransport*, *12*(2), 2–4.

Hietanen, S., & Sahala, S. (2014). Mobility as a Service. Can it be even better than owning a car. *The New Transport Model.*

Hu, X., Giang, N. K., Shen, J., Leung, V. C. M., & Li, X. (2015). Towards mobility-as-a-service to promote smart transportation. *2015 IEEE 82nd Vehicular Technology Conference (VTC2015-Fall)*, 1–5.

Ison, S., & Sagaris, L. (2016). Workshop 4 report: Developing inter-modal transport systems. *Research in Transportation Economics*, *59*, 175–178. doi:10.1016/j.retrec.2016.10.006

Jittrapirom, P., Caiati, V., Feneri, A.-M., Ebrahimigharehbaghi, S., González, M. J. A., & Narayan, J. (2017). Mobility as a Service: A Critical Review of Definitions, Assessments of Schemes, and Key Challenges. *Urban Planning*, *2*(2), 13–25. doi:10.17645/up.v2i2.931

Jittrapirom, P., Marchau, V., van der Heijden, R., & Meurs, H. (2018). *Future implementation of mobility as a service (MaaS): Results of an international Delphi study*. Travel Behaviour and Society. doi:10.1016/j.tbs.2018.12.004

Kamargianni, M., Li, W., Matyas, M., & Schäfer, A. (2016). A critical review of new mobility services for urban transport. *Transportation Research Procedia, 14*, 3294–3303. doi:10.1016/j.trpro.2016.05.277

Kamargianni, M., Matyas, M., Li, W., Muscat, J., & Yfantis, L. (2018a). *The MaaS Dictionary*. MaaSLab, Energy Institute, University College London. Available at: www.maaslab.org

Kamargianni, M., Yfantis, L., Muscat, J., Azevedo, C. L., & Ben-Akiva, M. (2018b). Incorporating the Mobility as a Service Concept into Transport Modelling and Simulation Frameworks. *Transportation Research Board 98th Annual Meeting*, (18), 1–23.

Karlsson, I. C. M., Sochor, J., & Strömberg, H. (2016). Developing the "Service" in Mobility as a Service: Experiences from a Field Trial of an Innovative Travel Brokerage. *Transportation Research Procedia, 14*, 3265–3273. doi:10.1016/j.trpro.2016.05.273

Kazhamiakin, R., Marconi, A., Perillo, M., Pistore, M., Valetto, G., Piras, L., . . . Perri, N. (2015). Using Gamification to Incentivize Sustainable Urban Mobility. *1st IEEE International Smart Cities Conference*. DOI:10.13140/RG.2.1.2622.2166

Li, Y., & Voege, T. (2017). Mobility as a Service (MaaS): Challenges of Implementation and Policy Required. *Journal of Transportation Technologies*, 7(02), 95–106. doi:10.4236/jtts.2017.72007

Litman, T. (2015). *Evaluating Accessibility for Transportation Planning - Measuring People' s Ability To Reach Desired Goods and Activities*. MaaS Alliance. Retrieved from http://maas-alliance.eu/

Litman, T. (2017). *Public Transportation's Impact on Rural and Small Towns: A Vital Mobility Link*. Community Transportation Association of America.

Masuch, N., Lützenberger, M., & Keiser, J. (2013). An open extensible platform for intermodal mobility assistance. In Procedia Computer Science (Ed.), *4th International Conference on Ambient Systems, Networks and Technologies* (Vol. 19, pp. 396–403). Elsevier. 10.1016/j.procs.2013.06.054

Meurs, H., Sharmeen, F., Marchau, V., & van der Heijden, R. (2019). Organizing the integration of firms in mobility-as-a-service systems: Principles of alliance formation applied to a MaaS-pilot in the Netherlands. *Transportation Research Part A, Policy and Practice*. doi:10.1016/j.tra.2019.09.036

Meurs, H., & Timmermans, H. (2017). *Mobility as a Service as a multi-sided market: Challenges for modeling* (No. 17-00830).

Mörch-Pierre, T. (2019). *Evaluating Payment Methods for MaaS*. Academic Press.

Mulley, C., Nelson, J. D., & Wright, S. (2018). Community transport meets mobility as a service: On the road to a new a flexible future. *Research in Transportation Economics*, *69*(March), 583–591. doi:10.1016/j.retrec.2018.02.004

Nuzzolo, A., Coppola, P., & Papa, E. (2014). Marginal Activity Access Cost (MAAC): A New Indicator for Sustainable Land Use/Transport (LUT) Planning. In *EWGT2013 – 16th Meeting of the EURO Working Group on Transportation* (Vol. 111, pp. 450–459). 10.1016/j.sbspro.2014.01.078

Pangbourne, K., Stead, D., Mladenovic, M., & Milakis, D. (2018). The case of Mobility as a Service: a critical reflection on challenges for urban transport and mobility governance. In Governance of the smart mobility transition, (pp. 33–48). Emerald Publishing Limited. doi:10.1108/978-1-78754-317-120181003

Polis Traffic Efficiency & Mobility Working Group. (2017). *Mobility as a Service: Implications for Urban and Regional Transport. Polis - European Cities and Regions Networking for Innovative Transport Solutions*. Retrieved from https://www.polisnetwork.eu/uploads/Modules/PublicDocuments/polis-maas-discussion-paper-2017---final_.pdf

Polydoropoulou, A., Pagoni, I., & Tsirimpa, A. (2018). *Ready for Mobility as a Service? Insights from stakeholders and end-users*. Travel Behaviour and Society. doi:10.1016/j.tbs.2018.11.003

Puhe, M. (2014). Integrated Urban E-ticketing Schemes - Conflicting Objectives of Corresponding Stakeholders. In Transportation Research Procedia (Ed.), Mobil. TUM 2014 "Sustainable Mobility in Metropolitan Regions" (Vol. 4, pp. 494–504). Elsevier B.V. doi:10.1016/j.trpro.2014.11.038

Shaheen, S., Cohen, A., & Zohdy, I. (2016). *Shared Mobility: Current Practices and Guiding Principles. Development*. Report No.FHWA-HOP-16-022.

Sherly, J., & Somasundareswari, D. (2015). Internet of things based smart transportation systems. *Interntional Raesearch Journal of Engineering and Technology*, *2*, 1207–1210.

Sochor, J., Arby, H., Karlsson, I. C. M. A., & Sarasini, S. (2018). A topological approach to Mobility as a Service: A proposed tool for understanding requirements and effects, and for aiding the integration of societal goals. *Research in Transportation Business & Management*, *27*(March), 3–14. doi:10.1016/j.rtbm.2018.12.003

Sochor, J., & Sarasini, S. (2017). Article. In *1st International Conference of Mobility as a Service (ICoMaaS)* (pp. 137–138). Tampere: Academic Press.

Stead, D. (2016). Key research themes on governance and sustainable urban mobility. *International Journal of Sustainable Transportation, 10*(1), 40-48.

Stein, M., Meurer, J., Boden, A., & Wulf, V. (2017). Mobility in Later Life - Appropriation of an Integrated Transportation Platform. In *Conference on Human Factors in Computing Systems - CHI '17* (pp. 5716–5729). 10.1145/3025453.3025672

Streeting, M., Chen, H., Teng, Y., Edgar, E., & Koh, J. (2016). *Mobility as a Service The next transport disruption — a perspective from China*. LEK.

Utriainen, R., & Pöllänen, M. (2018). Review on mobility as a service in scientific publications. *Research in Transportation Business & Management*, *27*, 15–23. doi:10.1016/j.rtbm.2018.10.005

Webb, J., Wilson, C., & Kularatne, T. (2018). Will people accept shared autonomous electric vehicles? A survey before and after receipt of the costs and benefits. *Economic Analysis and Policy*.

Chapter 2
Public-Private-People Partnership Networks and Stakeholder Roles Within MaaS Ecosystems

Jenni Eckhardt
https://orcid.org/0000-0002-6089-7986
VTT Technical Research Centre of Finland Ltd., Finland

Aki Aapaoja
https://orcid.org/0000-0002-6968-3191
VTT Technical Research Centre of Finland Ltd., Finland

Harri Haapasalo
https://orcid.org/0000-0001-5413-5638
University of Oulu, Finland

ABSTRACT

Mobility as a service (MaaS) is an emerging concept offering integrated mobility services. Combining different transport modes and services, as well as collaboration of stakeholders, are prerequisites for viable and attractive MaaS services. MaaS is expected to increase the sustainability and efficiency of transport. Public-private-people partnership (PPPP) networks are seen as a potential solution to meet these expectations, especially in rural areas. The purpose here is to present a PPPP network for MaaS, which integrates market-based mobility services and subsidized transportation. The chapter also describes the roles and responsibilities of primary and secondary MaaS stakeholders at different levels, including authority, service provider, MaaS operator, and user levels.

DOI: 10.4018/978-1-7998-1614-0.ch002

INTRODUCTION

Several trends and drivers are pushing demand for new mobility services. Targets to reduce greenhouse gas emissions require measures such as new sources of low-emission fuel, technological development of vehicles, and improving the energy efficiency of transport systems through services (Särkijärvi, Jääskeläinen, & Lohko-Soner, 2018). New mobility concepts are seen as a significant opportunity to bring about low-carbon mobility (European Commission, 2017a). As public funding decreases, pressure is growing to produce more efficient subsidized transport services. Urbanization results in increased congestion, as well as land use and parking problems, while causing more challenges in rural areas to provide efficient transport and viable mobility services. Advances in technology and digitalization enable integrated travel chains and shared services via smartphones and mobile applications (e.g. Hallock & Inglis, 2015). User behavior is also changing; younger generations are not as keen to own a car and customers are more environmentally oriented and open to circular economy services (Heikkilä, 2014). It has been recognized that technology-enabled transportation services have the potential to reduce driving and car ownership (Hallock & Inglis, 2015). In addition, reduction of car ownership may also relate to life events and income reductions (Clark, Chatterjee, & Melia, 2016).

Mobility as a Service (MaaS) is an emerging concept of integrated transport services that is seen as a solution to tackling future mobility challenges and meeting new customer needs. In rural areas, MaaS can increase efficiency of transport and accessibility of remote areas (Eckhardt, Nykänen, Aapaoja, & Niemi, 2018). Public-private partnerships (PPP) have been recommended for MaaS (National Academies of Sciences, Engineering, and Medicine, 2016; Eckhardt et al., 2017a; Polis, 2017; ERTICO, 2018). PPP refers to cooperation between public-private actors in which they jointly develop services and share risks, costs and resources which are connected with these services (Van Ham & Koppenjan, 2001).

The objective of this chapter is to present a potential public-private-people partnership (PPPP) network for MaaS, bringing in the 'people' aspect in terms of prosuming, where users both use and produce services. The aim is also to identify stakeholders within the MaaS ecosystem and define their roles. The chapter deals with the following research questions:

- What kind of PPPP networks can be created for rural MaaS?
- Who are the stakeholders involved in MaaS ecosystems?
- What are the roles and responsibilities of the identified stakeholders?

The chapter is structured as follows: A review of the literature describes the MaaS concept and relevant discussions on PPP and PPPP, as well as business and stakeholder networks. Next is a description of the adopted methodology (interviews and stakeholder workshops) and results for the PPPP network in the rural context and MaaS stakeholders, including roles and responsibilities. The results are discussed and followed by concluding remarks.

CONTEXT AND LITERATURE REVIEW

Mobility as a Service (MaaS) is an emerging concept of transport services. MaaS refers to "multimodal and sustainable mobility services addressing customers' transport needs by integrating planning and payment on a one-stop-shop principle" (MAASiFiE, 2016). Hietanen (2014) defines MaaS as "a mobility distribution model in which a customer's primary transportation needs are met over one interface and are offered by a service provider" (p. 2). MaaS could offer travelers easy, flexible, reliable, priceworthy and environmentally sustainable everyday travel, such as public transport, carsharing and car leasing, as well as more efficient goods shipping and delivery possibilities (European Commission, 2017b).

The main characteristics of different MaaS definitions and descriptions include customer-orientation and integrated services (e.g. Hietanen 2014; Burrows, Bradburn, & Cohen, 2015; Kamargianni, Li, & Matyas, 2016; König, Eckhardt, Aapaoja, Sochor, & Karlsson, 2016a; Transport Systems Catapult, 2016; MaaS Alliance, 2017; ERTICO, 2018). Integration also has several classifications. Shared mobility, Booking/Ticketing and Multimodal traveler information are the components of integration that König et al. (2016a) consider a MaaS service. Mobility services with fewer integration components could be considered MaaS-related services (König et al., 2016a). Kamargianni et al. (2016) have classified integration of MaaS in Ticket & Payment integration, Mobility package and ICT integration. Lyons, Hammond, and Mackay (2019) propose Levels of MaaS Integration (LMI) taxonomy including operational and cognitive user efforts and suggest that operational and informational integration should be principal determinants of scaling up MaaS. Sochor, Arby, Karlsson, and Sarasini (2018) have classified the following integration levels:

0. **No integration:** Single, separate services
1. **Integration of Information:** Multimodal travel planner, price info
2. **Integration of Booking and Payment**: Single trip – find, book and pay
3. **Integration of Service Offer:** Bundling/subscription, contracts, etc.
4. **Integration of Societal Goals**: Policies, incentives, etc.

The Netherlands Institute for Transport Policy Analysis proposes that integration level 2 of the classification by Sochor et al. (2018) would be the minimum level for MaaS (Harms, Durand, Hoogendoorn-Lanser, & Zijlstra, 2018). Even though mobility packages and subscriptions exist, such as Whim (n.d.) monthly packages, pay-as-you-go is a commonly offered tariff option of MaaS services (Jittrapirom et al., 2017).

The aspect of sustainability appears more rarely in definitions (e.g. MAASiFiE, 2016; König et al., 2016a; European Commission, 2017b; ERTICO, 2018), but new mobility concepts are seen as a major opportunity to bring about low-carbon mobility (European Commission, 2017a). Other attempts to mitigate environmental impact include zero or low-emission vehicles and optimizing the transport system, including measures such as multimodality and digital mobility (European Commission, 2016). Digital technologies can make transport more efficient and inclusive and enable seamless door-to-door mobility and value-added services (European Commission, 2016).

MaaS is expected to provide several benefits related to low-carbon mobility, such as a modal shift from car to public transport and sharing services, an increase in multimodal trips and resource efficiency, and decreased emissions (Karlsson, Sochor, Aapaoja, Eckhardt, & König, 2017). According to a global stakeholder survey, 80% of respondents expected private car use to be significantly reduced due to MaaS (König, Sochor, Eckhardt, & Böhm, 2016b). In Sweden, the UbiGo MaaS pilot in an urban context showed attitude changes towards various travel modes: participants became less positive towards the private car and more positive towards alternative modes such as carsharing, bus/tram and bikesharing (Sochor, Karlsson, & Strömberg, 2016). Particularly for younger generations, renting and sharing mobility services are increasingly popular and owning a private vehicle is becoming less important (European Commission, 2017a).

Goods deliveries are also connected to MaaS (e.g. Eckhardt et al., 2017a; European Commission, 2017b; MaaS Alliance, 2017; Jittrapirom et al., 2017). However, they are considered an additional feature and most of the MaaS services and pilots do not include freight services. A MaaS operator could, in addition to the mobility of people, include last-mile deliveries such as parcel, post, shopping, pharmacy products and meal deliveries. (Eckhardt et al., 2017a). Combining transport of goods is considered more relevant for rural areas due to long distances, sparse population and narrow flows of people and goods (Eckhardt et al., 2017a).

MaaS relies on existing public transport and is expanded with additional services, such as sharing services and taxis (Eckhardt et al., 2017a; Jittrapirom et al., 2017). However, rural areas have limited or even non-existing public transport, which challenges accessibility. Thus, rural solutions could include e.g. on-demand transport and sharing services (International Transport Forum, 2015; Eckhardt et al., 2018), as well as integrating social and health service and school transportation with commercial transportation (Eckhardt et al., 2018; The Finnish Innovation Fund Sitra, n.d.).

Shared mobility in its various forms supports the MaaS concept (e.g. Hallock & Inglis, 2015; Shaheen, Cohen, & Zohdy, 2016). Shared mobility refers to shared use of a motor vehicle, bicycle, or other low-speed transportation mode enabling users to obtain short-term access to transportation as needed, without requiring ownership (Shaheen et al., 2016). Several types of sharing services in the transportation sector already exist. There are commercial carsharing companies (e.g. DriveNow, ZipCar) and bike-sharing services, and some include both car and bike sharing (e.g. DB/DIMIS) (Eckhardt et al., 2017a). The sharing services that private people offer include peer-to-peer carsharing services (e.g. Blox Car) and ridesharing services (e.g. BlaBlaCar) (Eckhardt et al., 2017a). There are also organized hitchhiking services, and e.g. Rezo Pouce was developed to assist with rural mobility (Rezo Pouce, n.d.). Ridesourcing services, such as Lyft or Uber, connect drivers and customers via smartphone application, and are also used for trips when public transit is not available (National Academies of Sciences, Engineering, and Medicine, 2016). In this changing era of new mobility, the meaning of public transport is changing towards 'passenger transport services by shared resources' (Laurell, 2017).

The transport sector is the largest in value utilizing PPP (European PPP Expertise Centre, 2018). PPP is widely used for transport infrastructure projects but is also related to e.g. transport technology and services. New mobility services should be developed in collaboration with the private sector and local and transport authorities, to support city and regional transport priorities and policies (Polis, 2017). MaaS has the potential to improve the efficiency of existing transport services and public resources (Polis, 2017; Eckhardt et al., 2018; ERTICO, 2018). However, new models and tools for public-private collaboration are needed to effectively govern the development and spreading of sustainable MaaS (Smith, Sarasini, Karlsson, Mukhtar-Landgren, & Sochor, 2019a).

The generally recognized motivation for PPP is improved value for money (Organisation for Economic Cooperation and Development, 2012; Eadie, Millar, & Toner, 2013; European PPP Expertise Centre, 2015). PPP also enables combining public and private expertise efficiently (European Union, 2018). In addition, PPP may improve existing public sector practices and promote more innovative ways of delivering public services (European PPP Expertise Centre, 2015). Large-scale deployments of MaaS necessitate extensive modifications to current business models and business relations, making public-private open innovation challenging (Smith, Sochor, & Karlsson, 2019b). PPP contracts can be complex and resource intensive, and lack of public sector capacity and experience is one of the main challenges for PPPs (Organisation for Economic Cooperation and Development, 2012; European

PPP Expertise Centre, 2015). Potential negative factors of PPP also include delays in negotiation, time-consuming management and high participation costs (Li, Akintoye, Edwards, & Hardcastle, 2005).

Public-private-people partnership (PPPP, 4P) introducing the 'people' aspect has been studied related to e.g. urban development process, real estate industry and infrastructure building (e.g. Majamaa, 2008), as well as to the city resilience-building process (Marana, Labaka, & Sarriegi, 2018). PPPP is also related to Living Lab activities, including MaaS development (Jin & Qiu, 2019). However, in these examples, 'people' implies participating in planning through discussion and ideas. Here, 'people' also refers to an active stakeholder who would consume and produce mobility services as a 'prosumer'. Prosumers could integrate e.g. private cars into the MaaS ecosystem in the form of ridesharing and peer-to-peer carsharing services (Eckhardt et al., 2017a).

To understand the current state of the MaaS ecosystem, this study applies the approach of network theory. Ultimately an ecosystem is a network – and a system also – comprising various interconnected actors having relationships between each other and the surrounding environment (Rubenstein-Montano et al., 2001). Network thinking depends heavily on how a system is defined, since it studies interconnections between the different parts of the network. Boundaries must be set to separate what parts of the world are contained inside the system and what parts are considered the external environment (e.g. internal and external stakeholders) (Leviäkangas et al., 2014). An example of these network theories is a well-known ARA network approach defining networks as constellations of Activities Resources and Actors (Håkansson & Johanson, 1992). Studies on value-creating networks describe the operations and partially evolutionary nature of networks, giving insight also into the role of the public sector from the business point of view. Value networks open up possibilities to see benefits beyond organizational boundaries and improve efficiency through collaboration.

A business ecosystem is an economic community and a network supported by a set of interacting participants and organizations. Ecosystem participants can co-evolve and co-create competences and skills around innovations by collaborating to create and support new offerings, satisfy customer needs and eventually discover innovations (Moore, 1993). A business ecosystem is defined by Iansiti and Levien (2004) as "loose networks – of suppliers, distributors, outsourcing firms, makers of related products or services, technology providers, and a host of other organizations – that affect and are affected by the creation and delivery of a company's offerings" (p. 68). Ecosystems are not static and may include not only private companies, but also authorities and other public organizations (Peltoniemi & Vuori, 2004). Also, rival companies may be a part of the same ecosystem. One of the first publications in the ecosystems literature, "Predators and prey" by Moore (1993), descriptively presents the idea of evolution in ecosystems. The nature of this definition includes the assumption that ecosystems evolve and also hook in new participants, i.e. the ecosystem itself evolves according to its actors, which is also the basic element of a business network. In developing (or more likely guiding the evolution of) these networks, their former nature needs to be understood and the 'food chain' recognized in order to build economically viable ecosystems.

A key benefit of business ecosystems is that individual participants may gain or achieve more value within the ecosystem than they could by themselves (Adner, 2006), or incur fewer costs than operating alone. Typically, business ecosystems aim at extending the offering or co-developing it with others in various ways. Additionally, they can open up possibilities in new markets or enhance resource and asset utilization.

In this study, the participants or actors of ecosystems are called stakeholders. A stakeholder is any group or individual that affects or is affected by the organization's behavior (Freeman, 1984). Stakeholders are generally divided into primary and secondary (e.g. Freeman, 1984; Clarkson, 1995; Eesley & Lenox, 2006) or even tertiary stakeholders (Aapaoja & Haapasalo, 2014). The participation of primary stakeholders is required for the survival of an organization. These include e.g. customers, suppliers and public stakeholder groups (the governments and communities providing e.g. infrastructure, laws and regulations) (Clarkson, 1995). Stakeholders can also be seen as internal or external (Kinnunen, Aapaoja, & Haapasalo, 2014). Internal stakeholders are inside the ecosystem. External stakeholders are not formal members of the ecosystem but may have an impact on or be impacted by the ecosystem (Cleland, 1986; Winch & Bonke, 2004). Stakeholder identification and periodization of the most salient stakeholders is critical to achieve at least a reasonable amount of transparency and an economically viable ecosystem (Aapaoja & Haapasalo, 2014). MaaS involves an 'ecosystem' including several stakeholders from both the public and private sectors. A MaaS ecosystem combines different variations of PPP and consists of four levels: 1) public and regulatory level; 2) transport and logistics service provider level (i.e., supply-side); 3) mobility service level ('a MaaS operator'), and 4) end-user level (Aapaoja, Eckhardt, Nykänen, & Sochor, 2017).

METHODOLOGY

The research methodology (Figure 1) includes five steps: a literature review, interviews, workshops, results and consolidation (Eckhardt et al., 2017a; Eckhardt, Nykänen, Aapaoja, & Niemi, 2017b). The literature review was conducted first and consisted of the Mobility as a Service, public-private partnership, and business and stakeholder network theories.

Figure 1. Methodology

Qualitative data was then collected through interviews and workshops. The interviews consisted of 60 thematic interviews with 58 organizations (Table 1). The first set of interviews, half of the total, dealt with MaaS covering different geographical areas and were performed internationally. The second set focused on MaaS in rural areas and were with Finnish stakeholders only. The interviewees were selected based on their expertise in the MaaS field and to ensure coverage of widely different stakeholder groups and geographical areas; thus they were from different counties or parts of Finland. All the interviews were conducted similarly on a semi-structured basis (Yin, 2003) and included questions related to MaaS networks and stakeholders, e.g.:

- What kind of service combination could be created for different geographical areas? (First set of interviews) / What kind of service combination could be created for rural areas? (Second set of interviews)
- How could freight transport be combined with the mobility of people?
- Who would be the main player or MaaS operator?
- Who are the main stakeholders and what would their roles be?
- What are the enablers/barriers of/to MaaS?

MaaS operators or service providers were also asked about their current roles and responsibilities as well as their service and business model.

National workshops were organized in Finland, Sweden and Austria to define the roles and responsibilities of MaaS stakeholders. The participants varied in the workshops but all included a wide variety of key stakeholders (Table 2). The roles and responsibilities were based on a roadmap workshop defining short-term (1–3years) and medium-term (4–9 years) actions and transitions of MaaS. Actions and transitions were discussed in groups for drivers, markets, MaaS services and enablers, and the results were presented using Post-it notes on a large roadmap paper sheet. The Post-it notes were then organized by topic, and the groups discussed the responsible actor(s) for each topic and possibly other important stakeholders as well. Finally, the group work results on responsible actors were presented and set in order of priority by the participants.

Table 1. Interviewed organizations

Primary Stakeholders		
National authorities	Ministry of Transport and Communications, Finland	Finnish Transport Safety Agency, Finland
	Ministry of Agriculture and Forestry, Finland	Finnish Transport Agency, Finland
	ASFINAG, Austria	
Local and regional authorities	Centre for Economic Development, Transport and the Environment of Lapland, Finland	Centre for Economic Development, Transport and the Environment of South Ostrobothnia, Finland
	Centre for Economic Development, Transport and the Environment of Southeast Finland	Soite - Central Ostrobothnia joint municipal authority of social and health service, Finland
	Regional Council of Central Ostrobothnia, Finland	South Karelia Social and Health Care District, Finland
	Regional Council of Lapland, Finland	The city of Hämeenlinna, Finland
	Regional Council of South Karelia, Finland	The city of Imatra, Finland
	Trafikkontoret, Gothenburg, Sweden	The city of Rovaniemi, Finland
	Region of Västra Götaland, Sweden	
MaaS operators	MaaS Global, Finland	Ubigo Innovation, Sweden
	Telia Company, Finland	Sito, Finland
	Tuup, Finland	SNCF, France
Service providers (transport, logistics, digital)	Västtrafik public transport, Sweden	VR, Finland
	ÖBB, Austria	Uber, Finland
	Sunfleet carsharing, Sweden	Mobisoft, Finland
	PayiQ, Finland	Municipality logistics of Finland
	Siemens, Finland	Petri Pekkala (trade name), Finland
	Tuomi Logistics Ltd., Finland	Posti Group Corporation, Finland
	Tieto, Finland	
Secondary Stakeholders		
Academia	Gothenburg University, Sweden	DLR, Germany
	Trivector, Sweden	Victoria Swedish ICT, Sweden
	ÅF Consulting, Sweden	Tampere University of Technology, Finland
	Vinnova, Swedish innovation agency, Sweden	VTT Technical Research Centre of Finland Ltd.
Other stakeholders	Forum Virium Helsinki, Finland	The social insurance institution of Finland (Kela)
	Linna business development (Hämeenlinna), Finland	Association of People with Disabilities, Finland
	E-Mobility, Austria	The Association of Finnish Local and Regional Authorities, Finland
	The council of rural policy MANE, Finland	Visit Rovaniemi, Finland
	The hospital district of Lapland, Finland	Finnish Taxi, Finland
	The hospital district of North Ostrobothnia, Finland	Growth Corridor Finland
	Samtrafiken, Sweden	

Table 2. Stakeholders participating in the workshops

Primary Stakeholders		Secondary Stakeholders
Ministries	Transport agencies/ Road administrations	Consultancy
Infrastructure operators	Municipalities and cities	Academia
Regional administrations	Social and health care districts in Finland	The social insurance institution of Finland (Kela)
Public transport operators	Car sharing companies	Taxi union (Finland)
Railway operator	Logistics operators	Bus union (Finland)
MaaS operators	Technology providers	Funding agencies

Preliminary results were obtained by integrating and processing the collected data from the interviews and workshops in research group working meetings. Finally, the preliminary results were further developed, prioritized and validated in consolidation workshops. The participants in the consolidation workshops included representatives of national, regional and local authorities and institutions. Regarding roles and responsibilities, the results of the workshop and the first set of interviews were first integrated and validated in a European consolidation workshop. These results were further developed by the second set of interviews and then validated in a national consolidation workshop in Finland. The rural MaaS PPPP model was developed based on the interviews. The preliminary results including several potential rural MaaS network models were further developed based on group discussions in the Finnish consolidation workshop. The models were prioritized and the selected model validated.

RURAL MAAS EXPERIENCES AND IMPLICATIONS

So far, fully integrated MaaS services do not exist in Finnish rural areas. However, there are and have been pilots and services that include some aspects of integration (see Table 3).

Telia Company initiated the *Ylläs Around* PPP service pilot in the Ylläs ski resort area of Northern Finland in the spring of 2016. The second season of the pilot, winter 2016–17, was operated by Semel. Telia Company/Semel operated the service in cooperation with other main stakeholders comprising local transport operators, the Kolari local municipality, the Ylläs Travel Association and the Finnish Transport Agency. The main idea of the Ylläs Around pilot was to offer one-stop-shop transport services in the Ylläs ski resort area and between the Ylläs

Table 3. Characteristics of rural MaaS pilots and services

Service (S) / Pilot (P) Name	Multimodal Information	Shared Trip/ Vehicle	On-Demand	People and Goods	Integrated Modes or Services	Private and Subsidized Customers	Ticketing/ Mobile Payment
Ylläs Around (P)	x	x	x		x		x
Lapin reittiopas (P)	x						
Matkahuolto (S)		x		x	x		x
ALPIO (P)		x	x			x	x

area and the main local transport hubs Kittilä airport and Kolari railway station. The pilot included multimodal transport services available to end-users through a mobile application including payment and ticketing features. The transport modes on offer included buses, taxis, shared taxis and combinations of these. The original plan was to include freight/parcel deliveries, meal/food deliveries and snowmobiles in the future, but after financing for the pilot ended, Ylläs Around did not continue. Despite largely positive feedback from users and recognition of the pilot's future potential, the Ylläs Travel Association found it challenging to cover the expenses of the system with the achieved number of users (König et al., 2016a; Finnish Transport Agency, 2017).

In Finnish Lapland, the Open Arctic MaaS project (2018–2019) has developed a comprehensive digital search and trip planner, '*Lapin reittiopas*', in collaboration with several municipalities, ski resorts and other actors. The aim is to improve the accessibility of sparsely populated Northern Finland and support sustainable growth of tourism at the same time. The trip planner provides real-time information on service alerts, vehicle positions and trip updates. This development is seen as an enabler for future MaaS integrating different mobility services. (Open Arctic MaaS, n.d.)

Matkahuolto is a company promoting bus and coach services in Finland. Its main business areas include Travel and Parcel Services. Travel Services develop and maintain nationwide timetables and an online store including the services of over 100 bus companies. Also, a mobile application is available for purchasing tickets and reserving seats among other features. Matkahuolto operates 50 terminals and service points, and through its partners nearly 2000 service points are available. It also offers different Parcel Services including a bus parcel option. Matkahuolto thus combines the transport of people and goods on buses covering nearly the whole country. (Matkahuolto, n.d.)

The *ALPIO project* includes a PPP pilot in rural areas of the Tampere region, which started in the spring of 2019. The pilot consists of demand-responsive transport, shared transport of different user groups and combining trips that include customers paying themselves and those being publicly subsidized. Thus, elderly and mobility impaired people using demand-responsive service transport (minibus) and health and social service clients using subsidized taxi services are integrated with self-paying customers. Users paying for their own trips use a smartphone application to order a demand-responsive transport at a public transport fare. Primarily, self-paying customers are integrated with service transport minibus but may also be integrated with taxi rides when needed. A digital platform is created by companies participating in the project to optimize routes and integrate rides. A call center service is also available for subsidized customers. (The Finnish Innovation Fund Sitra, n.d.)

PPPP Network

Public-private collaboration is considered suitable especially for sparsely populated areas where public transport is limited or does not exist. Public actors bring in statutory and subsidized transport services that form the backbone of transport flows into which other transport services can be integrated. Public-private collaboration can take place in many forms and at many levels. Figure 2 proposes a conceptual model of a potential future PPPP network for MaaS based on interviews. It is targeted especially at rural areas, where integration of different transport solutions and private resources would increase efficiency and enable a wider variety of transport services.

The PPPP interface is based on a one-stop-shop principle. Customers would contact the same interface regardless of whether they are private customers paying for the services themselves or whether society is subsidizing the trips. Responsibility for operation of the interface could be assigned to a MaaS operator or split between a MaaS operator for market-based services and a transport service center for publicly subsidized services. In the latter case, tight collaboration would mean that for the customer there is still only a single interface.

Statutory transportation includes social and health service transportation, which refers e.g. to transportation of the elderly and handicapped, illness-related transportation, non-urgent medical transportation by hospital districts and school transportation. Homecare services refers to medical or non-medical services such as meal deliveries. The mobility solutions of homecare personnel are also relevant, not only product and meal deliveries. In addition to publicly subsidized transportation and homecare services, a transport service center could be responsible for logistics deliveries of the public sector, and for coordinating and combining these transportations.

A MaaS operator would be responsible for integrating market-based mobility services and potential additional services such as goods deliveries and activity services. Activity organizers could also offer mobility services as an additional service. Activity here refers to e.g. events such as concerts and sports games, tourist activities such as safaris and attraction visits, and leisure activities.

Figure 2. PPPP network for MaaS

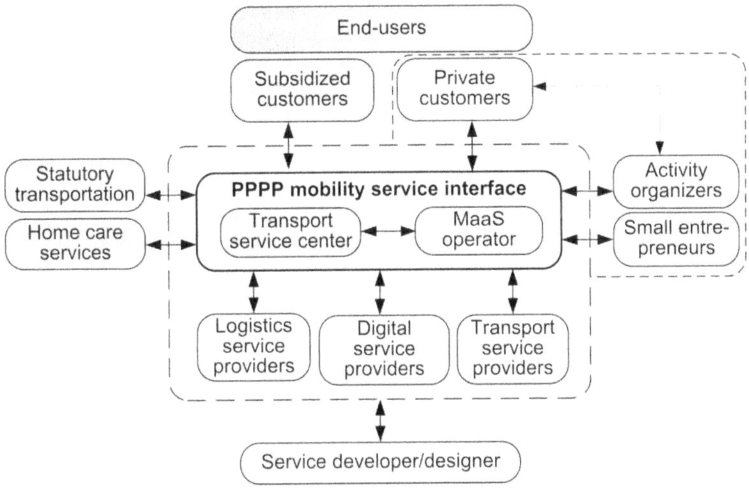

Transport service providers (TSP) would offer a variety of mobility services by different transport modes. These services could include public transport, sharing and rental services, ridesourcing, etc. The public transport offering in rural areas usually refers to buses. In cities it often includes also metros, trams, commuter trains and city bikes. In addition to traditional scheduled operation, demand-responsive transport (DRT) is increasingly receiving attention.

Logistics service providers (LSP) would offer transport services for both the public and private sectors. Small enterprises could use, but also offer, logistics/ mobility services in addition to their core business. Especially in rural areas with low utilization rates, free capacity could be offered through the PPPP mobility service interface. If national legislation allows, transportation of people and goods could be combined, for example, by taxis and buses. Also, vans and trucks could offer taxi services while transporting freight.

Private people, activity organizers and small enterprises can be considered as both customers and mobility service providers. Transport services could also be provided by shared private resources, bringing the 'people' aspect into this public-private-people partnership (PPPP or 4P) network. Shared private resources would integrate e.g. peer-to-peer rental services, carpooling and organized hitchhiking with the MaaS service offering. Private people could also provide ridesharing services for goods on a crowdsourcing basis.

Digital service providers (DSP) offer information and communication technology (ICT) solutions such as mobility applications for smartphones. ICT is also required for the demand-responsive transport service system, and to ensure interoperability and integration of travel chains and different user groups. Mobility applications would include multimodal journey planners, booking, mobile ticketing and payment. Also real-time data (e.g. real-time bus locations) and deviation and rerouting information could be provided. Although smartphones and mobile applications are becoming increasingly common, call services should also be available at least for special groups.

The service developer/designer would ensure that mobility services and user interfaces are user-friendly and customer-oriented. Optimally, the user interface would be easy to use and have the intelligence to guide the user to find fit-for-purpose solutions, avoid mistakes and minimize typing errors, e.g. by providing options to select.

Stakeholder Roles and Responsibilities

MaaS aims to integrate all transport modes and services. To achieve this at operational level, interoperability and collaboration must be achieved at higher levels also. Thus, the various roles and responsibilities of MaaS stakeholders within the emerging MaaS ecosystems are essential to ensuring this process. A MaaS ecosystem consists of primary stakeholders on four levels: 1) public and regulatory level; 2) transport and logistics service provider level (i.e. supply side, including mobile/ICT service providers); 3) mobility service level ('a MaaS operator') and 4) end-user level (Aapaoja et al., 2017). In addition, there are secondary stakeholders involved. Table 4 lists the primary and some secondary stakeholders and their roles and responsibilities. However, the roles may vary between countries, which should be considered when interpreting these for a specific country context. For example, some countries do not have tolling, and the division of responsibilities between different authorities may vary.

National road authorities, ministries and government are responsible for the operational environment for MaaS trough policy and regulation, and hence have a role as MaaS enablers. They create incentives and disincentives for MaaS, sustainable transport policy and new technological infrastructure, and allocate investments accordingly. They also prepare regulations and rules that can promote MaaS development. This may refer e.g. to reregulation of rules hindering MaaS implementation and opening up transport data. National regulations are partly based on EU regulations. European Commission Member States are setting up their National Access Points to "facilitate access, easy exchange and reuse of transport related data, in order to help support the provision of EU-wide interoperable travel and traffic services to end users" (European Commission, 2019). Regulations should

Table 4. Roles and responsibilities of stakeholders within MaaS ecosystems

Level	Stakeholder	Roles and Responsibilities
Primary Stakeholders		
National authorities	Government	National incentives and subsidies; legislation processes
	Ministries	Legislator; transport policy and strategies; infrastructure investments; funding
	Road authorities	Implementation of transport policy, strategy and investments; open transport data; permits; licenses; collaboration
	Infrastructure operator	Road network operations; tolling
Local and regional authorities	Regional transport agency	Regional public transport; locations of stations and stops
	Regional government	Land use planning; strategic planning and development
	Municipalities and cities	Transportation and traffic planning; municipal public transport; statutory transportation
Transport service providers	Municipal public transport	Schedules, ticketing and real-time information
	Long-haul public transport	Schedules, booking, ticketing and real-time information
	Bike sharing	Fares, locations and availability
	Car services	Fares, vehicle information, booking information, availability, locations
	Ride sharing	Drivers and rides platform
	Taxi	Fares, vehicle information, booking information
Logistics service providers	Logistics operators	Management of material flows between points of origin to end-use destination
Digital service providers	ICT and mobile service providers	Key enabling technology solutions, services and applications to MaaS operators and service providers
MaaS operators	MaaS company, public transport operator, PPP etc.	Combine existing transport services on the 'one-stop-shop' principle; customer service; service offering; service level; business model
Users	Citizens, customers	Communicate user needs and feedback; provide services ('prosuming')
Secondary Stakeholders		
Funding agencies	Private, public and R&D&I	Financial support and risk capital
OEMs and Resellers	Automotive industry	Fleet; technology; servitization
Academia	Universities, research institutes	Evaluation and impact assessment; new trends and knowledge; recommendations
Insurance companies	Private and business insurances	New insurances for MaaS

also ensure transparent and fair market conditions and competition. As indicated in Table 4, several stakeholders are involved in MaaS. Road authorities could promote collaboration between different stakeholders and transport modes towards common goals of MaaS.

Authorities also issue permits and licenses. These include basic permits and licenses for transport operators and drivers, but also for new trials of automated and robotized vehicles. The infrastructure operator is responsible for planning, constructing and maintaining the road infrastructure as well as tolling. Based on bilateral agreements, MaaS services may include special offers for tolling. In France, for example, SNCF's iDVROOM carpooling service offers a free automatic toll badge, no management costs, and a monthly downloadable invoice for their clients (iDVROOM, 2019).

National regulations and policy, including sustainability goals and incentives, are implemented by local and regional authorities at regional and municipal levels. Regions within a country may also be called e.g. counties, provinces and cantons. Local and regional authorities are responsible for transport planning, regional and municipal public transport and land use, as well as for creating preconditions for business. They are also responsible for organizing statutory transportation. This refers e.g. to school transportation, social and health service transportation and home care services. New and innovative public procurement processes enable organizing and combining different statutory transportation and public transport, which is especially important in sparsely populated areas. The role of local and regional authorities may strengthen in the future as they play a strategic role in a PPP MaaS ecosystem and even as a MaaS operator.

Local and regional authorities are responsible for opening public transport interfaces and data to enable travel chains and digital services. They provide locations of stations and stops, as well as stop names. Digital services use bus-stop names indicated by the authorities, which should therefore avoid misleading names and preferably use names that describe the location well.

Transport service providers include different transport modes and various services, including sharing and rental services. All TSPs are responsible for offering quality services, customer service and a good user experience of their own services. They also need to open data and Application Programming Interfaces (APIs) to enable reselling and integration of various services. In addition, every TSP is responsible for providing relevant information to its users; for e.g. public transport this includes schedules, fares and ticketing and real-time information. Long-haul buses and rail would also provide booking information. For car sharing and rental services, locations and availability of vehicles is important, as is docking station information for bike sharing. For ride sharing, a data platform for drivers and rides is essential to identify registered users and to match demand and supply. This also applies to on-demand transport. Taxis are increasingly using mobile applications and are part of the MaaS service offering. They provide fares, vehicle information and booking information.

Logistics service providers provide management of material flows. This may include, in addition to transportation, other logistics services such as warehousing and packaging. Logistics is seen as an add-on service to MaaS, not a primary operation. However, it is important to combine flows of people and goods when relevant to gain e.g. efficiency in rural areas.

Digital service providers are responsible for providing key enabling technology, services and applications to MaaS operators and transport service providers. This refers e.g. to mobile ticketing and payment, information and data services as well as different platforms, databases and open interfaces such as APIs. DSPs can also utilize intelligent and connected infrastructure (also known as C-ITS) as a MaaS element. DSPs offer technology services that enable MaaS and thus the integration of different services, travel chains, on-demand transport and the one-stop-shop principle. DSPs also provide the end-user interface, which is important for the user experience and perceived ease of the service.

The role of MaaS operator is a new take on traditional transport services and holds a key position. It is responsible for the service offering, business model, bilateral agreements with different service providers, and for combining services on a one-stop-shop principle. The MaaS operator is responsible for the service level of the entity of services offered. It can be a MaaS company for which MaaS is the main business – such MaaS Global which provides e.g. the Whim service in Helsinki – or a company for which it is a secondary business supporting the company's primary business. The public transport operator may also act as a MaaS operator. Thus, traditional public transport services are being complemented by new additional services. A good example is SNCF, the French national railway company, which proposes several first and last mile services to create the entire travel chain for their customers. In addition, municipal or regional public transport operators may complement their service offering with e.g. city bikes and shared vehicles. The PPP MaaS operator is considered suitable especially for sparsely populated areas where market-based solutions are not viable and different types of transport need to be combined to increase efficiency.

Users, aside from using MaaS services, are also expected to give feedback and communicate their needs in order to promote the development of the services. Users are also involved as 'prosumers', consuming and producing services and sharing their resources on a collaborative economy basis. Users could, for example, offer peer-to-peer carsharing and ride sharing of people and goods. Thus, PPP could be further developed into a PPPP model including the 'people' aspect. This is in line with the Swedish definition of public transport as 'passenger transport services by shared resources' (Laurell, 2017).

Secondary stakeholders support MaaS but not in a critical role. These include e.g. funding agencies, original equipment manufacturers (OEM) and resellers, academia and insurance companies. Funding agencies, including innovation and research funding, include both public and private agencies that provide financial support and risk capital for MaaS business solutions. As MaaS is a new concept requiring new business models, and consumer habits change slowly, time, persistence and funding are needed for an extensive MaaS implementation. Thus, funding opportunities are important for MaaS actors.

OEMs and resellers produce a sustainable fleet for MaaS, such as electric vehicles (EV). They also create new technology, for example keyless systems for shared vehicles. They promote new business models and servitization by participating in the development of new services such as car sharing, rental and leasing services.

Academia and research organizations provide neutral and valuable new information regarding e.g. trends and evaluation of MaaS. Impact assessment of MaaS services and pilots is needed to identify the potential impacts of MaaS on business, user and societal levels and the entire transport system. Research results and recommendations assist decision- and policy-makers and other stakeholders. Research also brings information and communication to the public about new developments and helps create trust in new transport services.

Insurance companies develop new insurance services for MaaS in collaboration with service providers, such as sharing, rental, peer-to-peer and ridesharing services. As an example, the peer-to-peer car rental service Blox Car offers, in collaboration with an insurance company, an insurance where there is no bonus loss in case of an accident during the rental period (Blox Car, n.d.). The French ride sharing company BlaBlaCar proposes a car insurance where clients can benefit from sharing rides (BlaBlaCar, n.d.).

DISCUSSION

MaaS is a new emerging concept with multiple stakeholders and a complex ecosystem. In these early stages, several pilots and some services already exist but tend to be restricted to limited geographical areas or transport modes or to a small number of mobility services. The level of integration is also generally quite low, and many existing services could be considered MaaS-related due to their low integration level. Broad ecosystems with PP(P)P involvement do not yet exist, thus there is no evidence of how they would work out, but the PP(P)P model is expected to have several benefits and pilots and development are underway towards this kind of collaboration.

The results of the conceptual PPPP network model and stakeholder roles and responsibilities are derived from qualitative data based on the knowledge and experience of experts representing different stakeholders of the MaaS ecosystem. The results are comprehensive, but may not cover all essential aspects, especially due to the early stages of MaaS and limited experience. Also, future MaaS development will show the applicability of the results. At the moment, PPP MaaS pilots are underway in Finland, and shared resources are seen as a vital part of future public transport in Sweden, supporting these results. Even though the results are based on work mainly with Finnish stakeholders and partly with experts from other European countries (e.g. Sweden and Austria), the results are also relevant and applicable to other countries. For example, in the U.S., similar stakeholders to those listed in Table 4 have been identified related to the Mobility on Demand (MOD) concept (Shaheen, Cohen, Yelchuru, & Sarkhili, 2017). However, countries and rural areas even within the same country have differences and specific characteristics, which should be considered when interpreting the results. For example, in Finland, data network coverage is comprehensive also in rural areas, enabling the widespread use of digital services. Internationally, Finland is leading the way in mobile data usage (see e.g. McCarthy, 2017). In some countries, inadequate data network coverage may limit digital service deployment, especially in rural areas.

The vision is to see the whole transport sector as a cooperative, interconnected ecosystem, providing services based on the needs of stakeholders, especially customers, and seamlessly combining different transport modes such as private vehicles, public and collective transport, biking and walking (Eckhardt et al., 2017a). This requires a general understanding of the roles and responsibilities of each ecosystem stakeholder. In addition to understanding, actions need to occur. Despite the definition of roles, it is not always obvious which stakeholder takes responsibility for holistic MaaS development and implementation. Public stakeholders might create better preconditions, for instance through regulation, and expect the business sector then to lead the development. However, in the early stages MaaS usually requires financial support even in cities. In addition, the responsibility for MaaS development might be shunted between authorities. If large and sustainable MaaS ecosystem development is pursued, the public sector has to take a strong role and lead the integration of public and private collaboration.

The public sector could start their mobility service development by digitalizing transport data and integrating and rationalizing public and statutory transport services. In sparsely populated areas, this might mean the use of demand-responsive transport and combining different user groups as well as people and goods. Later on, more services could be integrated, such as commercial and shared mobility services.

Municipalities could also collaborate to increase efficiency and learn from each other. Public procurement defines the preconditions for transport services. Procurement of new mobility services is challenging and needs further research in order to propose good practices and guidance regarding collaboration models, innovative procurement and ensuring technological openness and scalability.

The PPP and PPPP models can potentially reduce emissions, bring remarkable cost savings to the public sector and huge benefits to users. More personalized mobility services could help attract new customers by facilitating door-to-door transport. The benefits of a PP(P)P network could be more remarkable in rural areas where the transport service offering is usually limited. Rural areas could be more vital and accessible when integrating subsidized transport and shared private resources with market-based mobility services. This kind of integration would bring more variety to the service offering and better use of available vehicle capacity.

Shared and integrated mobility services require technology and digitalization. It is challenging currently to get all the relevant actors of the MaaS ecosystem to digitalize their transport data in order to enable interoperability. Also, as mentioned previously, data network coverage may hinder the development of MaaS especially in rural areas. Other possible challenges are dependency on the public sector's resources, skills and willingness to develop MaaS and procure new services. Bringing in the people aspect may also cause challenges. It is uncertain to what extent people are willing to adopt these services, both as producers and consumers. Also, if the service offering in rural areas depends a lot on prosumers offering e.g. ridesharing, peer-to-peer carsharing and organized hitchhiking, the provision of services might be occasional, and a sufficient service offering may not be guaranteed at all times.

Due to the potential benefits of the PPPP model, it will be likely introduced in rural areas in the near future. Like other MaaS pilots and services, it will probably start only with limited geographical areas and services. The speed and success of large-scale implementation is uncertain. Even though public-private collaboration is expected to become increasingly popular in cities too, in rural areas PPP is a necessity. In cities, it is possible for the business sector to offer viable services without public involvement. However, PPP is also beneficial in cities for the integration of city policies with mobility services.

More research on PPP and PPPP models is needed in relation to both MaaS in rural areas and cities. As users usually change their behavior slowly, it should be ensured that the duration of the pilots is long enough. Otherwise, it could be falsely concluded that there is not enough demand for new mobility services, as users have not had time to adopt new habits. Marketing of new pilots and services is important in order to raise awareness of new services. Marketing is also essential to have enough users to obtain reliable and possibly encouraging results for the continuation of services. Based on pilots and other MaaS experiences, good practices should be widely disseminated and scaled to other areas and more extensive service integration. Impact assessments of PP(P)P MaaS services should be performed to assist decision-making.

CONCLUSION

This chapter set out to explore potential PPPP networks for rural mobility and the stakeholders involved in MaaS ecosystems. A PPPP model was proposed which combines market-based and subsidized transport services on a one-stop-shop principle. Due to long distances, sparse population and low utilization and occupancy rates of vehicles in rural areas, *combining different mobility services of people and goods, as well as user groups, is crucial to improving efficiency and enabling viable services*. Public actors may combine various types of stakeholders and services in a MaaS system, which can rationalize the services on public actors' responsibility, e.g. statutory social and health service transportation and freight/delivery. *Public actor involvement is required where market-based public transport and other transport services are not viable*. Statutory transport services and shared private resources form a basis for mobility services in sparsely populated areas. Thus, *the PPPP model has huge potential for future rural mobility*. The involvement of public actors will likely lead to integration of societal goals and thus could take MaaS to the highest level of integration according to the classification of Sochor et al. (2018). This classification does not take freight into account. The mobility of goods can be considered similar to the mobility of people in the classification, except that rural MaaS can be more a matter of combining different services and user groups than modes, since accessibility is the key concern for people and services regarding MaaS. Rural MaaS could be defined as *mobility of people and goods on a one-stop-shop principle consisting of various user groups and integrated transport services*.

Twenty stakeholders were identified in MaaS ecosystems, of which sixteen are primary stakeholders and four are secondary stakeholders. As primary stakeholders are crucial for business, it can be concluded that *MaaS is a complex system requiring collaboration of a large number of stakeholders in order to create extensive and integrated MaaS*. Primary stakeholders are authorities, service providers, MaaS operators and users. They all have an important role and they should be aware of their roles and interaction between stakeholders. Authorities create preconditions for MaaS by setting legislation, regulations and incentives to better enable new mobility services. The MaaS operator is a new player, who in collaboration with service providers integrates a variety of services to offer them on a one-stop-shop principle. *The role of the MaaS operator is crucial, as it in the last resort determines how attractive MaaS services are for users*. The MaaS operator designs the service offering, e.g. services included, geographical coverage, pricing, payment system (subscription/pay-as-you-go) and user interface. It could be considered that good quality MaaS services are enough to boost MaaS. However, some curiosity and flexibility are required from users to change their behavior and try new services. Especially in rural areas where the mobility service offering is poor, *users can also have a remarkable role as prosumers* (producer and consumer) by offering e.g. peer-to-peer, carpooling and organized hitchhiking services. Secondary stakeholders are also important for promoting MaaS e.g. by enabling financing, developing new service components and providing information for decision- and policy-making.

The future of MaaS is uncertain, as MaaS is in its early stages. Regardless of how the MaaS concept will develop, features for future mobility include collaboration on various levels, implementing new mobility services and utilizing digitalization. These new services refer to e.g. travel chains, demand-responsive transport, carpooling/ridesharing, sharing services (car, bike, scooter…), peer-to-peer rental services, shared taxi rides, ridesourcing/ridehailing, organized hitchhiking, and combining the mobility of people and goods, as well as combining publicly subsidized and market-based transport services. *Digitalization is the key to integrating different transport services*, and the transport sector is actively promoting digital open data, open interfaces and other technological development. Above all, *integrated MaaS services require collaboration between relevant stakeholders in the MaaS ecosystem, even between rivals.*

ACKNOWLEDGMENT

This research was supported by the Conference of European Directors of Roads (CEDR) Transnational Road Research Programme and the Development Fund of the Ministry of Agriculture and Forestry of Finland. The authors gratefully acknowledge the funding organizations and researchers contributing to the projects.

REFERENCES

Aapaoja, A., Eckhardt, J., Nykänen, L., & Sochor, J. (2017). MaaS service combinations for different geographical areas. *Proceedings of the 24th World Congress on Intelligent Transportation Systems.*

Aapaoja, A., & Haapasalo, H. (2014). A framework for stakeholder identification and classification in construction projects. *Open Journal of Business and Management, 2*(1), 43–55. doi:10.4236/ojbm.2014.21007

Adner, R. (2006). Match your innovation strategy to your innovation ecosystem. *Harvard Business Review, 84*(4), 98–107. PMID:16579417

BlaBlaCar. (n.d.). *BlaBlaCar Une nouvelle façon d'économiser grâce à votre assurance auto.* Retrieved from https://assurance.blablacar.fr/

Blox Car. (n.d.). *Blox Car How it works.* Retrieved from https://bloxcar.fi/how-it-works

Burrows, A., Bradburn, J., & Cohen, T. (2015). *Journeys of the Future. Introducing Mobility as a Service.* London: Atkins.

Clark, B., Chatterjee, K., & Melia, S. (2016). Changes in level of household car ownership: The role of life events and spatial context. *Transportation, 43*(4), 565–599. doi:10.100711116-015-9589-y

Clarkson, M. B. E. (1995). A Stakeholder Framework for Analyzing and Evaluating Corporate Social Performance. *Academy of Management Review, 20*(1), 92–117. doi:10.5465/amr.1995.9503271994

Cleland, D. I. (1986). Project stakeholder management. *Project Management Journal, 17*(4), 36–44.

Eadie, R., Millar, P., & Toner, L. (2013). Public private partnerships, reevaluating value for money. *International Journal of Procurement Management, 6*(2), 152–169. doi:10.1504/IJPM.2013.052467

Eckhardt, J., Aapaoja, A., Nykänen, L., Sochor, J., Karlsson, M., & König, D. (2017a). *Deliverable 2: European MaaS Roadmap 2025.* MAASiFiE project funded by CEDR. Retrieved from https://www.vtt.fi/sites/maasifie/PublishingImages/results/cedr_mobility_MAASiFiE_deliverable_2_revised_final.pdf

Eckhardt, J., Nykänen, L., Aapaoja, A., & Niemi, P. (2017b). *Liikkumispalvelut maaseudun elinvoimaisuuden ja saavutettavuuden mahdollistajana* [Mobility as a Service as an enabler for vitality and accessibility of rural areas]. VTT research report VTT-R-03429-17. Retrieved from https://www.vtt.fi/sites/maaseutumaas/Documents/MaaseutuMaaS_raportti_revised_final2.pdf

Eckhardt, J., Nykänen, L., Aapaoja, A., & Niemi, P. (2018). MaaS in rural areas - case Finland. *Research in Transportation Business & Management, 27*, 75–83. doi:10.1016/j.rtbm.2018.09.005

Eesley, C., & Lenox, M. J. (2006). Firm Responses to Secondary Stakeholder Action. *Strategic Management Journal, 27*(8), 765–781. doi:10.1002mj.536

ERTICO. (2018). *Mobility as a Service - from modes to mobility*. Brussels: ERTICO. Retrieved from https://ertico.com/wp-content/uploads/2018/09/Ertico_MaaS-vision-paper-2018_www.pdf

European Commission. (2016). A European Strategy for Low-Emission Mobility (COM(2016) 501 final). Brussels: European Commission.

European Commission. (2017a). An agenda for a socially fair transition towards clean, competitive and connected mobility for all (COM(2017) 283 final). Brussels: European Commission.

European Commission. (2017b). *Smart, green and integrated transport, Horizon 2020 work programme 2016-2017*. Retrieved from https://ec.europa.eu/research/participants/data/ref/h2020/wp/2016_2017/main/h2020-wp1617-transport_en.pdf

European Commission. (2019). *National Access Points*. Retrieved from https://ec.europa.eu/transport/themes/its/road/action_plan/nap_en

European PPP Expertise Centre. (2015). *PPP Motivations and Challenges for the Public Sector*. Luxembourg: European Investment Bank.

European PPP Expertise Centre. (2018). *Market Update. Review of the European PPP Market in 2017*. Luxembourg: European Investment Bank.

European Union. (2018). *Public Private Partnerships in the EU: Widespread shortcomings and limited benefits* (Special report No 9/2018). Luxembourg: European Court of Auditors.

Finnish Transport Agency. (2017). Ylläs Around impact study (in Finnish). Research reports of the Finnish Transport Agency 31/2017. Helsinki: Finnish Transport Agency.

Freeman, E. R. (1984). *Strategic Management: A Stakeholder Approach*. New York, NY: Pitman Publishing Inc.

Håkansson, H., & Johanson, J. (1992). A model of industrial networks. In B. Axelsson & G. Easton (Eds.), *Industrial Networks: A New View of Reality* (pp. 28–34). London: Routledge.

Hallock, L., & Inglis, J. (2015). *The Innovative Transportation Index: The Cities Where New Technologies and Tools Can Reduce Your Need to Own a Car*. Frontier Group.

Harms, L., Durand, A., Hoogendoorn-Lanser, S., & Zijlstra, T. (2018). *Exploring Mobility-as-a-Service: Insights from Literature and Focus Group Meetings*. Den Haag: Ministry of Infrastructure and Water Management.

Heikkilä, S. (2014). *Mobility as a Service – A Proposal for Action for the Public Administration: Case Helsinki* (Master's Thesis). Retrieved from Aaltodoc database: http://urn.fi/URN:NBN:fi:aalto-201405221895

Hietanen, S. (2014). 'Mobility as a Service' – the new transport model? ITS & Transport Management Supplement. *Eurotransport*, *12*(2), 2–4.

Iansiti, M., & Levien, R. (2004). Strategy as ecology. *Harvard Business Review*, *82*(3), 68–78. PMID:15029791

International Transport Forum. (2015). *International Experiences on Public Transport Provision in Rural Areas*. Paris: International Transport Forum.

iVDROOM. (2019). *iVDROOM Télépéage offert*. Retrieved from https://lp.idvroom.com/telepeage-gratuit-covoiturage/

Jin, Z. R., & Qiu, A. Z. (2019). Mobility-as-a-Service (MaaS) Testbed as an Integrated Approach for New Mobility - A Living Lab Case Study in Singapore. In H. Krömker (Ed.), *HCI in Mobility, Transport, and Automotive Systems* (pp. 441–458). Cham: Springer; doi:10.1007/978-3-030-22666-4_32

Jittrapirom, P., Caiati, V., Feneri, A.-M., Ebrahimigharehbaghi, S., Alonso-González, M. J., & Narayan, J. (2017). Mobility as a Service: A Critical Review of Definitions, Assessments of Schemes, and Key Challenges. *Urban Planning*, *2*(2), 13–25. doi:10.17645/up.v2i2.931

Kamargianni, M., Li, W., & Matyas, M. (2016). A Comprehensive Review of "Mobility as a Service" Systems. *Transportation Research Board 95th Annual Meeting 2016*, 16-5401.

Karlsson, M., Sochor, J., Aapaoja, A., Eckhardt, J., & König, D. (2017). *Deliverable 4: Impact Assessment*. MAASiFiE project funded by CEDR.

Kinnunen, T., Aapaoja, A., & Haapasalo, H. (2014). Analysing internal stakeholders' salience in product development. *Technology and Investment*, *5*(2), 106–115. doi:10.4236/ti.2014.52011

König, D., Eckhardt, J., Aapaoja, A., Sochor, J., & Karlsson, M. (2016a). *Deliverable 3: Business and operator models for MaaS*. MAASiFiE project funded by CEDR. Retrieved from https://www.vtt.fi/sites/maasifie/PublishingImages/results/cedr_mobility_MAASiFiE_deliverable_3_revised_final.pdf

König, D., Sochor, J., Eckhardt, J., & Böhm, M. (2016b). State-of-the-art survey on stakeholders' expectations for Mobility-as-a-Service (MaaS). In *Proceedings of the 23rd World Congress on Intelligent Transport Systems*. Melbourne: ITS Australia.

Laurell, A. (2017). *Swedish Mobility Program (SMP): Den avslutande rapporten för projektet Vitt papper och samtidigt ett förarbete till Swedish Mobility Program (SMP)*. Samtrafiken.

Leviäkangas, P., Aapaoja, A., Kinnunen, T., Pilli-Sihvola, E., Hautala, R., & Zulkarnain, Z. (2014). The Finnish road weather business ecosystem: Turning societal benefits into business and the other way round. *Engineering Management Research*, *3*(1), 56–67. doi:10.5539/emr.v3n1p56

Li, B., Akintoye, A., Edwards, P., & Hardcastle, C. (2005). Perceptions of positive and negative factors influencing the attractiveness of PPP/PFI procurement for construction projects in the UK: Findings from a questionnaire survey. *Engineering, Construction, and Architectural Management*, *12*(2), 125–148. doi:10.1108/09699980510584485

Lyons, G., Hammond, P., & Mackay, K. (2019). The importance of user perspective in the evolution of MaaS. *Transportation Research A. Policy & Practice*, *121*, 22–36. doi:10.1016/j.tra.2018.12.010

MaaS Alliance. (2017). *White paper. Guidelines & Recommendations to Create the Foundations for a Thriving MaaS Ecosystem*. Brussels: MaaS Alliance.

MAASiFiE. (2016). *About MaaS concept and MAASiFiE project*. Retrieved from http://www.vtt.fi/sites/maasifie/

Majamaa, W. (2008). *The 4th P - People - in Urban Development Based on Public-Private-People Partnership* (Doctoral dissertation). Retrieved from Aaltodoc database: http://urn.fi/URN:ISBN:978-951-22-9585-2

Marana, P., Labaka, L., & Sarriegi, J. M. (2018). A framework for public-private-people partnerships in the city resilience-building process. *Safety Science Part C, 110*, 39–50. doi:10.1016/j.ssci.2017.12.011

Matkahuolto. (n.d.). Retrieved from: https://matkahuolto.fi/

McCarthy, N. (2017). Mobile Data Subscriptions: Which countries use the most gigabytes? *Forbes, 11*(10), 2017. Retrieved from https://www.forbes.com/sites/niallmccarthy/2017/10/11/mobile-data-subscriptions-which-countries-use-the-most-gigabytes-infographic/#2fd00e997220

Moore, J. F. (1993). Predators and prey: A new ecology of competition. *Harvard Business Review, 71*(3), 75–86. PMID:10126156

National Academies of Sciences, Engineering, and Medicine. (2016). *Shared Mobility and the Transformation of Public Transit*. Washington, DC: The National Academies Press. doi:. doi:10.17226/23578

Open Arctic MaaS. (n.d.). Retrieved from: https://www.arcticmaas.fi/

Organisation for Economic Cooperation and Development. (2012). *Recommendation of the Council on Principles for Public Governance of Public-Private Partnerships*. OECD.

Peltoniemi, M., & Vuori, E. (2004). Business ecosystem as the new approach to complex adaptive business environments. In *Proceedings of eBRF 2004*. Tampere, Finland: Tampere University of Technology and University of Tampere.

Polis. (2017). *Mobility as a Service: Implications for Urban and Regional Transport*. Brussels: Polis.

Rezo Pouce. (n.d.). *Rezo Pousse Médias et Presse*. Retrieved from https://www.rezopouce.fr/page/Medias_et_Presse

Rubenstein-Montano, B., Liebowitz, J., Buchwalter, J., McCaw, D., Newman, B., & Rebeck, K. (2001). A systems thinking framework for knowledge management. *Decision Support Systems, 31*(1), 5–16. doi:10.1016/S0167-9236(00)00116-0

Särkijärvi, J., Jääskeläinen, S., & Lohko-Soner, K. (Eds.). (2018). *Action programme for carbon-free transport 2045: Final report by the Transport Climate Policy working group (Publications of the Ministry of Transport and Communications, 13/2018)*. Helsinki: Ministry of Transport and Communications of Finland.

Shaheen, S., Cohen, A., Yelchuru, B., & Sarkhili, S. (2017). *Mobility on Demand Operational Concept Report*. Washington, DC: U.S. Department of Transportation. Retrieved from https://rosap.ntl.bts.gov/view/dot/34258

Shaheen, S., Cohen, A., & Zohdy, I. (2016). *Shared Mobility: Current Practices and Guiding Principles*. Washington, DC: U.S. Department of Transportation.

Smith, G., Sarasini, S., Karlsson, I. C. M., Mukhtar-Landgren, D., & Sochor, J. (2019a). Governing Mobility-as-a-Service: Insights from Sweden and Finland. In M. Finger & M. Audouin (Eds.), The Governance of Smart Transportation Systems. Cham: Springer. doi:10.1007/978-3-319-96526-0_9

Smith, G., Sochor, J., & Karlsson, I. C. M. (2019b). Public–private innovation: Barriers in the case of mobility as a service in West Sweden. *Public Management Review*, *21*(1), 116–137. doi:10.1080/14719037.2018.1462399

Sochor, J., Arby, H., Karlsson, M., & Sarasini, S. (2018). A topological approach to Mobility as a Service: A proposed tool for understanding requirements and effects, and for aiding the integration of societal goals. *Research in Transportation Business & Management*, *27*, 3–14. doi:10.1016/j.rtbm.2018.12.003

Sochor, J., Karlsson, M., & Strömberg, H. (2016). Trying Out Mobility as a Service: Experiences from a Field Trial and Implications for Understanding Demand. *Transportation Research Record: Journal of the Transportation Research Board*, *2542*(1), 57–64. doi:10.3141/2542-07

The Finnish Innovation Fund Sitra. (n.d.). *Public and private mobility services to be offered side by side*. Retrieved from https://www.sitra.fi/en/projects/public-private-mobility-services-offered-side-side/#what-is-it-about

Transport Systems Catapult. (2016). *Exploring the opportunity for Mobility as a Service in the UK*. Milton Keynes, UK: Transport Systems Catapult.

Van Ham, H., & Koppenjan, J. (2001). Building Public-Private Partnerships: Assessing and managing risks in port development. *Public Management Review*, *3*(4), 593–616. doi:10.1080/14616670110070622

Whim. (n.d.). *Whim Find your plan*. Retrieved from https://whimapp.com/plans/

Winch, G. M., & Bonke, S. (2002). Project stakeholder mapping: analyzing the interest of project stakeholders. In D. P. Slevi, D. I. Cleland, & J. K. Pinto (Eds.), *The frontiers of project management research. Newton Square Pennsylvania*. Project Management Institute.

Yin, R. K. (2003). *Case Study Research: Design and Methods* (3rd ed.). Beverly Hills, CA: SAGE Publications.

ADDITIONAL READING

Eckhardt, J., Aapaoja, A., Nykänen, L., Sochor, J., Karlsson, M., & König, D. (2018). The European Roadmap 2025 for Mobility as a Service. In Proceedings of 7th Transport Research Arena TRA 2018. Vienna.

International Association of Public Transport. (2019). Mobility as a Service. Brussels: International Association of Public Transport (UITP). Retrieved from https://www.uitp.org/sites/default/files/cck-focus-papers-files/Report_MaaS_final.pdf

Jeekel, H. (2017). Social Sustainability and Smart Mobility: Exploring the relationship. *Transportation Research Procedia*, *25*, 4296–4310. doi:10.1016/j.trpro.2017.05.254

Kamargianni, M., Li, W., Matyas, M., & Schäfer, A. (2016). A critical review of new mobility services for urban transport. *Transportation Research Procedia*, *14*, 3294–3303. doi:10.1016/j.trpro.2016.05.277

Mulley, C. (2017). Mobility as a Services (MaaS) – does it have critical mass? *Transport Reviews*, *37*(3), 247–251. doi:10.1080/01441647.2017.1280932

National Academies of Sciences, Engineering, and Medicine. (2016). *Between Public and Private Mobility: Examining the Rise of Technology-Enabled Transportation Services*. Washington, DC: The National Academies Press. doi:10.17226/21875

Sochor, J., Strömberg, H., & Karlsson, I. C. M. (2015). Implementing Mobility as a Service: Challenges in Integrating User, Commercial, and Societal Perspectives. *Transportation Research Record: Journal of the Transportation Research Board*, *4*(2536), 1–9. doi:10.3141/2536-01

Strömberg, H., Karlsson, I. C. M., & Sochor, J. (2018). Inviting travelers to the smorgasbord of sustainable urban transport: Evidence from a MaaS field trial. *Transportation*, *45*(6), 1655–1670. doi:10.100711116-018-9946-8

KEY TERMS AND DEFINITIONS

Collaborative Economy: Sharing, lending/borrowing, or renting products (e.g., vehicles) without the need to own them. Also participating in services together. A way to increase resource efficiency assisted by digital platforms.

Demand-Responsive Transport: A transport service that operates according to client needs in a specific zone and time slots without predefined routes or schedules.

Mobility as a Service: A concept that integrates different transport services, such as public transport, sharing services and taxi. Services are offered on a one-stop-shop principle utilizing digitalization and including e.g. multimodal route planning and mobile ticketing and payment. Additional services may also be included, such as freight.

Peer-to-Peer: Private people offering services to other private people.

Prosumer: A consumer who also produces services. Regarding transport services, prosuming may refer, for example, to peer-to-peer carsharing, ridesharing of people and goods, and organized hitchhiking.

Public-Private-People Partnership: A form of collaboration between public and private sectors as well as users ('people'). In the context of transport services, 'people' can be considered as prosumers.

Ridesharing: Private people offering rides or goods deliveries to other private people.

Ridesourcing: A transport service based on smartphone application that connects passengers and drivers. Instead of commercial vehicles, drivers typically use their own car.

Statutory Transportation: A transportation service that according to legislation has to be offered to all citizens fulfilling certain criteria. This may refer to reduced mobility capabilities of handicapped and elderly people.

Subsidized Transportation: A transportation service that is not fully market based and receives financial support from the public sector.

Chapter 3
Potential User Groups of Mobility as a Service in Finland

Timo Liljamo
Tampere University, Finland

Heikki Liimatainen
Tampere University, Finland

Markus Pöllänen
https://orcid.org/0000-0001-6534-8380
Tampere University, Finland

Roni Utriainen
Tampere University, Finland

Riku Viri
https://orcid.org/0000-0002-4931-2182
Tampere University, Finland

ABSTRACT

With a mobility as a service (MaaS) offering, the transport needs of the end customer may be fulfilled in a way, which may challenge the car dominant mobility practices of today. Up to now, there has been few studies focusing on the interest of end users towards MaaS. This chapter presents results from a survey in Finland (N=1,176) and focuses on the interest towards MaaS services among 18-64 year olds and especially the potential of MaaS in different user groups. The authors found that particularly the respondents who currently use public transport, the respondents living in households without a car, the respondents aged 25-34, and the respondents with higher education were more interested to adopt MaaS services. They also found that people living in urban areas and in apartment houses indicated a higher interest towards MaaS. The same groups had also more often no need or will for car ownership if a competitive MaaS service would be available. Continuing urbanisation offers increasing potential for MaaS as there are many mobility services, which can be bundled into a MaaS offerings in cities.

DOI: 10.4018/978-1-7998-1614-0.ch003

INTRODUCTION

In the developed countries, automobiles have grown to be the dominant mode of transport since their launch in late 1800s. During the 20th century, the number of passenger cars as well as passenger car mileage grew to support the lifestyles of the modern society. Together with increasing number of cars and kilometres travelled, problems associated with cars have increased to a state where options for this mode are asked for more and more. Could we own and use cars less, yet still fulfil our daily mobility needs? As one suggested model, Mobility as a Service, MaaS, offers an option by aiming to make travelling with different modes easy and fluent. This way MaaS could compete with the current car-based system. As Heikkilä (2014) in one of the first studies on MaaS described, Mobility as a Service can be seen as a system, where mobility operators provide a comprehensive range of mobility services to customers. MaaS can also be considered as a solution for more sustainable mobility, especially as it could replace car-based mobility. König et al. (2016) defined MaaS as "multimodal and sustainable mobility services addressing customers' transport needs by integrating planning and payment on a one-stop-shop principle".

The aim of this chapter is to analyse people's interest towards MaaS, and which user groups find MaaS offerings tempting and are the most potential MaaS customers. The potential is analysed based on a survey with a representative sample of 6,000 Finns. By combining the findings from this survey with previous studies and Finnish data on demographics and mobility patterns, the potential is further analysed. This chapter aims to answer the following questions: what is the interest towards MaaS among ordinary people, which are the most potential user groups of MaaS, and how do current trends in demographics and mobility affect the potential of MaaS. This paper strives to contribute to recognising current attitudes of ordinary people towards MaaS and the potential user groups of MaaS. Current attitudes towards MaaS and the potential user groups are in the interests of different stakeholders working on mobility services and MaaS related issues from companies to scientific community as well as authorities from local level to national level.

Since Mobility as a Service is a new concept, the number of studies on MaaS and its potential is still low. Kamargianni et al. (2015) analysed the potential of MaaS in the context of a large city as they identified supply and demand of transport services London, UK, and presented a MaaS-London concept for the future. Additionally, related to MaaS-London concept, Matyas & Kamargianni (2017) carried out stated preference study to understand people's mobility choices under MaaS, and Kamargianni et al. (2018) studied Londoners' attitudes towards car ownership and MaaS. Li & Voege (2017) studied the challenges of MaaS from the point of view of conditions in the city (availability of different modes, e-tickets, open data, and e-payment options).

Reflecting different aspects of MaaS potential, Giesecke et al. (2016) presented four key conceptual issues of MaaS, namely 1) nature of travel (e.g. trip purpose, trip length, mode, and means), 2) interoperability, 3) end-user behaviour related to e.g. different user groups, gender issues, and aging, and 4) sustainability. In this chapter, we will mostly focus on the end-users, but also discuss the nature of travel and sustainability related to MaaS. Giesecke et al. (2016) emphasised a strong focus on the end-user perspective, in which user group segments as well as user acceptance criteria, based on user attitudes and behaviour, are particularly important.

Next, in section 2, previous research findings are discussed related to the potential users and user groups of MaaS. In section 3, the original survey is presented, by which end-user perspectives on MaaS were asked. Thereafter, section 4 presents the findings of this survey, and in section 5, these findings will then be discussed in relation to previous studies and other data. Section 6 discusses the reliability of the results. Finally, section 7 presents the conclusions.

PREVIOUS RESEARCH ON THE POTENTIAL USER GROUPS OF MAAS

There are only a few publications in which the potential of MaaS has been studied from end-user perspective. We have not found similar MaaS-related studies as analysed in this paper with a large survey sample. Kamargianni et al. (2018) studied Londoners' attitudes towards MaaS and found that some demographic groups were more interested in MaaS. They also found that the characteristics of MaaS had an impact on people's willingness to adopt these services. Based on the survey responses of about 1,200 Londoners, 55% of the participants who were under 30 years old would utilise MaaS if it lowered mobility costs due to bundled mobility services, whereas 35% of over 49 years old would utilise MaaS for the same reason.

Thc study by Kamargianni et al. (2018) also found that MaaS could strengthen use of sustainable transport modes as 28% of regular public transport users would use more public transport. In addition, 35% of regular car users would use more public transport and 17% of car users would bicycle more. In conclusion, MaaS was expected to affect mostly people under 30 years old, while the effects on mobility patterns of older than 50 years old were expected to be the lowest.

ITS Australia (2018) also found out based on a MaaS-related survey with 4,000 Australian participants that younger people and especially men and educated people were most likely to become MaaS users. People living in rural areas, not college educated and rarely using public transport were the most likely people not to utilise MaaS.

Transport Systems Catapult (2016) stated that there will be no one MaaS model that fits all. This holds both from the business perspective, as there is a great number of different stakeholders, which make the models probably different, as well as from the user perspective as the service should align with the lifestyles of different customers. It may be challenging for the MaaS operators to introduce attractive value offerings for a significant proportion of the consumer market. However, the size of the mobility market and the diversified mobility needs offer potential for MaaS model. As the early adopters, new millennials are likely MaaS consumers.

A MaaS package can be a solution for household's mobility, and fulfil not only adults', but also children's mobility needs. Another important point in the MaaS business model is that the MaaS operator helps the transport operators improve their service by sharing the data on customers' mobility needs (Transport Systems Catapult 2016).

Men and high-income groups are estimated to have a positive attitude towards new technology (El Zarwi et al. 2017), which could also be a supporting issue for MaaS. Becker et al. (2017a, 2017b) studied car-sharing schemes in Switzerland and found that the most potential user group in car-sharing was young and highly educated people.

MaaS trials and commercial schemes are also a way to assess the potential of MaaS. Strömberg et al. (2016) found that trials provide a safe space for experimenting new behaviour, and this may lead to participants engaging behaviour that they would probably not do without the trial. Karlsson et al. (2016) studied how a MaaS trial, UbiGo, in Gothenburg, Sweden responded to user needs, and found that the customers were more satisfied during the trial compared with their previous travel solution. Up to now, the number of MaaS pilots has been low, and the findings related to commercial MaaS schemes have not yet been widely published.

In a survey carried out in April 2017, only 16 per cent of Finns had heard of MaaS. However, in the same survey, 80 per cent of respondents saw it necessary to able to travel the whole trip chain with the same ticket, which indicates a need for MaaS solutions. Over half of respondents under 30 years of age were willing to give up car ownership if transport services would be adequate. The willingness to pay for a monthly package covering all the mobility needs was relatively low as the respondents were willing to pay less than 150 euros a month on the average, although 10% of the respondents were willing to pay more than 400 euros (Solita 2017). From MaaS provider's perspective, this 10% is a very lucrative market segment. Additionally, the young willing to abstain from car ownership, is a highly potential customer group for MaaS services.

DATA AND METHODS

In April - June 2018, a survey was conducted in Finland on the attitudes of 18-64-year-old towards transport system and people's readiness and will to use new transport services like MaaS. The survey was conducted in Finnish. English translation of the questionnaire is presented in appendix A.

The questionnaire consisted of five parts. Total number of questions was 23 and it took about 10-15 minutes to complete the survey. The questions consisted of propositions with Likert scale 1 to 5, multiple choice, and open-ended questions. The survey also mapped the respondent's background information, such as age, gender, and whether the respondent had a driving licence, to help categorise the respondents.

In this study, survey's questions number 7 and 11c are used to identify the most potential user groups of MaaS services by cross tabulating answers to these questions with data related to questions 5, 12, 13, 14, 15, 16, 18, 19 and 21 on user background (age, gender, driving licence, education, car ownership, household size, type of house, the use of car and public transport). In question 7, an assumption was presented that a suitable mobility package (basically a MaaS offering) would be able to fulfil respondent's all transport needs, and related to this assumption, the respondents were asked to state a suitable price to start using such a package. The respondents had the option to answer 'I do not know' and 'I would not adopt the package'. In question 11c, the respondents were asked whether the respondent had the need or will to own an own private car in future, if the annual costs of the mobility package (including e.g. public transport, shared taxis, and shared cars) would be markedly lower and one could fulfil all mobility needs with the mobility package.

Additionally, we will present and analyse responses to question 1, which asked whether the respondent had heard or read of the concept 'Mobility as a service' before answering the survey. This question was only to gather interest, whether the term MaaS is widely known or not, or whether it has an effect to the interest of adoption of the packages. The actual questions about the person's willingness to use mobility offerings were formulated in a way that they could provide answer without prior knowledge about MaaS.

The sample group for the survey was randomly selected from the population register by age and gender group in proportion to population's age and gender distribution. Population Register Centre of Finland extracted the sample per procuration. For study purposes, the names, addresses, ages and gender of the people were available in the sample data. Throughout our study, we followed Population Register Centre's terms and conditions related to data use, and in analysing the survey responses, we followed Finnish legislation and good academic practices.

Table 1. Sample group, number of respondents and response rate by age and gender

Age Groups	Men			Women		
	Sample Size	Number of Respondents	Response Rate	Sample Size	Number of Respondents	Response Rate
18-24	415	50	12.0%	397	59	14.9%
25-34	666	86	12.9%	632	138	21.8%
35-44	662	107	16.2%	626	115	18.4%
45-54	635	111	17.5%	621	131	21.1%
55-64	663	176	26.5%	683	186	27.2%
In total	**3,041**	**530**	**17.4%**	**2,959**	**629**	**21.3%**

The population of the sample consisted of 18-64-year-olds living in Finland. The size of the population was 3,263,361 people, based on the population projection of 2018 (Statistics Finland 2015). The sampling fraction was 0.2%, and the size of the sample was 6,000 people. Table 1 presents the sample, number of respondents, and response rate in terms of the various age and gender groups.

All people, who were selected for the sample, were sent a postal questionnaire with a cover letter in April 2018. People had the option either to respond on the internet with an identification code or to send the questionnaire back by mail (postage paid). The last answers were received in June 2018. A total of 1,176 respondents participated in the survey, resulting to a total response rate of 19.6%. The mean age of all respondents was 45 years, for women 44 and men 45 years, respectively.

We present the survey results as the number of respondents and frequency percentages of different responses. We do not present the results with sampling factors, which would make the results representative of the Finnish 18-64-year-old population as our focus in the analysis is in the differences between user groups, and the parameters usually deployed in sampling factors (age and gender) are analysed separately.

Regarding the background information gathered with the survey, we found the respondents to represent the Finnish population relatively well e.g. in terms of driving licences. 94% of the respondents held a driving licence, which is slightly more than Finnish population of 18–64 years old in general, of which 89% held a driving licence on 1.1.2018, based on statistics from Finnish Transport Safety Agency (2018) and Statistics Finland (2018). 16% of the respondents stated to be from a household without a car, whereas 43% had one car, 33% two cars, 6% three cars, and 2% four or more cars in their household. These figures can be compared to average Finnish households, of which 26% were without a car, 54% had one car, 17% two cars, and 3% three or more cars in 2016 (Statistics Finland 2016b). It must be noted that the respondents of our survey did not include over 65 years old people, who typically live in small households with less cars.

The respondents' place of residence is also a key piece of background information. The respondents' addresses' postal codes were connected through Statistics Finland's (2016a) database with the seven-level urban-rural classification by Finland's environmental administration (2017). To better facilitate our analysis, the places of residence were divided into three categories, which are I) densely populated urban area (includes inner city and outer city areas of the original classification), II) sparsely populated urban area (includes exurbs, local rural centres areas, and countryside near cities), and III) sparsely populated area (includes rural heartland and sparsely populated countryside).

The survey results were analysed by using IBM SPSS Statistics software. Cross tabulations were used in the statistical analysis of the responses. These provided information on which factors affected the answers and how. Statistical significance of the cross tabulations was tested with the Chi-Square test, which is suitable for testing statistically significant differences of categorical variables (Taanila 2017).

RESULTS

20.6% of the survey respondents had heard or read about the concept Mobility as a service before the survey (N=1,169). Men were clearly more often familiar with MaaS compared to women, as 27.2% of men and 14.6% of women stated having heard or read about MaaS. Correspondingly, the respondents with higher level of education compared to respondents with lower level of education stated more often that MaaS concept was familiar to them. 34.3% of respondents with master's degree and 6.3% of the respondents with primary school education had heard or read about MaaS before the survey. From the different age groups, 24.4% of the respondents of the 25-54-year-old respondents had heard or read about MaaS before, whereas the share was 13.9% among the 18-24-year-old and 15.5% among the 55-64-year-old. As mentioned in the data and methods section, the survey was designed so that answering was not limited to prior knowledge of MaaS, and the term MaaS was not used in other survey questions.

There were two questions in the questionnaire, which depict the respondents' attitudes towards adopting MaaS in the form of a mobility package. In the first question, we found that 43.3% of the respondents were willing to adopt the mobility package, which would fulfil all mobility needs, at a price they themselves stated. The respondents, who informed a price, were willing to pay on average 139 euros a month for the package. The median value was 100 euros, and standard deviation was 114 euros. The highest price a respondent informed to pay was 700 euros and the lowest was 10 euros. 31.2% of the respondents did not know at which price they would adopt the mobility package, and 25.5% would not adopt the package. As the share of the respondents, who did not know was relatively high, and these respondents might be unsure not only of the price they would pay but also whether they would adopt the package at all, at any price, we will next refer to interest to adopt MaaS based on the respondents, who stated a price for the MaaS package.

To analyse the potential of MaaS among different user groups, the interest to adopt the mobility package was analysed against respondents' background information. We present the survey data (number of respondents in each analysed group and the share of answers in each option) and the statistical test results in appendix B. Results are statistically significant or very significant unless otherwise stated.

58.2% of the respondents living in households without a car were willing to adopt the mobility package, whereas the corresponding figure was only 24.0% for the respondents in households with four or more cars (appendix B, table 1). In households with cars, we found that the less cars there are, the more interested the respondents were to adopt the mobility package. Regarding different age groups, the 25 to 34 years old were more often willing to adopt the mobility package compared to other age groups (appendix B, table 2). We also found that the respondents with a higher educational level compared to lower level were more interested in adopting the mobility package (appendix B, table 3).

We also found that the respondents dwelling in densely-populated areas seemed to be more interested to adopt the mobility package, but the differences between areas were not statistically significant (appendix B, table 4). However, we found statistically significant differences regarding respondents' type of house. The respondents living in apartment houses were more interested in adopting the mobility package compered to others (appendix B, table 5). We found that that there was a moderate correlation ($r = 0.599$) between respondents' type of house and number of cars in a household, which can somewhat explain the higher interest in MaaS among dwellers in apartment houses, as the ones living in apartment houses usually own less cars compared to ones living in detached houses or row houses.

The respondents, who do not use passenger cars or use them only a little, were more willing to adopt the mobility package compared to the respondents, who use cars more (appendix B, table 6). We also found that the respondents, who use public transport more often, were more interested in adopting the mobility package compared to the ones using less public transport (appendix B, tables 7 and 8).

The respondents, who had heard or read about MaaS concept before the survey, were more often willing to adopt MaaS services compared to other respondents in general (appendix B, table 9). However, there is a correlation with educational level; the ones, who have a higher level of education, had more often heard or read about MaaS before the survey and they were also more interested in adopting the mobility package compared to other respondents.

The other question in the questionnaire related to the interest towards MaaS and MaaS potential asked whether the respondents needed or wanted to own a private car if the costs of a mobility package, including e.g. public transport, shared taxis and shared cars, were markedly lower than owning a private car. Additionally, it was assumed that the respondent could fulfil all one's mobility needs with the mobility package. In this hypothetical setting, 57.9% of the respondents (N=1,168) did not need or want to own a car. In households currently without a car, 76.8% of the respondents would not need or want to own private car, whereas the corresponding figure of respondents from one car households was 59.6%, and among respondents from households with two cars or more, the figure was 48.4% (appendix B, table 10).

We found that there is a correlation between current car usage and the need and will to own a car in future. Those who drove car less were not that eager to own a car, if a competitive mobility package would be available, compared with the ones using car more (appendix B, table 11).

The respondents, who dwell in densely-populated areas, do not express a need or will to own a car to a similar degree as to those who live in sparsely populated areas (appendix B, table 12). Similarly, the respondents living in apartment houses are not as eager to own a car as other respondents (appendix B, table 13). Regarding the different age groups, 25-34-year-olds stated more seldomly the need or will to car ownership compared to other age groups. Overall, younger age groups express less need or will to car ownership, if a competitive mobility package would be available, compared to older respondents (appendix B, table 14).

Respondents with a lower educational level express more often the need or will for car ownership compared to those with higher education (appendix B, table 15). Additionally, on average women express less need or will for car ownership, if a competitive mobility package would be available, compared to men (appendix B, table 16). Public transport usage correlates with the willingness to own a car, and the ones currently often using public transport express less need or will for car ownership compared to respondents who use public transport less often (appendix B, tables 17 and 18).

DISCUSSION

In this section, we will discuss the potential user groups of MaaS services in Finland by combining findings from previous studies and data on demographics and mobility patterns with survey's results presented in the previous section. This will further highlight the issues related to MaaS potential from end-user perspective and provide insight to how current trends in demographics and mobility affect the potential of MaaS. As different countries as well as areas within one country differ considerably from each other, the characteristics related to each area need to be considered in the analysis. Therefore, we start this discussion with a short introduction of Finland to put the results into context.

Finland is an EU member state and a Nordic country with a relatively small population, 5.5 million in 2017, especially compared to the large land area, resulting in a population density of 18.1 inhabitants per square kilometre (Statistics Finland 2017a). Most of Finland is sparsely populated. There is only one larger metropolitan area, Helsinki with 1.5 million inhabitants, and additionally six cities with above hundred thousand inhabitants in Finland (Statistics Finland 2017d). In these cities and the other urban areas, the population density is naturally significantly higher than the Finnish average.

Finnish people have typically a positive attitude towards digitalisation, and Finnish people considered themselves to have sufficient skills for digital technologies in everyday life more often than on average in EU countries. Yet, the majority of respondents in the survey in each EU member state in 2017 agreed they are sufficiently skilled in the use of digital technologies in their daily life, which indicates that despite differences among people's skills in different countries, most people consider themselves to be digitally competent (European Commission 2017). The level of digital skills should not therefore generally limit the use of MaaS solutions.

The available mobility services (e.g. public transport, city bike and electric scooter services, taxis, rental cars, etc.) and their quality differ around Finland. This reflects to the possible MaaS schemes, in which differences can be expected between urban and rural areas, between cities of different sizes, and between location in the urban structure, e.g. city centre vs. suburban area. There are great differences between cities in Europe and the USA in terms of population density and the non-auto and automobile dependent planning which reflect to the viable modes in different cities (Klinger et al. 2013). From this point of view, it is expected that the greatest MaaS potential is in the biggest cities offering widest selection of different transport modes and services whereas MaaS schemes in rural areas can be very limited in terms of different services bundled. However, in the survey, the same assumption regarding the availability of MaaS service and the mobility package's ability to fulfil all respondent's mobility needs, was presented.

The services in rural areas may need more public support whereas the areas with large transport demand can support commercial MaaS schemes. A MaaS pilot in Gothenburg, Sweden, proved that public support is vital for a successful and sustained service (Karlsson et al. 2016). In less populated rural areas MaaS can contribute to the livelihood of the area with lower costs with the help of more responsive transport services (Rantasila 2016).

In Finland, the first commercial MaaS scheme Whim was fully launched by MaaS Global in Helsinki in November 2017. With Whim, under one subscription and within a single app, the user is able to combine, plan, and pay for public transport, taxi, car rental, car sharing and city bike trips in Helsinki. Based on data set from Whim users in 2018, the results show that the early adopters have been younger generations compared to the population in Finland in general. Whim users have also used more public transport compared to the Helsinki Metropolitan Area in general, and they also combine taxis to public transport trips three times more often than typical Helsinki residents. Overall, the Whim users use taxis 2.1 times more often than a Helsinki resident in general. The overall number of trips per day for a Whim user is about the same as for typical Helsinki residents. The results related to Whim users should be interpreted with caution as the service has been developing during the analysed year (2018) and the number of Whim users and trips has increased considerable during the year. The early adopters may also be more adventurous, more experimental, and less fixed in their ways than public at large (Ramboll 2019).

Based on our survey responses, people living in cities are more interested in using MaaS services than people in general. We also found differences related to respondents' types of houses as those living in apartment houses were more willing to use MaaS services. We can state that in addition to cities providing more mobility options and more options for bundling services to mobility packages, people in cities are more interested to adopt MaaS services compared to others. Yet, it is noteworthy that the differences in attitudes between urban and rural dwellers are not that great. From this point of view, there is solid ground to offer MaaS services also in areas with sparser population if the service can be tailored according to the local circumstances. As an example, the MaaS pilot in Ylläs, a tourist attraction in Kolari with 3,800 inhabitants (Statistics Finland 2017d) in Lapland, Finland, included journey planning and payment using an app from train station and airport to Ylläs, and a skibus or taxi to travel with to ski slopes (Tekes 2017).

In Finland, the densely populated urban areas are very small portions of the land area, but together with the peri-urban areas, larger areas and transport corridors can be seen to offer a larger potential for MaaS in terms of geographical area. Considering future demographic development in Finland, the urban areas and especially the largest urban regions are expected to grow whereas the population in rural areas is expected to decrease because of internal migration, i.e. continuing urbanisation, and differences in birth and mortality rates.

In addition to the total amount of people and their residential areas, population's age distribution is important when considering the mobility practices and MaaS potential. From general mobility point of view and by using age group division, we can distinguish children and senior citizens that are outside the labour markets, young adults studying and entering the working life, and the working-age population.

Considering differences in mobility, even more detailed groups could be recognised, such as children before and in school age as well as younger and older adults and senior citizens.

The challenge of MaaS related to adults, which present the greatest group in terms of population, is that the adults are likely to be relatively locked to their mobility practices. As a contrast to this group, the young are especially interesting from MaaS point of view as they don't own cars to a high degree. Especially in the capital city, Helsinki, the 18-years-olds are getting driving licences much less than before and compared to other areas in Finland. In Helsinki, only 33% acquired the driving license when turning 18 whereas the share was 75% in rural areas in 2012 (Löytty 2014). Every year about 60,000 Finns reach the age of 18 and make decisions on whether to get a driving licence and a private car or not. Of these youngsters, two thirds live in urban areas and one third in rural areas (Statistics Finland 2017c). As the travel behaviour and patterns of younger generations are not fixed, the interest in MaaS services among youngsters could be strengthened by addressing knowledge enhancement and MaaS communication especially to this group.

The survey results clearly presented the 25-34-year-olds as the most potential age group to adopt MaaS services. In terms of MaaS potential, the second age group was the 18-24-year-olds, but the difference to 24-35-year-olds was very clear. This can be due to several factors, such as level of education. The 24-35-year-old have typically finished their studies and have more experience in mobility and different kinds of mobility services as well as life in general, whereas the 18-24-year-olds may still be benefiting from student discounts and their parents in terms of mobility, e.g. paying reduced charges and using parents' car. Also, in the study on attitudes towards automated vehicles in Finland, the 25-34-year-olds were found to be clearly more often expressing a positive attitude towards automated vehicles than other age groups (Liljamo et al. 2018). The 25-34-year-olds may have more comprehension of transport transformation than younger generations, which links with a more positive attitude towards new technologies and services.

From future point of view, it is worth to note that the greatest shift in population in Europe is forecasted to be in the number of senior citizens (European Commission 2014). In Finland, the relative share of senior population is especially high in rural areas. By 2030, the amount of people over 75 years of age will increase from 500,000 to over 800,000 (Statistics Finland 2017b). In all the other age groups, the population is expected to stay roughly at the same level. From MaaS point of view, it would be very interesting to consider this demographic development by thinking how a MaaS scheme could best serve this growing population. As people get older, they may prefer services in many occasions, e.g. as driving a car may get more difficult. Easy to use and access services, both physically and virtually, and perhaps also assisted, would benefit especially this group.

Table 2. The share of trips done by different transport modes in different age groups in Finland in 2016

	Age Groups								
	under 11	**12-17**	**18-24**	**25-34**	**35-44**	**45-54**	**55-64**	**Over 65**	**All**
Walking	27.4%	23.4%	21.1%	21.2%	16.1%	15.8%	18.7%	25.2%	20.4%
Cycling	16.5%	17.1%	6.9%	6.1%	5.3%	4.3%	5.0%	4.8%	6.8%
Passenger car as driver	0.2%	0.3%	37.9%	42.6%	54.5%	55.2%	50.9%	42.9%	41.5%
Passenger car as passenger	44.5%	35.9%	14.7%	12.9%	10.3%	9.7%	11.5%	16.3%	16.3%
Public transport	5.1%	14.4%	15.7%	13.4%	8.5%	8.0%	7.6%	5.8%	9.3%
Taxi	1.8%	1.0%	0.5%	0.7%	0.6%	0.6%	0.6%	0.8%	0.8%
Other	4.4%	7.8%	3.1%	3.0%	4.8%	6.3%	5.7%	4.2%	4.8%
N/A	0.0%	0.0%	0.1%	0.1%	0.0%	0.0%	0.0%	0.0%	0.0%
Grand Total	**100%**	**100%**	**100%**	**100%**	**100%**	**100%**	**100%**	**100%**	**100%**

Source: (Data from Finnish Transport Agency 2018)

The modal shares in different age groups in Finland are presented in table 2. Under 18 years old do most of their trips as car passengers, pedestrians and cyclists. The relatively high share of children's taxi use is explained by school rides, especially typical in rural areas with long distances. It's worth to note that many of children's trips are done together with their parents. The MaaS solution should acknowledge this and for instance offer a mobility package including services for the whole family. For children under 18 years of age, over 40% of the trips is done by walking and cycling, and considering this especially from sustainability point of view, the MaaS scheme could include bikes (e.g. shared bikes, city bikes or rental bikes), and encourage children to cycle even more, maybe replacing some the current car trips by biking.

As can be seen in table 2, the 18-24-year-olds are in a transition phase where the main mode is already passenger car as a driver (38%), but not as clearly as for the 35-64-year-olds, who do more than 50% of their trips as car drivers. Walking has an important role in every age group, but it is worth to note that these trips are short, thus representing only a small portion of the passenger mileage. In cycling, we can recognise a diminishing role the older generations are. Like walking trips, cycling trips are short in general, and they are not very likely replaced by mobility services. The share of public transport is the highest in age group 18-24, which also indicates high potential from MaaS perspective as the youngsters are already more familiar to using mobility services compared to other age groups.

Based on national travel survey data from Finnish Transport Agency (2018), we can find that most of the trips are done by single transport mode in Finland. In 2016, multimodal trips accounted for 16% of all trips. Most typical travel chains are passenger car and walking (6,5% of all trips) and public transport (without changes between public transport means) and walking (5,6%). Other combinations, e.g. public transport, cycling and walking, are clearly much rarer (Pastinen 2018). With a currently low share of all trips being multimodal, this offers simultaneously a challenge and potential for MaaS as MaaS solutions are aiming to offer competitive multimodal trip chains.

Compared to the share of trips in Finland, the passenger mileage is even more car dominant reflecting longer trips done by car than on foot or by bike. Clearly the smallest passenger car mileage is in the inner urban area. The mileage grows in the outer and peri-urban areas likely as a result of sparser population density and longer distances to attractions such as workplaces and commercial activities. The highest mileage is in the sparsely populated rural areas where the distances are usually long and therefore passenger car is the most viable option for travelling (Finnish Transport Agency 2018). From MaaS perspective, the difference between passenger car driver and passenger is that drivers are less eager to change mode and give up driving compared to passengers who are already more familiar to the role of passenger. It has been explicitly noted, that drivers are very attached to using their vehicles (Tertoolen et al. 1998). Based on the responses to the survey, we can state that the ones currently using passenger car to great extent were not as ready to alter their travel behaviour and adopt MaaS services as the respondents, who use cars less or not at all. The ones driving a lot are used to using their own car and thus giving up car ownership and adopting mobility services is considered difficult.

Looking at mobility from a household perspective, it is usual that the passengers in passenger cars are from the same household as the driver. This presents at the same time a challenge and a potential for MaaS as a MaaS scheme for a household could change the mobility practices for the whole household to other modes than passenger cars, or at least the person(s) usually being the car passenger(s) could adopt different mobility services in the role of a passenger. Yet, it requires a big shift, and is a major challenge to overcome for MaaS. Based on the survey results, the type of household and number of household members are not essential when considering the MaaS potential. Instead, other issues influence the interestedness towards MaaS services. However, the survey was directed to individuals, and the situations, where MaaS services could fulfil the mobility needs of the entire household, were not considered.

In Finland, there is a clear difference in the number of cars in use in households of different sizes. Almost half of the one-person households have not a car in use, whereas over half of four+-person households have more than one car in use

(Statistics Finland 2012). Considering this from MaaS point of view, there is clear potential related to the households not owning a car. The challenge is to make the MaaS scheme tempting for the households with one or more cars. For multicar-households, MaaS can offer the possibility to give up car ownership, and even if not totally, mobility services could replace the second or third car. The survey results indicate that people in households without a car are considerably more often interested in MaaS services compared to people in households with a car. We also found that the respondents living in households with one car are more often interested in adopting MaaS services compared to respondents in households with two or more cars. In the survey, MaaS was considered as the primary transport option and as the alternative to passenger car. If the MaaS service would have been considered as a secondary mobility option, replacing the possible second car in the household, the results would have probably been different.

In addition to positive outcomes from transport policy point of view, MaaS could also lead to a mode-shift away from public transport and increasing number of journeys (Transport Systems Catapult 2016). If MaaS offers improved individual mobility, this could mean more traffic and eventually more congestion (Rantasila 2016). People could e.g. replace conventional timetabled public transport with flexible car-sharing services during the peak hours (Henscher 2017).

Based on the survey results, the respondents who use public transport are more interested in adopting MaaS services. Additionally, the respondents, who currently drove less, were more interested in MaaS services. These results indicate that MaaS may increase traffic and may lead to a shift towards car-oriented mobility, if MaaS offerings include e.g. shared cars and taxi services. In this case, the share of sustainable modes, i.e. walking, cycling, and public transport, may decrease as trips are done more often by car. The current public transport users as well as the ones, who currently travel less by car, live mostly in urban areas. Therefore, the increasing traffic may lead to notable problems e.g. due to congestion in cities.

THE RELIABILITY OF THE RESULTS

This study is based on findings from previous research and analysing the results of an original survey on the potential user groups of MaaS. The survey with a large sample size of 6,000, was carried out as a postal questionnaire with the option for the respondent to answer on the internet. The response rate of the survey was rather low, 19.6%. Compared to a somewhat similar, recent Finnish survey on the attitudes towards automated vehicles, we can find the response rate to be nearly the same, as the percentage in that study (Liljamo et al. 2018) was 20.4%. Low response rate makes it possible that the respondents do not present the actual view of the whole population. The low response rate can be partly a result of people not being interested in the topic, and the ones responding may then be more interested than people in general.

Generally, older generations had a higher response rate compared to younger generations, and women compared to men. We found that these groups, which were more active to respond to the survey, were less interested in adopting MaaS. It can be speculated that if we had had more men or young to respond to the survey, and a higher response rate among these respondent groups, would the results have been different.

In the survey, we tried to be as clear as possible with the questions and statements. Still, it can be that the respondents have had difficulties to comprehend some of the issues in the questionnaire as most respondents were not that familiar with integrated mobility services. Additionally, some might have experienced it to be difficult to answer the hypothetical questions, e.g. if there is service available that can fulfil all mobility needs, and with a competitive price, is there a need or will for car ownership. We found that about 30% of the respondents chose "I do not know" to survey's question 7, which might be a result of experienced difficulties. In the survey feedback, the questionnaire was mostly considered explicit, but some questions were found difficult to answer. The greatest concern of the respondents was related to the feasibility of Maas in rural context, and some of the rural respondents found the survey somewhat irrelevant. The difficulties in grasping the issues of the survey may also have an influence on the low response rate.

Related to the validity of the study, we next discuss the connections between the research questions and the questions in the survey. Three research questions presented in this chapter were 1) what is the interest towards MaaS among ordinary people, 2) which are the most potential user groups of MaaS, and 3) how do current trends in demographics and mobility affect the potential of MaaS. The first and second research question is mostly approached through analysing responses to the survey's questions 7 and 11c. Of these question 7 (would the respondent adopt the mobility package at a price the respondent was asked to define) was focusing merely on the willingness to adopt MaaS, whereas question 11c questioned the need and will to own a private car if there was a competitive mobility package available. It's noteworthy that both these survey questions were hypothetical and presented assumptions of a high quality and price competitive MaaS solution, which might be difficult to realise in real life setting (in question 7 the mobility package is stated to able to fulfil all transport needs, and in question 11c the mobility package has additionally markedly lower annual costs compared to private car). The research question 3 was approached in section 5 by discussing the findings related to the survey in connection with previous studies and data related to Finnish demographics and mobility.

Related to reliability, the statistical significance of survey results was analysed. Even though the response rate was rather low, the differences between the analysed user groups were usually statistically significant or very significant. As the response rate and the number of responses in different groups was relatively low (e.g. 50 responses from 18-24-year-old males), we chose to analyse the findings related to different background information separately (not e.g. analysing specific age groups in densely-populated areas or related to educational level).

As there is currently only a small number of previous studies, and especially research published in peer-reviewed scientific journals, on potential user groups of MaaS that are based on end-user surveys, a direct comparison cannot be made between previous studies and the results of this study. However, there are some MaaS-related reports presenting survey results. In the survey by ITS Australia (2018), same groups (young, men, highly educated, urban dwellers, using public transport more often) were identified as the most potential MaaS users. The Finnish survey by Solita (2017) found similar low recognition of MaaS concept among the respondents, as well as similar level of willingness to pay. The reliability of this paper is also supported by the findings from MaaS-London project (Kamargianni et al. 2018), in which especially the young have been identified as potential utiliser of MaaS. Also, some of the services which are expected to be bundled to MaaS offering, e.g. car-sharing, are found to attract especially young and highly educated people (Becker et al. 2017a, 2017b), which again highlight these groups as especially potential user groups.

CONCLUSION

MaaS is still a new concept, and to ordinary people MaaS as a term and concept remains to be generally unknown. In Finland, 20.4% of the survey respondents in April - June 2018 had heard or read about Mobility as a Service before the survey. This is about 4 percentage points higher than in the survey by Solita (2017) a year earlier in Finland. This low awareness of MaaS is notable and may also affect the results as higher awareness of MaaS could lead to better recognition of MaaS potential. In the design of survey, it was recognised the MaaS is a new concept and likely unfamiliar to most respondents and therefore the survey questions were formulated in a way that the respondents could answer without prior knowledge about MaaS.

MaaS has potential to change the mobility practices of people. The potential varies anyhow in different geographical areas and for different population groups. To be able to assess and address the potential and challenges of MaaS it is important to recognise the differences between user groups. In this paper, we found many differences in the interest towards MaaS adoption. We found that particularly the ones who currently use public transport, live in households without a car, are aged 25-34, and are highly educated were more interested to adopt MaaS services. We also found that people living in urban areas and in apartment houses indicated a higher interest towards MaaS adoption. In general, we found that MaaS has the greatest potential on urban areas and the younger people are especially interesting from MaaS point of view. Based on the results, the MaaS offerings can be targeted to the areas and user groups with the greatest potential. This is especially recommended for commercial MaaS operators who seek to gather enough users to operate a commercially successful MaaS scheme. For authorities, this means that if they want to support MaaS, most support is needed outside these areas with highest potential. The interests of business and authorities can also be aligned especially in urban areas, where public transport could attract more users when being offered also as a part of a MaaS scheme.

For all stakeholders, it is recommended to invest in making MaaS a concept which would be generally recognised. Even if the term MaaS would not become familiar to average Joe, people should be aware of systems, which integrate mobility services and offer an option for private car ownership and use. Additionally for MaaS operators, this study highlights that the MaaS offerings should be reasonably priced as people in the survey were generally willing to pay on average 139 euros a month for a mobility package, which would fulfil all mobility needs. For authorities, if they see MaaS as a sensible medium for more sustainable transport, it should be considered to increase the costs related to car ownership and use in order to make MaaS a more tempting option especially for those, who currently use car to a high degree.

Even though this chapter draws especially on the findings from a Finnish survey, we can note that the potential user groups are similar to those found in other countries, e.g. Australia. In future research, it would be interesting to undertake comparisons between countries, and to analyse, e.g., to which extent the people in general are interested in MaaS and especially what type of MaaS offerings are of interest to different users, e.g. what are the mobility services people want to include in the mobility package. In addition to focusing on the individual user's point of view, also the household perspective of MaaS should be addressed, e.g. by a MaaS survey focusing on the mobility needs of entire households, including children, too. Also, the MaaS potential among senior citizens should be studied in future.

There are also many other important MaaS related issues, which require further research. Since there is not much data of current MaaS users available as MaaS services have just recently been launched and the companies providing MaaS services have not openly shared the data, the research in this chapter is based on alternative sources. In future, if the data of current users can be used for research, it should be studied whether the identified user groups match results of this study. The findings from different MaaS pilots and commercial schemes are very interesting as these may be combined to reveal on the potential user groups (e.g., who are using the services, how are the services used) and the potential of MaaS overall. Additionally, specific user groups such as the elderly, who were not surveyed in this study, or people in living rural areas should be studied more precisely in the future. It should be also studied, how people without smartphones or with lacking skills to use mobile applications could benefit from MaaS. Finally, the willingness to adopt MaaS services, e.g. willingness to pay for different types of services and in different user groups, is of great research interest.

REFERENCES

ITS Australia. (2018). *Mobility as a Service in Australia. Customer insights and opportunities*. Author.

Becker, H., Ciari, F., & Axhausen, K. W. (2017a). Comparing car-sharing schemes in Switzerland: User groups and usage patterns. *Transportation Research Part A, Policy and Practice*, *97*, 17–29. doi:10.1016/j.tra.2017.01.004

Becker, H., Loder, A., Schmid, B., & Axhausen, K. W. (2017b). Modeling car-sharing membership as a mobility tool: A multivariate Probit approach with latent variables. *Travel Behaviour and Society.*, *8*, 26–36. doi:10.1016/j.tbs.2017.04.006

El Zarwi, F., Vij, A., & Walker, J. L. (2017). A discrete choice framework for modeling and forecasting the adoption and diffusion of new transportation services. *Transportation Research Part C, Emerging Technologies*, *79*, 207–223. doi:10.1016/j.trc.2017.03.004

European Commission. (2014). Population ageing in Europe – Facts, implications and policies. Author.

European Commission. (2017). Special Eurobarometer 460, Attitudes towards the impact of digitisation and automation on daily life. Author.

Finland's Environmental Administration. (2017). *Urban-rural classification*. Available at http://www.ymparisto.fi/en-US/Living_environment_and_planning/Community_structure/Information_about_the_community_structure/Urbanrural_classification

Finnish Transport Agency. (2018). *Finnish National Travel Survey*. Retrieved from https://www.liikennevirasto.fi/web/en/statistics/national-travel-survey

Giesecke, R., Surakka, T., & Hakonen, M. (2016). Conceptualising Mobility as a Service. *2016 11th International Conference on Ecological Vehicles and Renewable Energies.*

Heikkilä, S. (2014). Reorganization of the mobility service provision - Public governance as a contributor. *2014 21st World Congress on Intelligent Transport Systems.*

Hensher, D. A. (2017). Future bus transport contracts under a mobility as a service (MaaS) regime in the digital age: Are they likely to change? *Transportation Research Part A, Policy and Practice, 98,* 86–96. doi:10.1016/j.tra.2017.02.006

Kamargianni, M., Matays, M., Weibo, L., & Schäfer, A. (2015). *Feasibility study for "Mobility as a Service" concept in London.* UCL Energy Institute & Department for Transport.

Kamargianni, M., Matyas, M., Li, W., & Muscat, J. (2018). *Londoners' attitudes towards car-ownership and Mobility-as-a-Service: Impact assessment and opportunities that lie ahead.* MaaSLab – UCL Energy Institute Report.

Karlsson, I. C. M., Sochor, J., & Strömberg, H. (2016). Developing the 'Service' in Mobility as a Service: Experiences from a field trial of an innovative travel brokerage. *Transportation Research Procedia, 14,* 3265–3273. doi:10.1016/j.trpro.2016.05.273

Klinger, T., Kenworthy, J. R., & Lanzendorf, M. (2013). Dimensions of urban mobility cultures – a comparison of German cities. *Journal of Transport Geography, 31,* 18–29. doi:10.1016/j.jtrangeo.2013.05.002

König, D., Eckhardt, J., Aapaoja, A., Sochor, J. & Karlsson, M. (2016). *Deliverable 3: Business and operator models for MaaS.* MAASiFiE project funded by CEDR.

Li, Y., & Voege, T. (2017). Mobility as a Service (MaaS): Challenges of Implementation and Policy Required. *Journal of Transportation Technologies,* 7(02), 95–106. doi:10.4236/jtts.2017.72007

Liljamo, T., Liimatainen, H., & Pöllänen, M. (2018). Attitudes and concerns on automated vehicles. *Transportation Research Part F, 59*(A), 24–44.

Löytty, M. (2014). *18–20-vuotiaat nuoret henkilöauton kuljettajina* (in Finnish). Trafi Publications 2/2014.

Matyas, M., & Kamargianni, M. (2017). A stated preference experiments for mobility-as-a-service plans. *5th IEEE International Conference on Models and Technologies for Intelligent Transportation Systems.* 10.1109/MTITS.2017.8005610

Pastinen, V. (2018). *Matkaketjut ja liikkuminen palveluna. Henkilöliikennetutkimus 2016. Faktakortta – laadittu helmikuussa 2018* [Travel chains and mobility as a service. National travel survey 2016. Fact card - February 2018]. Available at https://www.traficom.fi/sites/default/files/media/file/Faktakortti-HLT2016-MaaS.pdf

Ramboll. (2019). *Whimpact. Insights from the world's first Mobility-as-a-Service (MaaS) system.* Available at https://ramboll.com/-/media/files/rfi/publications/Ramboll_whimpact-2019

Rantasila, K. (2016). The impact of Mobility as a Service concept to land use in Finnish context. *International Conference on Sustainable Mobility Applications, Renewables and Technology (SMART).*

Solita. (2017). *Suomalaiset janoavat uusia liikennepalveluita* (in Finnish). Available at https://www.solita.fi/blogit/suomalaiset-janoavat-uusia-liikennepalveluita/

Statistics Finland. (2017a). *Population density by area 1.1.2017. Statistics Finland's PX-Web databases.* Retrieved from http://pxnet2.stat.fi/PXWeb/pxweb/en/StatFin/

Statistics Finland. (2017b). *Population projection 2015 according to age and sex 2015 - 2065, whole country. Statistics Finland's PX-Web databases.* Retrieved from http://pxnet2.stat.fi/PXWeb/pxweb/en/StatFin/

Statistics Finland. (2017c). *Population 31 Dec by Area, Urban-rural classification, Sex, Age and Year. Statistics Finland's PX-Web databases.* Retrieved from http://pxnet2.stat.fi/PXWeb/pxweb/en/StatFin/

Statistics Finland. (2017d). *Population by sex and area 31.12.2016 and increase of population. Statistics Finland's PX-Web databases.* Retrieved from http://pxnet2.stat.fi/PXWeb/pxweb/en/StatFin/

Statistics Finland. (2018). *Preliminary data on vital statistics by month of occurence 2018.* Available at: https://www.stat.fi/til/vamuu/2018/08/vamuu_2018_08_2018-09-25_tau_001_en.html

Statistics Finland. (2012). *Kotitalouksien jätteiden kierrättäminen on lisääntynyt* (in Finnish). Available at http://www.stat.fi/til/ktutk/2012/ktutk_2012_2012-11-05_tie_001_fi.html

Statistics Finland. (2015). *Population projection 2015 according to age and sex in 2015 to 2065, Statistics Finland's PX-Web database.* Available at http://pxnet2.stat.fi/PXWeb/pxweb/en/StatFin/StatFin__vrm__vaenn/statfin_vaenn_pxt_001.px/?rxid=b389490b-3693-4f46-95f2-3186ea600029

Statistics Finland. (2016a). *PAAVO – Open data by postal code area, Statistics Finland's PX-Web database*. Available at https://www.stat.fi/tup/paavo/index_en.html

Statistics Finland. (2016b). *Share of households with cars has stabilised, number of pet owners is increasing*. Available at http://www.stat.fi/til/ktutk/2016/ktutk_2016_2016-11-03_tie_001_en.html

Strömberg, H., Rexfelt, O., Karlsson, I. C. M., & Sochor, J. (2016). Trying on change - Trialability as a change moderator for sustainable travel behavior. *Travel Behaviour and Society.*, *4*, 60–68. doi:10.1016/j.tbs.2016.01.002

Taanila, A. (2017). *Kirjoituksia Aki Taanilan kvantitatiivisesta menetelmäpajasta – 6 Ristiintaulukointi ja Khiin neliö –testi* (in Finnish). Available at https://tilastoapu.wordpress.com/2011/10/14/6-ristiintaulukointi-ja-khiin-nelio-testi/

Tekes. (2017). *MaaS toimii jo kolmessa kaupungissa ja Ylläksellä* (in Finnish). Available at https://www.tekes.fi/nyt/uutiset-2017/maas-toimii-jo-kolmessa-kaupungissa-ja-yllaksella/

Tertoolen, G., van Kreveld, D., & Verstraten, B. (1998). Psychological resistance against attempts to reduce private car use. *Transportation Research Part A, Policy and Practice*, *32*(3), 171–181. doi:10.1016/S0965-8564(97)00006-2

Transport Systems Catapult. (2016). *Mobility as a Service. Exploring the opportunity for mobility as a service in the UK*. Author.

APPENDIX A: QUESTIONNAIRE

Figure 1.

PART I

1. Have you ever heard or read about the concept of MaaS (Mobility as a Service)?

☐ Yes ☐ No

2. What are your views on the following mobility pricing options?

a) **Package pricing**: a fixed monthly fee is paid for mobility. Mobility package can include for example unlimited local public transport and a certain amount of taxi rides or use of a rental car.

☐ Very positive ☐ Somewhat positive ☐ Neutral ☐ Somewhat negative ☐ Very negative

b) **Current form mixed pricing**: both fixed fees and taxes (e.g. vehicle insurance and vehicle tax) and fees based on use (e.g. fuels and single tickets in public transport)

☐ Very positive ☐ Somewhat positive ☐ Neutral ☐ Somewhat negative ☐ Very negative

c) **Usage based travel or kilometer-based pricing**: fee is solely based on the amount of use of mobility service or private car

☐ Very positive ☐ Somewhat positive ☐ Neutral ☐ Somewhat negative ☐ Very negative

3. What do you think of the following statements?	Strongly agree	Somewhat agree	Neither agree nor disagree	Somewhat disagree	Strongly disagree
It should be possible to combine all mobility services and use them with single ticket and application (e.g. taxi, train and bus with the same ticket).	☐	☐	☐	☐	☐
I would like to have all costs of my travelling to be included in a single monthly payment.	☐	☐	☐	☐	☐
I think that the cost of my mobility is currently low enough so I can fulfil my daily mobility needs.	☐	☐	☐	☐	☐
I think that it is difficult or cumbersome to use public transport (including long-distance public transport)	☐	☐	☐	☐	☐
I think that personal space, time and peace in private car are important criteria for me when choosing the mode of transport.	☐	☐	☐	☐	☐
I think that I can fulfil my daily mobility needs by using other modes of transport than private car.	☐	☐	☐	☐	☐
Transport transformation from using private cars towards mobility services would be a desirable trend.	☐	☐	☐	☐	☐

Figure 2.

PART II

4. How interested would you be in the following mobility services if they were available for use in your daily life or on holiday/business trips?

Shared taxi Works like a conventional taxi, but there may also be strangers as passengers so travel time can be slightly longer but the cost of travel is reduced			**Shared car** Shared cars would be available nearby to be booked and claimed with a mobile device. The cost would be approximately 1 € / km		
☐ Very interested	☐ Somewhat interested	☐ Not at all interested	☐ Very interested	☐ Somewhat interested	☐ Not at all interested

5. Which of the following describes the best your current mobility habits and mobility costs? <u>Choose one option for each row</u> that best describes your current mobility habits. In the parenthesis is the estimate of average monthly cost of each option (see the cover letter for more instructions).

Owning a car (incl. taxes, insurances, decrease in value, parking)	☐ No car (0 €/mo.)	☐ Car, value less than 10 000 € (150 €/mo.)	☐ Car, value 10 000 – 25 000 € (300 €/mo.)	☐ Car, value more than 25 000 € (450 €/mo.)
Use of a car as a driver or passenger (incl. fuel, maintenance, tires, washes etc.)	☐ No usage (0 €/mo.)	☐ Less than 10 000 km per year (50 €/mo.)	☐ 10 000 – 25 000 km per year (150 €/mo.)	☐ Over 25 000 km per year (300 €/mo.)
Use of local public transport	☐ No usage (0 €/mo.)	☐ Low usage (10 €/mo.)	☐ Weekly usage (30 €/mo.)	☐ Almost daily usage (50 €/mo.)
Use of long-distance public transport	☐ No usage (0 €/mo.)	☐ Usage a few times a year (20 €/mo.)	☐ Monthly usage (50 €/mo.)	☐ Weekly usage (100 €/mo.)
Use of other mobility services (taxi, rental car)	☐ No usage (0 €/mo.)	☐ Usage a few times a year (20 €/mo.)	☐ Monthly usage (50 €/mo.)	☐ Weekly usage (100 €/mo.)

6. Using the table above, you can calculate an estimate of the monthly cost of your mobility. Do you think that this sum corresponds to the real costs of your mobility?

☐ Yes

☐ No, the real costs of my mobility are approximately ________ euros per month.

☐ I do not know much money I use for my mobility

7. <u>Assume</u> that a suitable mobility package could cover all your mobility needs. What price should the mobility package have so that you would buy it?

The mobility package would include e.g. public transport, taxi and shared or rented cars.

________ euros per month

☐ I do not know

☐ I would not adopt the package

Figure 3.

PART III

8. Assume that you would buy a mobility package and you would not own a car at all. How much would you use the following mobility services? Choose one option for each row!

Shared or rented car	☐ Over 15 d/mo. (over 20 000 km per year)	☐ Approx. 10 d/mo. (10 000–20 000 km per year)	☐ Approx. 5 d/mo. (5 000–10 000 km per year)	☐ Less than 2 d/mo. (less than 5 000 km per year)
Use of local public transport	☐ Usage almost daily	☐ Weekly usage	☐ Monthly usage	☐ Usage a few times a year
Use of long-distance public transport	☐ Usage almost daily	☐ Weekly usage	☐ Monthly usage	☐ Usage a few times a year
Use of other mobility services (e.g. taxi, city bikes)	☐ Usage almost daily	☐ Weekly usage	☐ Monthly usage	☐ Usage a few times a year

9. Would you travel more often or longer trips, if

a) The mobility costs were cheaper than currently? ☐ Yes ☐ No

b) You had a fixed price monthly package that included an unlimited number of trips? ☐ Yes ☐ No

The national energy and climate strategy of Finland aims to promote mobility services so that travelling alone by a car reduces and the growth of car traffic in urban areas stops despite population growth.

10. What do you think of the following statements regarding the climate strategy?	Strongly agree	Somewhat agree	Neither agree nor disagree	Somewhat disagree	Strongly disagree
Private car traffic should be limited to reduce emissions	☐	☐	☐	☐	☐
Public transport should be funded more with tax revenues to decrease headways	☐	☐	☐	☐	☐
Transport investments should be allocated more to the construction of railways and tramways instead of road and street construction	☐	☐	☐	☐	☐
The obligation of building parking spaces should be waived when building new apartments	☐	☐	☐	☐	☐
In city centres, there should be more pedestrian and public transport streets and limited private car traffic.	☐	☐	☐	☐	☐
Shared cars should be exempt from car taxes and parking fees	☐	☐	☐	☐	☐
Congestion charges should be implemented in the metropolitan area	☐	☐	☐	☐	☐
The amount of the tax deductibility of commuting should be same for all mode of transport	☐	☐	☐	☐	☐

Figure 4.

PART IV

11. Would you want or need to own a private car, if

a) Public transport connections were good enough to fulfil your daily mobility needs ?

☐ Yes ☐ No

b) Rental or shared car were always available within 500 meters and the annual cost of using it was less than the annual cost of an own car?

☐ Yes ☐ No

c) The annual cost of the mobility service (incl. e.g. public transport, shared taxis and shared cars) was significantly lower than your own car and you could fulfil all your mobility needs with it.

☐ Yes ☐ No

Background information and open feedback

12. How old are you?

_______ years

13. What is your gender?

☐ Woman ☐ Man ☐ Other

14. Do you have a driver's license?

☐ Yes ☐ No

15. How many people are in your household, yourself included?

_______ person(s)

16. What kind of house do you live in?

☐ Apartment house ☐ Row house
☐ Detached house ☐ Other

17. Do you have any permanent injuries or illnesses that affect your driving?

☐ Yes ☐ No

18. What is your household gross income per month?

☐ Less than 2 000 € ☐ 4 000–6 000 €
☐ 2 000–4 000 € ☐ Over 6 000 €

19. How many cars do you have in your household?

_______ car(s)

20. Have you registered for one of the following mobility services: Whim, Kyyti, Kätevä Seinäjoki, DriveNow tai OP Kulku? (MaaS operators in Finland**)**

☐ Yes ☐ No

21. What is your highest level of education you have graduated or studying at the moment?

☐ Primary school ☐ High school
☐ Bachelor's degree ☐ Master's degree

22. Open feedback regarding mobility services

23. Open feedback regarding this survey

THANK YOU FOR THE PARTICIPATION!

APPENDIX B" RESULT TABLES

Question 7: Assume that a suitable mobility package could cover all your mobility needs. What price should the mobility package have so that you would buy it?

Table 3. Question 7 cross tabulated with cars in a household (N=1,155, Chi-Square p<0.001)

	0 (N=184)	**1 (N=504)**	**2 (N=378)**	**3 (N=64)**	**4 or More (N=25)**
Could buy	58.2%	43.7%	37.6%	35.9%	24.0%
Can not say	29.9%	31.0%	33.9%	26.6%	20.0%
Would not buy	12.0%	25.4%	28.6%	37.5%	56.0%

Table 4. Question 7 cross tabulated with age group (N=1,171, Chi-Square p<0.001)

	18-24 (N=109)	**25-34 (N=225)**	**35-44 (N=224)**	**45-54 (N=244)**	**55-64 (N=369)**
Could buy	45.9%	55.1%	46.9%	43.9%	32.8%
Can not say	33.0%	27.6%	29.5%	25.4%	37.7%
Would not buy	21.1%	17.3%	23.7%	30.7%	29.5%

Table 5. Question 7 cross tabulated with level of education (N=1,167, Chi-Square p=0.001)

	Primary School (N=80)	**High School (N= 377)**	**Bachelor's Degree (N=444)**	**Master's or PhD Degree (N=266)**
Could buy	26.3%	39.5%	44.4%	52.6%
Can not say	37.5%	34.5%	30.6%	25.9%
Would not buy	36.3%	26.0%	25.0%	21.4%

Table 6. Question 7 cross tabulated with place of residence in (N=1,171, Chi-Square p=0.119)

	Densely Populated Urban Area (N=592)	Sparsely Populated Urban Area (N=368)	Sparsely Populated Area (N=211)
Could buy	47.0%	39.4%	39.8%
Can not say	30.1%	32.1%	32.7%
Would not buy	23.0%	28.5%	27.5%

Table 7. Question 7 cross tabulated with type of the house (N=1,167, Chi-Square p<0.001)

	Apartment House (N=409)	Row House (N=192)	Detached House (N=557)
Could buy	49.4%	42.7%	39.5%
Can not say	32.8%	31.3%	29.8%
Would not buy	17.8%	26.0%	30.7%

Table 8. Question 7 cross tabulated with use of a car as a driver or passenger per year (N=1,167, Chi-Square p<0.001)

	No Usage (N=117)	Less Than 10,000 km (N=398)	10,000–25,000 km (N=467)	Over 25,000 km (N=185)
Could buy	47.0%	50.0%	37.5%	40.5%
Can not say	39.3%	30.7%	33.4%	22.7%
Would not buy	13.7%	19.3%	29.1%	36.8%

Table 9. Question 7 cross tabulated with use of local public transport (N=1,168, Chi-Square p<0.001)

	No Usage (N=519)	Low Usage (N=394)	Weekly Usage (N=121)	Almost Daily (N=134)
Could buy	36.4%	43.4%	50.4%	64.2%
Can not say	28.5%	36.0%	33.1%	25.4%
Would not buy	35.1%	20.6%	16.5%	10.4%

Table 10. Question 7 cross tabulated with use of long-distance public transport (N=1,168, Chi-Square p<0.001)

	No usage (N=472)	Usage a Few Times a Year (N=574)	Monthly Use (N=88)	Weekly Use (N=34)
Could buy	32.2%	47.0%	67.0%	76.5%
Can not say	32.8%	33.1%	19.3%	5.9%
Would not buy	35.0%	19.9%	13.6%	17.6%

Table 11. Question 7 cross tabulated with heard or read about concept of MaaS before (N=1,166, Chi-Square p<0.001)

	Yes (N=241)	No (N=925)
Could buy	52.7%	41.0%
Can not say	18.7%	34.2%
Would not buy	28.6%	24.9%

Question 11c: Would you want or need to own a private car, if the annual cost of the mobility service (incl. e.g. public transport, shared taxis and shared cars) was significantly lower than your own car and you could fulfil all your mobility needs with it.

Table 12. Question 11c cross tabulated with cars in a household (N=1,151, Chi-Square p<0.001)

	0 (N=185)	1 (N=503)	2 (N=376)	3 (N=63)	4 or More (N=24)
Yes	23.2%	40.4%	50.5%	49.2%	75.0%
No	76.8%	59.6%	49.5%	50.8%	25.0%

Table 13. Question 11c cross tabulated with use of car as a driver or passenger per year (N=1,163, Chi-Square p<0.001)

	No usage (N=117)	Less Than 10,000 km (N=397)	10,000–25,000 km (N=465)	Over 25,000 km (N=184)
Yes	20.5%	35.3%	50.1%	50.0%
No	79.5%	64.7%	49.9%	50.0%

Table 14. Question 11c cross tabulated with place of residence in (N=1,167, Chi-Square p=0.004)

	Densely Populated Urban Area (N=590)	Sparsely Populated Urban Area (N=366)	Sparsely Populated Area (N=211)
Yes	38.1%	43.2%	51.2%
No	61.9%	56.8%	48.8%

Table 15. Question 11c cross tabulated with type of the house (N=1,163, Chi-Square p<0.001)

	Apartment House (N=409)	Row House (N=191)	Detached House (N=554)
Yes	32.3%	41.9%	49.1%
No	67.7%	58.1%	50.9%

Table 16. Question 11c cross tabulated with age group (N=1,167, Chi-Square p<0.001)

	18-24 (N=108)	25-34 (N=222)	35-44 (N=223)	45-54 (N=245)	55-64 (N=369)
Yes	35.2%	29.7%	39.0%	47.3%	50.1%
No	64.8%	70.3%	61.0%	52.7%	49.9%

Table 17. Question 11c cross tabulated with level of education (N=1,163, Chi-Square p<0.001)

	Primary School (N=80)	High School (N= 374)	Bachelor's Degree (N=444)	Master's or PhD Degree (N=266)
Yes	53.8%	47.3%	43.0%	29.1%
No	46.3%	52.7%	57.0%	70.9%

Table 18. Question 11c cross tabulated with gender (N=1,168, Chi-Square p<0.001)

	Woman (N=630)	Man (N=539)
Yes	34.6%	51.0%
No	65.4%	49.0%

Table 19. Question 11c cross tabulated with use of local public transport (N=1,164, Chi-Square p<0.001)

	No Usage (N=517)	Low Usage (N=393)	Weekly Usage (N=121)	Almost Daily (N=133)
Yes	52.4%	41.2%	22.3%	22.6%
No	47.6%	58.8%	77.7%	77.4%

Table 20. Question 11c cross tabulated with use of long-distance public transport (N=1,164, Chi-Square p<0.001)

	No Usage (N=471)	Usage a Few Times a Year (N=571)	Monthly Use (N=88)	Weekly Use (N=34)
Yes	52.4%	37.1%	28.4%	17.6%
No	47.6%	62.9%	71.6%	82.4%

Section 2

MaaS as a Differentiating Mobility System in Rural and Urban Contexts

The second section gives an overview of some interesting and recent examples of MaaS applications developed for a wide variety of scenarios going from rural to urban spaces. Multiple aspects were analyzed that must be taken in account to ensure the feasibility and success in the implementation phase and that can be useful and serve as a reference point for other MaaS initiatives. In addition, some case studies, based on real-world experience using samples gathered from different geographical realities, are also presented.

Chapter 4
Challenge, Coordination, and Collaboration for Effective Rural Mobility Solutions

Kate Pangbourne
https://orcid.org/0000-0003-2100-1961
Institute for Transport Studies, University of Leeds, UK

ABSTRACT

In this chapter, the author argues that mobility as a service (or MaaS) as a principal means of accessing transport may make it difficult to meet existing policy aims for sustainable mobility that addresses the climate change crisis and demographic trends. Strong governance will be needed to design MaaS for these challenges whilst addressing existing and future social injustices. Furthermore, without determined coordination, an inherent individualism at the heart of many current formulations of MaaS privileges urban areas with existing multimodal options and risks further excluding rural environments. Drawing on examples from Europe and beyond, concerns regarding the ability to assess unanticipated undesirable impacts of transport innovations that are MaaS-like in character are highlighted and discussed. How might authorities collaborate with each other and MaaS brokers to provide low carbon accessibility in rural areas?

DOI: 10.4018/978-1-7998-1614-0.ch004

INTRODUCTION

An integrated approach to transport has been an objective of transport policy for many decades but it has yet to be fully realized. The emergence of Mobility as a Service or MaaS (a concept describing a set of characteristics for tools to access mobility) has been promoted as being the long-awaited enabler for such integration. MaaS itself is in turn facilitated by the increasing pace of digitalization of daily life. However, we are still in the early part of a hype cycle for MaaS, as is suggested by Eckhardt et al., (2018) for example. Therefore, it is timely to critically interrogate claims made for MaaS proposals before committing ourselves to a future reliance on MaaS for delivering public benefits.

In this chapter the author critically examines two of the promises made by promoters of MaaS which are important to supporting accessibility and reducing transport's environmental impacts, namely integration and efficiency. The author goes on to consider how these promised visions might play out when applied to rural settings. The objective is to show that MaaS tends to be presented as a universal remedy, but that there are different potential impacts for urban versus rural contexts. Some of these potential impacts have social and distributional implications, even though it is clear that there is evidence for the potential benefits of more flexible approaches to rural public transport. Without clarity of thinking, these differences pose a significant challenge for 'rural MaaS' because a key aspect of the development of MaaS is founded on a value proposition designed for urban conditions. Each aspect of the challenge, and alignment with current policy trends in relation to MaaS is discussed in section four and some solutions and recommendations are offered in section five. Firstly however, section two sets out the theory and method used to develop this argument, and section three provides a background to MaaS, particularly its definitions, evolution and the context of the main rural challenges. The chapter ends with an assessment of future research directions (section six), and a closing summary.

THEORETICAL PERSPECTIVE AND METHOD

In this chapter the author argues that the challenge to be addressed by MaaS differs between rural and urban settings. The objective is to show whether the potential for MaaS to benefit rural areas could be undermined by unanticipated outcomes from MaaS if too much trust is placed on the current claims regarding MaaS. These claims tend to either be aspatial (ignoring geographic factors) or they specifically refer to urban issues. Furthermore, these claims tend to be biased to being pro-MaaS, as rhetorical persuasion is identified as an important mechanism in how new technologies are taken up in society, by controlling the public discussion around what challenges technological solutions such as MaaS are intended to address (Berkhout, 2006).

Pangbourne, Stead, Mladenovic and Stead (in press) have closely examined this type of rhetoric in relation to the governance challenges posed by MaaS. This chapter carries out a similar exercise in relation to issues specific to rural settings. The author asked the research question "What are the priorities that need to be addressed in defining objectives for rural MaaS?" This question is addressed through a synthesis of facts about the current state of MaaS development, with a qualitative analysis of MaaS promotional and information documents to understand the degree to which rural areas are specifically addressed, followed by a consideration of rural need in light of seven core characteristics of MaaS (Jittrapirom et al., (2017) see below). This argument is underpinned by the background section which surveys academic literature about rural issues, including transport.

The method uses purposive and inductive documentary analysis. Grant (2018) defines documents as texts that can exist in either electronic or physical form. Document types include tweets, social media posts, interview and focus group transcripts, archive material and published documents of many types. Such documents can be created as part of the research process (primary data) or can be already existing (secondary data). This chapter uses existing documents about MaaS sourced using the internet (search terms Mobility as a Service, MaaS, rural AND MaaS, rural AND Mobility as a Service), and in the English language. Academic and factual sources are reviewed in the usual way, before a set of important MaaS promotional documents are analysed for their content relating to the application of MaaS to rural areas.

BACKGROUND

This section discusses the definitions and origins of MaaS. It then briefly covers the current state of deployment and the role of coalitions created to promote the interests of those with something to gain from promoting MaaS as a universal solution for integrating transport services. The different transport and societal characteristics of rural compared to urban settings are also outlined in order to inform the argument that is put forward in the rest of the chapter. Whilst there has been a rapid increase in the quantity of material published on MaaS, this is not an overarching review, as others have already done this (e.g. Utriainen & Pöllänen, 2018). Therefore, work included here has been selected specifically to address questions raised by the position put forward in this chapter.

Genealogy and Definition of Mobility as a Service

MaaS is a relatively new term in the world of transport, which is often credited in part to a Masters' thesis in Finland (Heikkilä, 2014). Aapaoja, Eckhardt, Nykänen, and Sochor (2017) point out that the typical characteristics of products described as MaaS are that they have a one-stop-shop principle. In other words they provide for mobile ticketing and payment, and include information to allow users to plan multimodal journeys and rerouting. However, others have shown that some key aspects of MaaS are very similar to earlier transport ideas, such as Advanced Traveller Information Systems (Lyons, Hammond & Mackay, 2018), which was promoted in the 1990s as an essential element in the integration of different transport modes which empowered users in planning and re-planning journeys themselves. Clearly the intermodal and multi-service provider aspect of completing an end-to-end journey is not entirely novel (Utriainen & Pöllänen, 2018; Hensher, 2017).

However, there is some novelty in personalization to individual preferences (such as desired travel time, pro-environmental values, cost, and accessibility requirements) as well as in the use of mobile apps as the primary user interface for booking and payment. In the servitized model first set out by Heikkilä and since promoted heavily by MaaS Global there is a user registration requirement for full access to the product features, and there is very often a monthly subscription package. This aspect of the business model delivers an additional value proposition and a new source of profit, as registered users generate a lot of data that has the potential to be used for service planning, traffic forecasting and urban management. The type of data generated by a MaaS platform is therefore an important commodity when combined with new data analysis capabilities.

Many of these features (such as the development of new sources of data), were found pre-2014 in schemes and projects that aimed to either increase the use of sustainable mobility modes, such as UbiGo (Karlsson, Sochor & Strömberg 2016) and Superhub (Forbes, Wells & Masthoff 2012) or to increase the efficiency of routing and mode choice such as the personalized multi-modal journey planner and interface built for MyWay (Pou, 2015). These projects which sought to achieve more sustainable *urban* mobility through the use of digitized coordination can be viewed as prototypes on the way to the current form of MaaS (see for example Sochor, Strömberg & Karlsson, 2015).

Digital innovation is generally seen as an opportunity for economic development. As a result MaaS has come to be seen as a growth area, and many public-private coalitions are emerging to build the case for MaaS and unlock government action and investment. Table 1 surveys a range of definitions of MaaS that are in current use by such coalitions. Most of these definitions owe much to Heikkilä (2014) and

Table 1. A selection of MaaS definitions in current use in Europe and beyond

Where	Who	Definition
Australia	Mobility as a Service Australia	*Mobility as a Service (MaaS) is all about you – providing you with personalised mobility solutions that make use of this wide range of mobility options. Mobility as a Service brings every kind of transport together into a single intuitive mobile app. It seamlessly combines transport options from different providers, handling everything from travel planning to payments. (*http://maasaustralia.com/*)*
Catalonia	MaaS Catalonia	*Mobility as a Service is a user-centric, intelligent mobility distribution model in which all mobility service providers' offerings are aggregated and supplied to users through a single digital platform.*
Europe	MaaS Alliance	*the integration of various forms of transport services into a single mobility service accessible on demand. To meet a customer's request, a MaaS operator facilitates a diverse menu of transport options, be they public transport, ride-, car- or bike-sharing, taxi or car rental/lease, or a combination thereof. (*https://maas-alliance.eu/homepage/what-is-maas/*)*
Finland	ITS Finland	*A mobility distribution model in which all of a customer's major transportation needs are met from a single platform by a single service provider that orchestrates each individual transport service component to meet a customer's end-to-end service expectations. Typically, services would be bundled in to a package similar to current mobile operator services. (*https://www.its-finland.fi/index.php/en/palvelut/mobility-as-a-service.html*)*
The Netherlands	Ministerie van Infrastructuur en Milieu, citing the White Paper 'Mobility as a Service' by MuConsult (2017)	*The provision of multimodal, demand-driven mobility services, whereby customers are offered tailor-made travel options via a digital platform (e.g. Mobile app) based on real-time information, including payment and the handling of transactions.*
UK	Parliamentary Transport Committee	*MaaS is the term for the digital platforms (often smartphone apps) through which people can access a range of public, shared and private transport, using a system that integrates the planning, booking and paying for travel.* (House of Commons, 2018, p3)

the subsequent promotional efforts of a Finnish company (MaaS Global) that has captured the policy imagination, particularly in Europe (Pangbourne et al., 2019).

There are differences between these definitions which can cause confusion. Jittrapirom et al., (2017) have identified a dozen conceptualisations of MaaS. They extrapolated from these a set of nine core characteristics: integration of transport modes; tariff option; a single platform; multiple actors; use of technologies; demand orientation; registration requirement; personalization; and customization, the first seven of which are used in this chapter to frame a set of requirements for rural MaaS.

MaaS Focal Points: Urban Areas, Efficiencies and a Commoditized Value Proposition

In relation to MaaS, most government and commercial focus has concentrated on urban developments, something that fits well with the Smart Cities paradigm which attracts a lot of investment. Whilst it is beyond the scope of this chapter to explore Smart Cities, it is useful to note that Hollands (2008) describes Smart Cities as consisting of using "networked infrastructures to improve economic and political efficiency and enable socio, cultural and urban development" (p. 307). Transport is one of the central pillars of the movement but there is nothing comparable for rural areas, even though rural areas are equally deserving of social and cultural development and would benefit from networked infrastructures.

This and other urban transport issues are important policy drivers that are harnessed to much rhetoric around transport innovation, including the claim of MaaS promoters that their model provides an efficient solution to congestion and sustainability, by enabling public transport to be supported by other traditional and innovative transport modes. At the same time, they promote MaaS as a benefit for end-users by making all forms of transport easier to use through a single interface (i.e. reducing the cognitive load involved in travel mode choice (Stradling, Hine & Wardman, 2000).

This idealized claim that MaaS improves the efficiency of transport leads to it being presented as the epitome of 'shared mobility'. Shared mobility is a term covering a number of forms of vehicle or ride-sharing, which is promoted as improving efficiency. It is not necessary to go further into it for the purposes of this chapter, though Le Vine and Polak (2015) is a place to start for interested readers. They also link shared mobility to the sharing economy, with both ideas generally perceived as underpinned by Information and Communications technology (ICT). Rhetorically of course 'sharing' is perceived as a virtue, so to link it to MaaS is to make MaaS sound virtuous by association. However, it has been pointed out by Bardhi and Eckhardt (2012) that the 'sharing economy', is not communitarian but commercial, with the promoted value proposition being more driven by commodifying and monetizing access to transport than by pro-social values of sharing systems supporting reciprocity and social connection. Thus the profit-motive will seek either to increase mobility when it more properly ought to be reduced for many people if transport's negative impacts are to be addressed (particularly climate change) or to make current levels of mobility more expensive once a market is captured, which would be undesirable for those who already suffer a mobility deficit due to poor affordability, or where transport costs are already higher, such as in rural areas. Therefore there are grounds for not assuming that MaaS will be a) virtuous or b) efficient in addressing congestion and other negative impacts associated with excess or lack of transport, even in urban settings.

Rural Contexts

Having established that the focus of innovation attention is on larger cities, this potentially leaves substantial land areas, and their residents, neglected. To take the United Kingdom (UK) as an example, there are several areas that are classed as rural. Within Scotland, for example, most of the Highlands and Islands region is classified as Remote Rural, which are defined as areas with a population of less than 3,000 people, and with a drive time of between 30 and 60 minutes to a Settlement of 10,000 people or more (Scottish Government, 2018). Other areas with similar characteristics within the UK include Wales, Cumbria and Cornwall. English, Scottish and Welsh rural areas are facing the challenge of an aging population - 60% of the English and Welsh rural population are over 45 years old (ONS, 2018). In Wales, the number of people aged 65 and over is projected to increase by 36.6 per cent between 2016 and 2041 (ONS, 2017) and about one-third of the Welsh population live in rural areas (Gartner, Gibbon & Riley, 2007). In a wider context, most European countries have substantial areas that may be classed as some variety of rural. The next section explores the characteristics of rural areas that make them distinctive in terms of transport provision.

CHARACTERISTICS OF RURAL AREAS

What many rural areas in developed nations have in common is that communities are generally facing an older but smaller population. The shrinking and ageing of rural populations is related both to general demographic trends with increased longevity combined with lower birth rates as well as to a continued loss of young people who 'self-urbanise' to seek higher education or employment (Amcoff & Westholm, 2007). The effect is to lead to difficulties in maintaining public services, bringing into question the long-term sustainability of rural areas. Whilst living in rural areas is sometimes framed by policy makers as a lifestyle choice that carries with it the need to trade-off certain expectations, this is a skewed perspective that is drawn upon to justify the closure and centralization of public services (e.g. Woods, 2006; Van Steen & Pellenbarg, 2010). For many countries rural regions will remain important for their primary industries, as settings for leisure and tourism and as managed (to greater and lesser extent) habitats providing critical ecosystem services. Therefore, there will remain a need for a rural workforce, which will live, work, raise families and retire there. This population will need mobility and their requirements will

Figure 1. The rural MaaS challenge

vary over the course of a life, yet the challenge of organizing public transport in rural areas is a long-recognised and increasing issue that is compounded by highly individualized solutions based on an increased reliance on personal vehicles.

Against this background rural MaaS needs to address specifically rural challenges arising from the combination of lower population densities with sparser transport networks, patchy ICT infrastructure, the accessibility requirements of older people, and supporting the continued societal need for the products of rural industries. This is itself a challenge as the requirements for supplying the mobility needs of rural livelihoods are quite different in many ways from the needs of the retired rural population. These challenges are summarized and represented visually in Figure 1.

In this section, each of these challenges and their rationale are described in turn, and the section ends with a discussion of how these issues are addressed in the typically positive MaaS rhetoric.

The Challenge of Sparse Populations and Transport Networks

Large areas and low-density populations are defining features of rurality. Mobility-related consequences of this are sparse transport networks and the need to cover long distances. Employment related consequences are a preponderance of primary industry or jobs based on the natural environment in some way.

Liimatainen and Mladenović (2018) highlight that public transport is the backbone of every MaaS offering thus far, so there is clearly a challenge to resolve for rural MaaS, where public transport has been tending to contract for several decades. Surakka, Härri, Haahtela, Horila and Michl (2018) highlight that Finland, for example, has a very low level of rural public transport provision which has led to a high degree of dependence on private cars, and they contrast this with the provision in Switzerland, in which an integrated system is delivered between the railways and the Postbus network which have closely coordinated timetables. However, the rural geographies and infrastructure of these two countries are not very comparable.

Whilst it is unlikely that all the developments that have been afforded to urban areas can be transferred to rural areas, delivering better accessibility more sustainably is essential in order to support the continued attractiveness of these areas as working landscapes where people will need to live in the future. Eckhardt et al., (2018) highlight that it is recognized that *mobility challenges are not only restricted to cities, as they also concern social inclusion and the connectivity of rural areas* (European Commission, 2017), yet most of the investment and attention thus far is centred on high density cities with a plurality of transport modes potentially available for integration within a MaaS ecosystem. It is also more challenging to address transport decarbonization in rural areas (Sovacool, Noel, Kester & de Rubens, 2018).

Additionally, something that is often left out of MaaS visions is non-passenger mobility. Some disruptive innovators impacting on the transport sector and urban environment certainly do keep aspects of logistics in their sights through short-lived experimental pilots or more established services (for example Uber, Deliveroo, or JustEat), but it is not a strong element in most material on MaaS, as passenger and freight have been largely separated in recent decades, though postbuses and parcels sent by passenger rail were standard in the past and the MaaSiFie project included rural logistics in its study (Eckhardt et al., 2018).

Under-Served and Ageing Demographics

It is well documented that there is an undeniable need for socially disadvantaged groups (e.g., older adults, young, and disabled) in rural and remotely located areas to be better served in relation to their mobility requirements (Currie, 2010). Delivering better rural accessibility more sustainably is a difficult challenge in relation to the ageing population. Health and gerontology research has established that out-of-home mobility is a significant component in maintaining physical health and mental well-being in old age. As having a driving licence and access to a car has become a significant predictor of higher overall mobility in our automobile-oriented society, loss of that access, through impairment to driving ability can have significant ill effects (Zeitler & Buys, 2015). For example, Marottoli et al., (2000) found a strong association between driving cessation and decreased out-of-home activity.

In a highly car dependent society, a lack of alternatives to driving leads to transport disadvantage amongst older adults (Engels & Liu, 2011), yet older adults now expect increased mobility compared to previous generations (Alsnih & Hensher, 2003) and may not plan for the possibility of driving cessation in retirement (Musselwhite & Shergold, 2013). Most focus has been on technological means of extending independent safe driving in private cars for longer (Nordbakke & Schwanen, 2015), rather than on providing other means of transport, such as demand responsive public transport (DRT), even though DRT models for public transport were demonstrated as feasible for rural areas several decades ago.

For rural areas, there is the added issue of the centralization of essential services, as discussed in the previous sub-section above, resulting in a need to consider the demand for all people, including older citizens, to travel beyond their local area for essential purposes (Ruyser & Halseth, 2012). A non-transport response to that has been an exploration by the health sector and health service researchers of care-at-a-distance via ICT, but that has not yet matured, and acceptance by the public of the need to engage with technology in order to access services has been proceeding more slowly outside metropolitan areas (Currie, Philip & Roberts, 2015). This is partly due to ICT connectivity issues in rural areas, but the characteristics of older demographics are undoubtedly a factor.

Whilst there is a cohort effect in that subsequent generations of older people will have more ICT skills and positive attitudes, these are modified by cognitive and physical impairments resulting in interface accessibility issues as well as the continual advance of ICT beyond existing users' abilities (e.g. Pangbourne, Aditjandra & Nelson, 2010). There is also a valid critique arising from the Smart Cities debate about the 'disciplining' of individuals through the increased 'forcing' of technological adoption (e.g. Vanolo, 2014), that would be applicable to rural MaaS as much as to the Scottish Government's insistence on a wholesale shift to smartcard ticketing or the Swedish move to become a cashless society. A more detailed discussion of this aspect of MaaS is needed but outside the scope of this chapter.

However, the issue of a lack of coordination between different policy sectors is a significant barrier to providing a better level of mobility in rural areas. Pangbourne et al. (2010) highlight the lack of coordination between the health service and public transport realities in setting appointment times for example, and there is also no coordination between transport organized and provided by the healthcare system with other socially necessary transport services, probably resulting in more vehicle movements than strictly necessary for the level of population mobility supplied. Better coordination between *sectors* is similarly a key recommendation of Eckhardt et al. (2018), and rural MaaS is potentially an ideal candidate for this type of coordination.

Limitations of Rural Information and Communications Technology Connectivity

Given the role of ICT in enabling MaaS, which is sometimes portrayed as the ultimate form of demand responsive mobility, the questions of adoption and acceptability cannot arise until there is availability. As noted above, ICT is not ubiquitous in rural areas, yet it is a key enabler for MaaS, and adoption of MaaS applications requires familiarity with mobile interfaces. The potential for MaaS to open a new front in digital exclusion has been highlighted by others (for example Sochor and Nikitas (2016)). For example, Polydoropoulou, Pagoni and Tsirimpa (in press) carried out stakeholder and end-user research in the Greater Manchester region (UK), and found that low levels of digital take up and low trust in digital payments are social barriers to MaaS implementation, and that poor internet coverage is both a social and operational barrier. For example, in terms of socially disadvantaged groups, it was recognized that the 'unbanked' would not be able to access MaaS services, and this was rated as a strong operational barrier. The results of their workshops in Budapest (Hungary) were substantially the same.

If these are barriers in highly urbanized areas, then the prospect for rural areas is unlikely to be more favourable. Rural areas have long been a focus of attention for academics interested in social inclusion due to transport connectivity issues and more recently ICT has been added to the list of issues. See Nutley (1980) for example in relation to transport-related disadvantage. Ashmore, Farrington and Skerratt (2017) look at community-led broadband in rural digital infrastructure to assess its impact on the resilience of rural communities. They find that community-led broadband featuring a "localism" development approach can strengthen local rural individual identity and rural resilience, but it may also entrench existing inequalities and feelings of exclusion. Wallace, Vincent, Luguzan, Townsend and Beel (2017) investigate how ICT influences social integration in rural communities. They conclude that ICT is an integral part of rural social relations but can play very different roles for different social and cultural groups as well as for communities in different locations. Ashmore, Farrington and Skerratt (2015) look at the relationship between rural dwellers and high speed broadband, highlight that the relationship is complex and can be contradictory at times, and that more research needs to be done.

MAAS AND TECHNOLOGICAL OPTIMISM

Despite what is known about the inequalities in technology diffusion and adoption, technology optimism, and even determinism, is dominant in the promotion of MaaS. Positive rhetoric which praises the perceived benefits of MaaS is widespread and a number of umbrella bodies have been created to bring together stakeholders, particularly commercial stakeholders, whether working at the centre or on the fringes of the concept. Several large claims have been made for the potential benefits of MaaS, such as *reduced road congestion, improved air quality, healthier travel choices, increased efficiency in transport networks and more effective management of transport demand.* (House of Commons, 2018 p. 3), but it is also recognized (beyond the walled gardens of the promotional alliances) that uncontrolled deployment of MaaS could have negative consequences (Pangbourne et al., 2018).

Examining MaaS Through a Rural Lens

Firstly and significantly, much of this rhetoric is spatially blind. Few of the claims made for MaaS focus on addressing rural problems. For example, the alliance organization, MaaS Catalonia, has developed a MaaS Manifesto containing a set of principles that should guide how MaaS should be deployed in Catalonia. This does not include any reference to addressing geographical inequalities, though one might interpret the principles as implicitly covering non-urban settings (*MaaS should foster social inclusion and promote socially responsible behaviour. Social inclusion and inequality concerns need to be a fundamental part of the service for it to be socially sustainable. For governments, MaaS is the opportunity to address a wide range of policy goals from environmental sustainability, to smart mobility and integration.* Principle 2, MaaS Manifesto, MaaS Catalonia). Given that the Generalitat de Catalunya was a project partner in MyWay (EU FP7 Grant number 609023) and MyWay was required to cover the whole of Catalonia, not just the metropolitan area of Barcelona, it would suggest that governance actors intend that rural areas should be served equitably through MaaS initiatives, but this is not always made sufficiently explicit.

The MaaS documents from promoters and stakeholders that were collected for this study have been surveyed for the attention they pay to deploying MaaS in a rural setting. The results are shown in Table 2, and with only a very few exceptions, the focus is clearly 'urban by default', with hardly any material specifically focused on requirements for rural MaaS. MaaS Scotland stands out, if only because their document contains a lot of summaries of pilots and projects under development in rural areas, which perhaps reflects the Scottish Government's commitment to the economic and social interests of its substantial rural areas, something that subsequently

Table 2. Mentions of 'rural' in non-academic MaaS-related documents, in date order from 2015-2018

Title	Source and Date	No. of Pages	Mentions of Rural	Comments
Mobility as a Service Can it be even better than owning a car?	Sampo Heitanen and Sami Sahala, slide set, no date	56	0	
Exploring the opportunity for Mobility as a Service in the UK	Transport Systems Catapult, July 2016	52	1	
Mobility as a Service: the next transport disruption	LEK Consulting sponsored by UITP & TTF Australia, no date but not before 2016	14	0	
Mobility as a Service The End of Car Ownership?	Sampo Heitenan, CEO, MaaS Global, slide set, no date but not before 2016	43	0	
The rise of mobility as a service	The Deloitte Review, issue 20, 2017	20	0	
Reimagine Places: Mobility as a Service	KPMG, August 2017	32	3	Developed an illustrative scenario for a rural setting
MaaS Roundtable	Foresight Future of Mobility Project, Government Office for Science (UK)	4	0	
Guidelines and recommendations to create the foundations for a thriving MaaS ecosystem	MaaS Alliance White Paper, September 2017	22	1	
EPOMM Project e-update	EPOMM Project, December 2017	4	0	
Moving Together	MaaS Alliance, leaflet, no date	2	0	
Mobility as a Service: Positioning Scotland for an Emerging Global Market	MaaS Scotland, January 2018	24	19	Rural is mentioned in a number of pilot project summaries.
Mobility-as-a-Service The value proposition for the public and our urban systems	ARUP, March 2018	52	4	On the basis of little evidence, the conclusion comments "Current evidence suggests that service provision in rural and low density suburban locations is a strong proposition" p. 41
Mobility as a Service Putting Transit Front and Center of the Conversation	Cubic, March 2018	24	1	
Data makes MaaS happen - MaaS Alliance Vision Paper on Data	MaaS Alliance, November 2018	9	0	
Exploring Mobility-as-a-Service: insights from literature and focus group meetings	Netherlands Institute for Transport Policy Analysis (KiM) for Ministry of Infrastructure and Water Management, November 2018	28	12	Most mentions indicate the geographic location of participant quotes rather than a substantive treatment of requirements or opportunities for rural MaaS. The conclusions are not promising for rural MaaS: "In order for MaaS to be successful, the service must, at any rate, offer autonomy and flexibility, be reliable, and ideally be available anytime and anywhere. Presumably, the latter is not a realistic point of departure in rural areas." p. 26
Mobility as a Service	Intelligent Transport In-Depth Focus, volume 2, issue 3, no date	13	0	
Mobility-as-a-Service and changes in travel preferences and travel behaviour: a literature review	KiM Netherlands Institute for Transport Policy Analysis for Ministry of Infrastructure and Water Management, September 2018	57	2	The mentions are in the context of a review of DRT in rural areas, with MaaS covered as an URBAN concept
A Business Case for Mobility as a Service	SkedGo, Business brochure, 2018	7	0	
Problem Statement: Mobility as a service business models	New South Wales Government, 2018	4	1	

is reflected in a recent funding call for innovation projects (Transport Scotland, 2019) (at the time of writing the successful projects had not been announced). These sources are listed in date order, but there is no apparent trend to increase or decrease focus on rurality.

Requirements for Developing Rural MaaS

It remains the case that most MaaS proposals, pilots and deployments have been initiated in highly urbanized settings. In order to understand what is necessary to apply MaaS to rural transport issues, we can look at the problem through the first seven core characteristics of MaaS identified by Jittrapirom et al., (2017). The author does not consider personalization and customization further here.

- **Integration of Transport Modes:** This characteristic presupposes that there is a choice of existing modes that should be integrated, to enable users to undertake multimodal trips more easily. Jittrapirom's model does extend beyond urban settings by encompassing long-distance services (bus/coach or train) as well as flights and ferries, but this overlooks the daily mobility practices of rural residents, which may be intra-rural or between rural and urban environments. There is therefore a need to understand the trip-making needs of rural residents and comparing that to the available transport modes before attempting design a rural MaaS.
- **Tariff Option:** A key characteristic of new MaaS models is that they offer users a tariff choice between a) paying a monthly subscription which buys access to a range of transport options with caps on use based on distance or time or numbers of trips or b) paying per use (Pay-as-you-Go, PAYG). Typically, the PAYG option is more expensive per trip than a bundled package, requiring the user to guestimate which option will provide them with better value. Caps based on distance or time are likely to work out very expensive in rural areas. Therefore, there is a need to design fair pricing models for rural areas.
- **One Platform:** This characteristic of MaaS is based on users having a single point of access (though it could have several interfaces, e.g. web-site or mobile app) for accessing mobility, potentially alongside other services, such as weather forecasts, synchronization with personal calendars, a trip record and other feedback. A potential issue for rural areas is that rural dwellers may fall between different MaaS providers' regions or need to travel alternately between the areas covered by different MaaS providers if these radiate outwards from urban centres, and this aspect of interoperability needs to be considered from the perspective of people outside urban settings.

- **Multiple Actors:** The digital MaaS platform is multi-sided, providing benefits for multiple actors in the transport ecosystem, both on the demand and supply side, as well the platform owners. Cooperation is also needed by additional non-transport actors – handling payments, providing data management services, telecoms services, other information services (e.g. weather forecasts). On the face of it, this aspect does not have peculiarly rural facets, except that there may be fewer other information services and telecoms services may depend on availability of infrastructure. In terms of achieving economies of scale to support public transport, in the past rural areas often relied on hybrid passenger/small logistics services such as Post Buses. ICT provides opportunities to recreate this approach for 21st century rural areas: this potential needs more in-depth evaluation and scoping (and could be linked to the tariff characteristic above).
- **Use of Technologies:** A key aspect of technology that underpins MaaS is the presence of a reliable mobile internet network. This is an area of weakness in many rural areas, that needs to be overcome if rural MaaS is to provide the quality of service to mobility users and suppliers alike. Suitable ICT infrastructure is a foundational determinant that precedes any consideration of user adoption and acceptability.
- **Demand Orientation:** The MaaS paradigm is described as user-centric, in its offer of 'the best' option(s) for the individual user, with demand-responsive modes being available. In a rural setting, there are generally far fewer alternative modes available from which to select a 'best' option, and the economies of scale for providing more mobility are likely to be reliant on concurrent sharing of rides, whereas sharing in urban settings can be more reliant on sharing vehicles sequentially. There is a public ambivalence about sharing that needs to be overcome (Marsden, Anable, Bray, Seagriff & Spurling, 2019); this is an aspect where rural areas might have an advantage of urban areas, as there is more of a culture of informal ride-sharing (Gray, Shaw & Farrington, 2006).
- **Registration Requirement:** The requirement to register with the MaaS platform is fundamental to the personalization elements (and data generation) that are central to the value proposition for MaaS. This can be as an individual, or via a group (such as a business or a household). Whilst the MaaS product might utilize public transport modes (such as bus, tram, metro or train), it is not itself public transport. The requirement to register is a barrier to use as it requires the adoption of one technology to use another, and it also requires users to have certain other privileges such as a bank account. Adoption of some of the technologies around this is affected by both rural infrastructure

issues and acceptance. For example, there may also be acceptance issues relating to registration with older demographics found in rural areas.

SOLUTIONS AND RECOMMENDATION

Rural mobility has been a topic of research for a many years, but transformation in rural connectivity is as patchy as ICT for many. In the UK as elsewhere, 'austerity' policies have resulted in dramatic contractions to the bus network. Considering this, there has been renewed interest in finding solutions. The problems of rural areas that are exacerbated by the characteristics of rural communications networks (both transport and ICT) are well recognized, as discussed above. It is also the case that MaaS solutions have been developed almost exclusively for urban settings, where the issues are congestion and air pollution, with social and distributional issues in accessibility. Consequently, the rhetoric of MaaS proponents is focused on addressing urban problems, omitting the opportunities offered by addressing rural needs, as is demonstrated in the analysis of the documents that promote and discuss MaaS (see Table 2).

Whilst MaaS is now being reconceptualized for rural environments, see for example the rural MaaS principles developed by Eckhardt et al., 2018, but also the guidance issued by Transport Scotland in relation to its MaaS Innovation Fund (Transport Scotland, 2019b), this chapter has sought to demonstrate how the problems that need to be addressed are different to those in urban areas. This is step is important for authorities to be able to take appropriate action. Accessibility and inclusion issues arise from poor connectivity and consequent functional car dependence is linked to high greenhouse gas emissions from rural transport – there are generally no public transport options beyond buses and providing bus public transport in rural areas is very expensive.

As most rural residents have increasingly provided for their own mobility by acquiring access to private cars, the existence of any kind of bus network becomes less and less viable. There has been awareness of this issue for many years. The EU-funded FAMS (Flexible Agency for collective demand-responsive Mobility Services) project achieved proof-of-concept for a technology that would support the provision of rural demand responsive public transport in 2004 (CORDIS, no date). The FAMS project ran successful trials in Tuscany (Italy) and Angus (Scotland, UK). This was clearly an early example of what we now refer to as MaaS. In relation to taking an existing DRT model and digitizing it, thereby bringing DRT closer to Intelligent Transport Systems (ITS), the FAMS project made some important advances. However, this model has yet to be widely adopted, with regulatory and institutional barriers identified by the project still existing.

However, there is also a fourth question that is relevant to many rural areas:

4. How might rural MaaS enable or constrain the mobility of day visitors and tourists to rural areas?

Given the economic significance of rural tourism, and its high environmental impact, it is important to direct tourists to services with lower impacts. Registration can be a barrier to that, and it is hard to imagine that rural residents will be willing to offer peer-to-peer sharing with non-resident strangers in sufficiently large numbers to have a meaningful environmental impact. Furthermore, given the substantial dependence of most rural populations on travel in private cars, there are some significant issues in developing a multi-modal MaaS provision that will be adopted by the rural population, is it possible for a rural MaaS to provide a comprehensive transport service without a substantial reduction in individual flexibility? Adopting a social enterprise model for a rural MaaS may be one appropriate way forward capable of addressing the four posed questions, and there could be technological solutions to support the establishment of interpersonal trust in privacy protective ways using agent-based software models.

FUTURE RESEARCH DIRECTIONS

In considering how rural MaaS needs to be designed to address specific spatial and social context of rural areas, it is helpful to understand that whilst it needs to contribute to solving a range of problems, it is unlikely to be the sole solution. This is a helpful perspective from which to consider how rural MaaS needs to be developed, and the four guiding questions above provide a framework for systematically interrogating the degree to which a rural MaaS proposition addresses those questions.

More attention needs to be paid to the range of perspectives on feasibility and adoption that exist amongst rural stakeholders and potential end-users. In relation to feasibility this relates strongly to what would constitute a suitable business model for rural MaaS. A true 'sharing' model may sit better with the traditional ideas of rural reciprocity and mutual support, suggesting that social enterprise models, and user-centric design, should be a priority for focused investigation.

Furthermore, there needs to be more research into how rural MaaS platforms can enable peer-to-peer sharing alongside both community and commercial provision (perhaps through a social enterprise model), how co-distribution and shared small logistics/passenger transport can be revived and stimulated and also how the continued need for personal-car based transport can be brought into the ecosystem beyond lift-sharing, to stimulate the uptake of low-carbon vehicles.

There are also specific and unique opportunities in rural areas: one transport issue in many rural areas is the long distances for parcel delivery (which in the case of some parts of the UK results in extra charges for customers), there are institutional barriers to overcome to integrate between logistics and passenger transport. Whilst many levers for the provision of passenger transport and delivery of transport policy lie at local level for example, this is not the case for logistics. There needs to be space made for testing innovative ideas, which creates a collaboration requirement.

The Scottish Government (Scotland is a constituent nation of the UK with its own devolved administration) has recognized that smart mobility poses an innovation issue for Scotland – whilst more than half the population live in the urbanized central belt, considerably more than half the land area is remote rural, and poorly connected to the railway network. Scotland is a highly car dependent nation outside the main urban centres. The Scottish Government is investing £2m over three years to stimulate near-market innovation in MaaS (the MaaS Scotland Investment Fund) (Transport Scotland, 2019a). There is an expectation that the proposals will contribute to practical testing of the viability of MaaS in Scotland in a manner that also addresses the strategic agenda of the Scottish Government. The aims are specified as being modal shift to sustainable public transport alternatives, reducing carbon emissions; healthier lifestyles, by incorporating active travel modes (traditional modes of transport such as bus or rail are considered active as they have an element of walking involved to access, as well as more regular active options such as bike schemes); but also supporting digital innovation within Scotland, specifically relating to public transport. In year 1 proposals must fit with one or more of three themes – rural, islands and communities; tourism; and Tackling Inequality, Accessibility & Mobility barriers (Transport Scotland, 2019b).

However, it remains to be seen whether proposals will be fit for rural areas, as the limitations of the fund restrict heavily the ability to undertake research, and thus are looking for pilot projects that already have a Minimum Viable Product, with a need for minimal development, existing evidence for a market, a commercial strategy for a sustainable business model and to have already undertaken co-creation or stakeholder engagement. There is also a technical requirement that pilot projects should be compatible with the existing smart mobility information infrastructures (cEMV, mobile and ITSO). This could strongly limit any community MaaS proposal and will hamper new ideas from outside the traditional transport sector. Notwithstanding that MaaS is not really a new concept, the fundamental features of what we now call MaaS has potential to provide some solutions to existing problems in transport provision is clearly recognized by policy-makers, transport authorities and academics.

However, there is a risk that the digitizing and 'subscriptionizing' of access to mobility through a single point might have dangers in relation to exclusion, particularly for those wishing to travel in rural areas – part of this relates to the existence of less well off or older individuals without bank accounts, or from a strong tendency not to use mobile technology at all, or at least not to use it for payment. However, a structural issue relates to the ICT connectivity and mobility infrastructure constraints of rural areas. Demand responsive transport, bookable by telephone, has long been an answer to providing socially necessary transport solutions in rural areas, and these have also been digitized (as with the FAMS discussed above), but as most DRT solutions are community transport services, often run by charity groups, this existing solution has not yet been systematically incorporated into a rural MaaS solution. However, if this is to make use of the benefits of mobile ICT, the rural connectivity issue must be addressed, which requires collaboration and coalition building between the *community* transport sector (to provide a use for the ICT infrastructure), government (to provide incentives) and telecommunications (to build the infrastructure).

Serving a collective good is a core ethos of community transport providers, and thus these actors may not naturally find common ground with commercial MaaS brokers. Focusing on individual preferences in MaaS product design (such a self-centred focus is clear in the MaaS Australia articulation of the concept, see Table 1) draws us away from the collective good. However, clarity is needed around how exactly MaaS can avoid exacerbating known future issues and this means that influencers and decision-makers need to ask three questions:

1. What are the local transport problems, and would MaaS address these?
2. Does the MaaS proposition reduce or worsen local social problems where unequal accessibility/mobility is a contributory factor?
3. Can MaaS genuinely make users' transport choices more environmentally sustainable by reducing GHG and other negative impacts in the specific context under consideration?

The answer to question one has been so far seen as most significant for urban areas which are beset by significant air quality problems and congestion. However, in this environment, the transport ecosystem is such that MaaS providers can be both regulated and incentivized to encourage their users to be less reliant on the car-based options in the multi-modal mix. i.e. the bulk of trips made by MaaS users need to be by public transport (bus, light and heavy rail) and to involve much more walking and cycling (potentially also e-scooters where these are legal).

There are some obvious limits to active travel's suitability in rural areas, where trip distances are typically much longer and lacking in safe infrastructure for walkers and cyclists. Furthermore, as highlighted in the foregoing sections, the contraction of rural public transport has proceeded in a vicious cycle with rising private car use

over many decades. Thus, the issues that MaaS propositions should focus on in ru areas are those relating to social inclusion and greenhouse gas emissions reductio

However, in rural areas cross-benefits might be achieved through co-distributio where freight vehicles are used by more than one company to improve loading and reduce overall vehicle mileage, or combining some logistics activities with passenger mobility, as with the example of the Swiss Postbus (which used to exist in the UK for example, at least until the 1980s in some places). Such a move would strongly synergise with another widespread trend: the increase in internet-facilitated home-shopping and associated rises in light goods vehicle-based deliveries. Better coordination of this kind of traffic, including co-distribution and the potential to combine with passenger movements, offers GHG emissions reductions for rural areas which are otherwise 'hard to treat'.

MaaS can also be designed to address question two. This is particularly pertinent to rural areas, and the community transport DRT services are a case in point. Rural traditions of informal lift-giving within families and communities might also be leveraged through a true 'sharing economy' facilitated with MaaS-style mobile apps that reward sharers in some way. However, the public nature of rural transport services might be undermined. For example, can we really call transport public if citizens must register before they can use it?

In relation to question three, there is a need to measure the true impacts of solutions devised to address questions one and two, especially as the sharing of vehicles and services is seen as particularly crucial to making transport more sustainable in rural areas, and such sharing is intuitively seen as more environmentally sustainable. If it can be shown that a rural MaaS can address the transport issues, and that the social needs of rural residents can be equitably supported by the product at the same time, can it also be shown that there is a net reduction in environmental impacts?

It appears to the author that one thing that has not been found in MaaS to date, even in urban settings, is the inclusion of vehicle fueling (though there are fuel card account systems used in the logistics industry or by businesses). For rural areas, where functional car dependence is almost certain to remain for some considerable time, the need to address climate change by switching to low carbon vehicles could be supported by including fueling for any fuel type or vehicle-to-grid contracts in rural MaaS business models. How this might work is that customers could book time on an EV point at a Park and Ride system (or wherever they need to go, be it their workplace, shopping centre, place of education, or healthcare provider) via the MaaS, and that the energy required by their vehicle is either PAYG or an allowance included in their monthly package; if their vehicle is supplying energy to the grid the user is paid instead. Some elements of this approach would be resilient to different forms of low-carbon vehicle, should EVs not be the sole or longer-term solution.

There are also specific and unique opportunities in rural areas: one transport issue in many rural areas is the long distances for parcel delivery (which in the case of some parts of the UK results in extra charges for customers), there are institutional barriers to overcome to integrate between logistics and passenger transport. Whilst many levers for the provision of passenger transport and delivery of transport policy lie at local level for example, this is not the case for logistics. There needs to be space made for testing innovative ideas, which creates a collaboration requirement.

The Scottish Government (Scotland is a constituent nation of the UK with its own devolved administration) has recognized that smart mobility poses an innovation issue for Scotland – whilst more than half the population live in the urbanized central belt, considerably more than half the land area is remote rural, and poorly connected to the railway network. Scotland is a highly car dependent nation outside the main urban centres. The Scottish Government is investing £2m over three years to stimulate near-market innovation in MaaS (the MaaS Scotland Investment Fund) (Transport Scotland, 2019a). There is an expectation that the proposals will contribute to practical testing of the viability of MaaS in Scotland in a manner that also addresses the strategic agenda of the Scottish Government. The aims are specified as being modal shift to sustainable public transport alternatives, reducing carbon emissions; healthier lifestyles, by incorporating active travel modes (traditional modes of transport such as bus or rail are considered active as they have an element of walking involved to access, as well as more regular active options such as bike schemes); but also supporting digital innovation within Scotland, specifically relating to public transport. In year 1 proposals must fit with one or more of three themes – rural, islands and communities; tourism; and Tackling Inequality, Accessibility & Mobility barriers (Transport Scotland, 2019b).

However, it remains to be seen whether proposals will be fit for rural areas, as the limitations of the fund restrict heavily the ability to undertake research, and thus are looking for pilot projects that already have a Minimum Viable Product, with a need for minimal development, existing evidence for a market, a commercial strategy for a sustainable business model and to havc already undertaken co-creation or stakeholder engagement. There is also a technical requirement that pilot projects should be compatible with the existing smart mobility information infrastructures (cEMV, mobile and ITSO). This could strongly limit any community MaaS proposal and will hamper new ideas from outside the traditional transport sector. Notwithstanding that MaaS is not really a new concept, the fundamental features of what we now call MaaS has potential to provide some solutions to existing problems in transport provision is clearly recognized by policy-makers, transport authorities and academics.

However, there is a risk that the digitizing and 'subscriptionizing' of access to mobility through a single point might have dangers in relation to exclusion, particularly for those wishing to travel in rural areas – part of this relates to the existence of less well off or older individuals without bank accounts, or from a strong tendency not to use mobile technology at all, or at least not to use it for payment. However, a structural issue relates to the ICT connectivity and mobility infrastructure constraints of rural areas. Demand responsive transport, bookable by telephone, has long been an answer to providing socially necessary transport solutions in rural areas, and these have also been digitized (as with the FAMS discussed above), but as most DRT solutions are community transport services, often run by charity groups, this existing solution has not yet been systematically incorporated into a rural MaaS solution. However, if this is to make use of the benefits of mobile ICT, the rural connectivity issue must be addressed, which requires collaboration and coalition building between the *community* transport sector (to provide a use for the ICT infrastructure), government (to provide incentives) and telecommunications (to build the infrastructure).

Serving a collective good is a core ethos of community transport providers, and thus these actors may not naturally find common ground with commercial MaaS brokers. Focusing on individual preferences in MaaS product design (such a self-centred focus is clear in the MaaS Australia articulation of the concept, see Table 1) draws us away from the collective good. However, clarity is needed around how exactly MaaS can avoid exacerbating known future issues and this means that influencers and decision-makers need to ask three questions:

1. What are the local transport problems, and would MaaS address these?
2. Does the MaaS proposition reduce or worsen local social problems where unequal accessibility/mobility is a contributory factor?
3. Can MaaS genuinely make users' transport choices more environmentally sustainable by reducing GHG and other negative impacts in the specific context under consideration?

The answer to question one has been so far seen as most significant for urban areas which are beset by significant air quality problems and congestion. However, in this environment, the transport ecosystem is such that MaaS providers can be both regulated and incentivized to encourage their users to be less reliant on the car-based options in the multi-modal mix. i.e. the bulk of trips made by MaaS users need to be by public transport (bus, light and heavy rail) and to involve much more walking and cycling (potentially also e-scooters where these are legal).

There are some obvious limits to active travel's suitability in rural areas, where trip distances are typically much longer and lacking in safe infrastructure for walkers and cyclists. Furthermore, as highlighted in the foregoing sections, the contraction of rural public transport has proceeded in a vicious cycle with rising private car use over many decades. Thus, the issues that MaaS propositions should focus on in rural areas are those relating to social inclusion and greenhouse gas emissions reduction.

However, in rural areas cross-benefits might be achieved through co-distribution where freight vehicles are used by more than one company to improve loadings and reduce overall vehicle mileage, or combining some logistics activities with passenger mobility, as with the example of the Swiss Postbus (which used to exist in the UK for example, at least until the 1980s in some places). Such a move would strongly synergise with another widespread trend: the increase in internet-facilitated home-shopping and associated rises in light goods vehicle-based deliveries. Better coordination of this kind of traffic, including co-distribution and the potential to combine with passenger movements, offers GHG emissions reductions for rural areas which are otherwise 'hard to treat'.

MaaS can also be designed to address question two. This is particularly pertinent to rural areas, and the community transport DRT services are a case in point. Rural traditions of informal lift-giving within families and communities might also be leveraged through a true 'sharing economy' facilitated with MaaS-style mobile apps that reward sharers in some way. However, the public nature of rural transport services might be undermined. For example, can we really call transport public if citizens must register before they can use it?

In relation to question three, there is a need to measure the true impacts of solutions devised to address questions one and two, especially as the sharing of vehicles and services is seen as particularly crucial to making transport more sustainable in rural areas, and such sharing is intuitively seen as more environmentally sustainable. If it can be shown that a rural MaaS can address the transport issues, and that the social needs of rural residents can be equitably supported by the product at the same time, can it also be shown that there is a net reduction in environmental impacts?

It appears to the author that one thing that has not been found in MaaS to date, even in urban settings, is the inclusion of vehicle fueling (though there are fuel card account systems used in the logistics industry or by businesses). For rural areas, where functional car dependence is almost certain to remain for some considerable time, the need to address climate change by switching to low carbon vehicles could be supported by including fueling for any fuel type or vehicle-to-grid contracts in rural MaaS business models. How this might work is that customers could book time on an EV point at a Park and Ride system (or wherever they need to go, be it their workplace, shopping centre, place of education, or healthcare provider) via the MaaS, and that the energy required by their vehicle is either PAYG or an allowance included in their monthly package; if their vehicle is supplying energy to the grid the user is paid instead. Some elements of this approach would be resilient to different forms of low-carbon vehicle, should EVs not be the sole or longer-term solution.

However, there is also a fourth question that is relevant to many rural areas:

4. How might rural MaaS enable or constrain the mobility of day visitors and tourists to rural areas?

Given the economic significance of rural tourism, and its high environmental impact, it is important to direct tourists to services with lower impacts. Registration can be a barrier to that, and it is hard to imagine that rural residents will be willing to offer peer-to-peer sharing with non-resident strangers in sufficiently large numbers to have a meaningful environmental impact. Furthermore, given the substantial dependence of most rural populations on travel in private cars, there are some significant issues in developing a multi-modal MaaS provision that will be adopted by the rural population, is it possible for a rural MaaS to provide a comprehensive transport service without a substantial reduction in individual flexibility? Adopting a social enterprise model for a rural MaaS may be one appropriate way forward capable of addressing the four posed questions, and there could be technological solutions to support the establishment of interpersonal trust in privacy protective ways using agent-based software models.

FUTURE RESEARCH DIRECTIONS

In considering how rural MaaS needs to be designed to address specific spatial and social context of rural areas, it is helpful to understand that whilst it needs to contribute to solving a range of problems, it is unlikely to be the sole solution. This is a helpful perspective from which to consider how rural MaaS needs to be developed, and the four guiding questions above provide a framework for systematically interrogating the degree to which a rural MaaS proposition addresses those questions.

More attention needs to be paid to the range of perspectives on feasibility and adoption that exist amongst rural stakeholders and potential end-users. In relation to feasibility this relates strongly to what would constitute a suitable business model for rural MaaS. A true ‘sharing’ model may sit better with the traditional ideas of rural reciprocity and mutual support, suggesting that social enterprise models, and user-centric design, should be a priority for focused investigation.

Furthermore, there needs to be more research into how rural MaaS platforms can enable peer-to-peer sharing alongside both community and commercial provision (perhaps through a social enterprise model), how co-distribution and shared small logistics/passenger transport can be revived and stimulated and also how the continued need for personal-car based transport can be brought into the ecosystem beyond lift-sharing, to stimulate the uptake of low-carbon vehicles.

CONCLUSION

This chapter has built on the argument that MaaS is not guaranteed to live up to the overly optimistic rhetoric that is employed by its proponents. Evidence is provided that the rhetoric is spatially blind in its neglect of the different socio-spatial conditions affecting rural transport versus urban transport. The reasons for this are presented, but it is a significant issue for rural areas, which have a unique set of trends, in relation to an ageing demographic, contracting population, and poor ICT and public transport service levels. In theory MaaS has a lot to offer in solving some of the social and environmental consequences of this, but more needs to be known about how to build reduced and zero-carbon rural transport options in to rural MaaS, a challenge that should not be underestimated. New collaborations could be achieved in rural areas that would create more viable MaaS offering, if better coordination can be achieved between currently separate areas of provision, such as health and social transport, as well as micro-logistics.

Secondly, rural areas perhaps need a more communitarian MaaS offer, to reflect the socio-cultural differences of rural areas and to reduce the need to make a profit where population density reduces the commercial opportunity. A more 'modern' image for a social MaaS rural transport service offer might also support the maintenance of population levels, that would help stem the outward migration and balance the age profile of rural areas which would further support the maintenance of sustainable rural livelihoods and ecosystem services.

A key conclusion is that addressing spatially specific issues needs to be made more prominent in the discourse around MaaS, with a recommendation that effort is made to include meaningful discussion on the potential for rural MaaS to be designed and operated for the collective good. This effort can and should draw on the wealth of knowledge about social need for mobility in rural areas, and on the abundant work exploring more flexible options that has been done over the years before MaaS took the spotlight. As the profitability of a commercial approach to rural MaaS is clearly in question, a social enterprise model should be a serious consideration.

ACKNOWLEDGMENT

The author would like to thank the editors of this volume for the invitation to contribute and their support, the valuable feedback from the anonymous reviewers, as well as co-authors from other outputs and her wider network of academics for the intellectual discussions that led to this sole-authored analysis. The author also acknowledges the funding support from the Engineering and Physical Sciences

Council of the UK for her personal fellowship that has enabled her to pursue this area of research as a spin-off from her research project (grant reference EP/N030524/1).

REFERENCES

Aapaoja, A., Eckhardt, J., Nykänen, L., & Sochor, J. (2017). MaaS service combinations for different geographical areas. *ITS World Congress*, Montreal, Canada.

Alsnih, R., & Hensher, D. A. (2003). The mobility and accessibility expectations of seniors in an aging population. *Transportation Research Part A, Policy and Practice, 37*(10), 903–916. doi:10.1016/S0965-8564(03)00073-9

Amcoff, J., & Westholm, E. (2007). Understanding rural change: Demography as a key to the future. *Futures, 39*(4), 363–379. doi:10.1016/j.futures.2006.08.009

Ashmore, F. H., Farrington, J. H., & Skerratt, S. (2015). Superfast Broadband and Rural Community Resilience: Examining the Rural Need for Speed. *Scottish Geographical Journal, 131*(3-4), 265–278. doi:10.1080/14702541.2014.978808

Ashmore, F. H., Farrington, J. H., & Skerratt, S. (2017). Community-led broadband in rural digital infrastructure development: Implications for resilience. *Journal of Rural Studies, 54*, 408–425. doi:10.1016/j.jrurstud.2016.09.004

Bardhi, F., & Eckhardt, G. M. (2012). Access-based consumption: The case of car sharing. *The Journal of Consumer Research, 39*(4), 881–898. doi:10.1086/666376

Berkhout, F. (2006). Normative expectations in systems innovation. *Technology Analysis and Strategic Management, 18*(3-4), 299–311. doi:10.1080/09537320600777010

CORDIS. (n.d.). *Flexible Agency for collective demand-responsive Mobility Services.* Retrieved from https://cordis.europa.eu/project/rcn/61514/results/en

Currie, G. (2010). Quantifying spatial gaps in public transport supply based on social needs. *Journal of Transport Geography, 18*(1), 31–41. doi:10.1016/j.jtrangeo.2008.12.002

Currie, M., Philip, L. J., & Roberts, A. (2015). Attitudes towards the use and acceptance of eHealth technologies: A case study of older adults living with chronic pain and implications for rural healthcare. *BMC Health Services Research, 15*(1), 162. doi:10.118612913-015-0825-0 PMID:25888988

Eckhardt, J., Aapaoja, A., Nykänen, L., Sochor, J., Karlsson, M. A., & König, D. (2018). The European Roadmap 2025 for Mobility as a Service. Proceedings of 7th Transport Research Arena TRA 2018.

Engels, B., & Liu, G. J. (2011). Social exclusion, location and transport disadvantage amongst non-driving seniors in a Melbourne municipality, Australia. *Journal of Transport Geography*, *19*(4), 984–996. doi:10.1016/j.jtrangeo.2011.03.007

European Commission. (2017). An agenda for a socially fair transition towards clean, competitive and connected mobility for all. COM (2017) 283 final. Brussels, Belgium: Author.

Forbes, P. J., Wells, S., & Masthoff, J. (2012). SUPERHUB: Integrating behaviour change theories into a sustainable urban-mobility platform. Proceedings of Using Technology to Facilitate Behaviour Change and Support Healthy. doi:10.14236/ewic/HCI2012.99

Gartner, A., Gibbon, R., & Riley, N. (2007). *A Profile of Rural Health in Wale*. Cardiff, UK: Wales Centre for Health.

Grant, A. (2018). *Doing EXCELLENT social research with documents: Practical examples and guidance for qualitative researchers*. Taylor and Francis. doi:10.4324/9781315177274

Gray, D., Shaw, J., & Farrington, J. (2006). Community transport, social capital and social inclusion in rural areas. *Area*, *38*(1), 89–98. doi:10.1111/j.1475-4762.2006.00662.x

Heikkilä, S. (2014). *Mobility as a Service – A Proposal for Action for the Public Administration, Case Helsinki (MSc dissertation)*. Aalto University. Retrieved from https://aaltodoc.aalto.fi/handle/123456789/13133

Hensher, D. A. (2017). Future bus transport contracts under a mobility as a service (MaaS) regime in the digital age: Are they likely to change? *Transportation Research Part A, Policy and Practice*, *98*, 86–96. doi:10.1016/j.tra.2017.02.006

Hollands, R. G. (2008). Will the real smart city please stand up? *City: Analysis of urban trends, culture, theory, policy, action, 12*, 303-320.

House of Commons. (2018). *Mobility as a Service*. Transport Committee, Eighth Report of Session 2017-19. HC950, London, UK.

Jittrapirom, P., Caiati, V., Feneri, A.-M., Ebrahimigharehbaghi, S., González, M. J. A., & Narayan, J. (2017). Mobility as a Service: A Critical Review of Definitions, Assessments of Schemes, and Key Challenges. *Urban Planning*, *2*(2), 13–25. doi:10.17645/up.v2i2.931

Karlsson, M., Sochor, J., & Strömberg, H. (2016). Developing the "Service" in Mobility as a Service: Experiences from a field trial of an innovative travel brokerage. *Transportation Research Procedia, 14*, 3265–3273. doi:10.1016/j.trpro.2016.05.273

Le Vine, S., & Polak, J. (2015). Introduction to special issue: New directions in shared-mobility research. *Transportation, 42*(3), 407–411. doi:10.100711116-015-9603-4

Liimatainen, H., & Mladenović, M. N. (2018). Understanding the complexity of mobility as a service. *Research in Transportation Business & Management, 27*, 1–2. doi:10.1016/j.rtbm.2018.12.004

Lyons, G., Hammond, P., & Mackay, K. (2018). The importance of user perspective in the evolution of MaaS. *Transportation Research Part A, Policy and Practice, 121*, 22–36. doi:10.1016/j.tra.2018.12.010

Marottoli, R. A., de Leon, C. F. M., Glass, T. A., Williams, C. S., Cooney, L. M., & Berkman, L. F. (2000). Consequences of driving cessation decreased out-of-home activity levels. *The Journals of Gerontology. Series B, Psychological Sciences and Social Sciences, 55*(6), S334–S340. doi:10.1093/geronb/55.6.S334 PMID:11078110

Marsden, G., Anable, J., Bray, J., Seagriff, E., & Spurling, N. (2019). *Shared mobility: where now? where next? In The second report of the Commission on Travel Demand.* Oxford, UK: Centre for Research into Energy Demand Solutions.

Musselwhite, C. B., & Shergold, I. (2013). Examining the process of driving cessation in later life. *European Journal of Ageing, 10*(2), 89–100. doi:10.100710433-012-0252-6 PMID:28804286

Nordbakke, S., & Schwanen, T. (2015). Transport, unmet activity needs and wellbeing in later life: Exploring the links. *Transportation, 42*(6), 1129–1151. doi:10.100711116-014-9558-x

Nutley, S. D. (1980). Accessibility, mobility and transport-related welfare: The case of rural Wales. *Geoforum, 11*(4), 335–352. doi:10.1016/0016-7185(80)90022-6

Pangbourne, K., Aditjandra, P. T., & Nelson, J. D. (2010). New technology and quality of life for older people: Exploring health and transport dimensions in the UK context. *IET Intelligent Transport Systems, 4*(4), 318. doi:10.1049/iet-its.2009.0106

Pangbourne, K., Mladenovic, M., Stead, D., & Milakis, D. (in press). Questioning Mobility as a Service: Unanticipated implications for society and governance. *Transportation Research A*. doi:10.1016/j.tra.2019.09.033

Pangbourne, K., Stead, D., Mladenovic, M., & Milakis, D. (2018). The case of Mobility as a Service: A critical reflection on challenges for urban transport and mobility governance. In G. Marsden & L. Reardon (Eds.), *Governance of the smart mobility transition* (pp. 33–48). Emerald Points. doi:10.1108/978-1-78754-317-120181003

Polydoropoulou, A., Pagoni, I. & Tsirimpa, A. (2018). Ready for Mobility as a Service? Insights from Stakeholders and End-users. *Journal of Travel Behaviour and Society*.

Pou, C. (2014). *MyWay Project: European Smart Mobility Resource Manager.* Presentation for Annual Polis Conference, Madrid, Spain. Retrieved from http://myway-project.eu/www.myway-project.eu/index.php/publications/presentations/index.html

Scottish Government. (2018). *Rural Scotland Key Facts. A National Statistics publication for Scotland.* The Scottish Government.

Sochor, J., & Nikitas, A. (2016). Vulnerable users' perceptions of transport technologies. Proceedings of the Institution of Civil Engineers. *Urban Design and Planning*, *169*(3), 54-162.

Sochor, J., Strömberg, H., & Karlsson, I. C. M. A. (2015). Implementing Mobility as a Service. Challenges in integrating user, commercial, and societal perspectives. *Transportation Research Record: Journal of the Transportation Research Board*, *2536*(1), 1–9. doi:10.3141/2536-01

Sovacool, B. K., Noel, L., Kester, J., & de Rubens, G. Z. (2018). Reviewing Nordic transport challenges and climate policy priorities: Expert perceptions of decarbonisation in Denmark, Finland, Iceland, Norway, Sweden. *Energy*, *165*, 532–542. doi:10.1016/j.energy.2018.09.110

Stradling, S. G., Hine, J., & Wardman, M. (2000). *Physical, cognitive and affective effort in travel mode choices.* Paper presented at the Symposium on Travel Mode Choice, Second International Conference on Traffic & Transport Psychology, Bern, Switzerland.

Surakka, T., Härri, F., Haahtela, T., Horila, A., & Michl, T. (2018). Regulation and governance supporting systemic MaaS innovations. *Research in Transportation Business & Management*, *27*, 56–66. doi:10.1016/j.rtbm.2018.12.001

Transport Scotland. (2019a). Retrieved from https://www.transport.gov.scot/our-approach/mobility-as-a-service

Transport Scotland. (2019b). *Mobility as a Service Investment Fund Guidance*. Retrieved from https://www.transport.gov.scot/media/45027/maas-mif-application-guidance-notes.pdf

UK Office of National Statistics. (2017). *National population projections: 2016-based*. Retrieved from https://www.ons.gov.uk/releases/nationalpopulationprojections2016basedstatisticalbulletin

Utriainen, R., & Pöllänen, M. (2018). Review on mobility as a service in scientific publications. *Research in Transportation Business & Management*, *27*, 15–23. doi:10.1016/j.rtbm.2018.10.005

Van Steen, P. J. M., & Pellenbarg, P. H. (2010). Population change and spatial transformation in the Netherlands. *Tijdschrift voor Economische en Sociale Geografie*, *101*(5), 612–618. doi:10.1111/j.1467-9663.2010.00636.x

Vanolo, A. (2014). Smartmentality: The Smart City as disciplinary strategy. *Urban Studies (Edinburgh, Scotland)*, *51*(5), 883–898. doi:10.1177/0042098013494427

Wallace, C., Vincent, K., Luguzan, C., Townsend, L., & Beel, D. (2017). Information technology and social cohesion: A tale of two villages. *Journal of Rural Studies*, *54*, 426–434. doi:10.1016/j.jrurstud.2016.06.005

Woods, M. (2006). Redefining the "rural question": The new "politics of the rural" and social policy. *Social Policy and Administration*, *40*(6), 579–595. doi:10.1111/j.1467-9515.2006.00521.x

Zeitler, E., & Buys, L. (2015). Mobility and out-of-home activities of older people living in suburban environments: 'Because I'm a driver, I don't have a problem'. *Ageing and Society*, *35*(04), 785–808. doi:10.1017/S0144686X13001086

Chapter 5
Public Transport as a Provider of MaaS in Rural Areas:
Mission (Im)Possible?

Lars E. Olsson
https://orcid.org/0000-0002-6570-6181
Karlstad University, Sweden

Margareta Friman
https://orcid.org/0000-0002-7475-680X
Karlstad University, Sweden

ABSTRACT

Mobility as a service (MaaS), in this case study, involves the integration of different travel modes into a mutual service that handles bookings and payments for individual trips. In this chapter, the authors discuss how a public transport authority has developed a MaaS for rural areas by integrating a public transport service with carpooling. The project's development, the platform's functionality, ideas for future development, and experiences of the service are analyzed from both a management and a user perspective. One overall conclusion reached is that public transport may very well develop and offer MaaS; however, there are a number of barriers to overcome concerning the legal aspects, changed travel behaviors, and density.

DOI: 10.4018/978-1-7998-1614-0.ch005

INTRODUCTION

Mobility as a Service (MaaS) is a relatively new and innovative concept whose background is the ongoing development of information technology. The starting point here is the potential to create new services by linking different travel modes together into a completely new type of mobility service. Since the area is new, many different definitions have been provided regarding the characteristics of Mobility as a Service. A recently published review of definitions, assessments of schemes, and key challenges (Jittrapirom et al. 2017) points out that Mobility as a Service can be thought of as a concept, a phenomenon, or a new transport solution. It is concluded that the novelty and fuzzy nature of MaaS makes defining what it really is a challenge.

However, one early definition of MaaS describes it as a system whereby a transport operator offers a number of mobility services (like a bundle of travel offers) to the user (Heikkilä, 2014). Within the MaaS-Alliance network, which works toward spreading MaaS around the world, a broader definition is used, which focuses more on users and their demands for a well-functioning mobility service. The MaaS-Alliance describes how Mobility as a Service puts the user in the center and offers customized solutions based on his/her need for mobility. Although Jittrapirom et al. (2017) speak of the fuzziness surrounding the MaaS concept, they nevertheless present a definition that focuses on new services based on mobility needs that are combined with new technology and merged in a new transport solution. The common denominator of the different definitions is the bundling of different mobility offerings into one mutual service.

If MaaS becomes a reality, this will certainly lead to changed practice in public transport. How this change will look is difficult to predict; however, by following the initiatives that are taken, it will be possible to gain more knowledge of the difficulties, challenges and best practices that public transport can use to prepare for the future. One regional public transport authority in Sweden (Blekingetrafiken) decided, early on, to learn more about the development of new service concept and to implement a MaaS-like service (the Hämta [pick-up] platform) in rural areas. The original idea was to include carpooling in the public transport offering, with the aim of helping users to find each other so that they could travel together, in their own private cars, either to their final destinations or to public transport stops. Thus, the user of the service could either choose to use the carpooling service alone or the carpooling service in combination with public transport. This offer was to be developed and offered by the public transport authority as an owner of the service. This chapter describes Blekingetrafiken's experiences of offering MaaS as one of the first regional public transport authorities in Sweden to do so. In this study, the project's development, the platform's functionality, ideas for future development, and experiences of the service will be analyzed from both a management and a user perspective.

Meeting the mobility needs of people living in rural areas is a challenge facing public transport. In addition to increased car use, sparsely-populated areas and an aging population are important explanations for this difficulty. The frequent use of cars in sparsely-populated areas leads to fewer people using public transport, which makes it difficult to justify the existence of a public transport service, both from an economic and a sustainable perspective. Offering a service that suits different groups at a reasonable cost is very difficult without heavy subsidization. At the same time, both researchers and politicians often agree that public transport is important for the continuing existence and further development of rural areas, providing access to education, work, and leisure activities. Furthermore, for the elderly, public transport is often a prerequisite for living in a rural area when driving is no longer an option. Several studies show that certain groups are particularly vulnerable, from a mobility point of view, in rural areas. This particularly applies to people who do not have a driver's license and/or access to a car and to people who, for health reasons, find it difficult to travel by public transport. Brake and Nelson (2007) argue that the increased use of the car leads to public transport being abandoned, with those affected being the elderly, the young, the disabled and households with low incomes.

Various solutions have been put forward regarding the supply of rural public transport. Examples include different forms of car sharing or carpooling services (e.g., Parker, Walker and Johnson. 2011), on-demand systems (Wang et al. 2015; Davison et al. 2012), different types of flexible, dynamic, transport solutions (Brake and Nelson, 2007; Nourbakhsh and Ouyangref, 2012) or solutions that are to a very great extent based on information technologies (Velaga, et al. 2012). Mobility as a Service is usually discussed as a solution in urban environments. Developing Mobility as a Service for the residents of rural areas, however, could have several benefits, e.g. increased accessibility to workplaces and schools/services, and the service could also be a positive service for people at great risk of social exclusion. Thus, MaaS could potentially be an option for people without either a license or access to a car, as well as for the elderly, those with a low income, and the young.

Various types of Mobility as a Service have been made available via digital platforms (Sochor, Karlsson and Strömberg, 2016). Reviews classifying MaaS using different characteristics have been presented in the research literature; however, the level of general knowledge of how such innovative services can meet the mobility needs of different groups in society, both in urban and rural areas, remains low. There are few projects testing Mobility as a Service in reality, providing limited knowledge of how such services are experienced by prospective users. There is even some uncertainty regarding who is going to use MaaS. Furthermore, there is also limited knowledge of the institutional prerequisites, expectations, development and implementation of MaaS.

MAIN FOCUS OF THE CHAPTER

The Hämta platform was an initiative developed by Blekingetrafiken, a regional public transport authority in Sweden. The goal was to achieve increased mobility and accessibility in rural areas, increased public transport use, and a reduced environmental impact. The Hämta platform is a type of Mobility as a Service. It is being offered by a regional public transport authority which tested including a carpooling service as part of the public transport offer. The Hämta platform (a web application) was designed to suit people seeking transport ("travelers") and people driving to or through the requested destination ("drivers"). The basic function of the Hämta platform is that the person wanting to travel requests a trip via the web application. Without a voluntary driver, there will be no trips to choose from. Anyone who has a mobility need can use the application, request a trip, and be notified as soon as a driver accepts or rejects a trip. When a driver starts out on a trip, he or she activates that trip in the application which then sends a signal to the traveler. The Hämta platform merges the carpooling service with the public transport service by providing maps of all the public transport stops in Blekinge. In this way, the user of the application can choose to either car pool from start to finish or to go to a stop and from there switch to public transport. If the user chooses to press on a bus stop on the map displayed in the application, then he/she will be updated with real-time public transport information and will be able to continue to a Travel Planner where it is possible to buy a ticket for the public transport leg of the trip (see Figures 1 and 2 in the User perspective section). In order to gain more knowledge of the development of MaaS, the platform's functionality, ideas for future development, and experiences will be analyzed, from both the management and user perspectives.

This chapter use multiple data sources which is a common methodology in case studies (Baxter and Jack, 2008). We have used data from a user test involving 61 unique users, complemented with semi-structured interviews with users registered for the platform; one interview conducted by ourselves, and results from multiple interviews conducted by the project organization. The interviews focused on use of the service, experiences, and thoughts related to the benefits and difficulties of this kind of MaaS in a rural context. We also conducted similar interviews with three actors within the project-owner organization to capture possibilities and difficulties from a regional public transport authority perspective. In addition, documentation provided by the project-owner organization were included. The various strands of data add strengths to our understanding of the case as the multiple sources are converged in the analysis. The results of the management and user perspectives provide an indication of the possibilities and prerequisites of using MaaS in rural areas, and are discussed in light of recent meta-analytical results regarding factors influencing carpooling. We conclude by answering whether or not it is an (im)possible mission for public transport authorities to act as the provider of MaaS in rural areas.

Management Perspective

Blekinge County is located in the south of Sweden (159,362 residents). Blekinge County Council is a self-governing local authority. Its most important responsibilities are public healthcare and public transportation. By law, each county in Sweden has its own public transport authority. Each authority is responsible for strategic decisions about public transport. In Blekinge County, it is Blekingetrafiken which acts as the regional public transport authority. This regional transport authority has to document and follow a regional Transport Provision Program. The director is the head of Blekingetrafiken, which manages and develops the public transport system within the county, as well as to and from Blekinge. Public transport is publicly procured and primarily financed via county council taxes and revenues from tickets.

This project was initially initiated by the director who saw the need to better understand current trends and the future role of public transport within the county. Management wanted to better understand the possible consequences of current trends (e.g., increased automation, electrification, digitalization, shared economy) and to initiate the development of new mobility solutions for rural areas where the supply of public transport is low. The director described how this regional public transport authority wanted to make the necessary adjustments to the public transport offer in order to be prepared for the future.

Work was divided into stages. During the first stage, a development team was appointed and tasked with focusing on dynamic and flexible transport services to create a better understanding of the residents' transport needs, and to focus on how travel by public transport could be made more satisfying. This assignment was complex and future-oriented and the development team decided to collaborate with a local innovation company.

A Design Thinking approach was important during the first stage of the project and design sessions were used as a working method whereby the development team interacted with potential travelers in order to develop ideas for possible future transport services. Members of the development team said that, by reframing the problem in a user-centric way, they better came to understand the human needs involved in travel.

A number of conclusions were formulated based on the material collected during the design sessions. First, the team members concluded that safety is an essential factor for the traveler in all forms of transport. It was realized that safety is not something that is enjoyed by everyone, or to the same extent, when traveling. It was clear that what would initially be needed in order to develop dynamic and flexible public transport was not more buses, but a service which the travelers could feel safe and secure with. During the discussions between the development team and the innovation company, the idea was put forward to exploit the enormous transport capacity that existed in the region in the form of private vehicles. Private vehicles were seen as an underutilized resource that could be used to develop a new mobility service for people living in areas with a sparse public transport service. Based on the initial design sessions jointly with potential travelers, and on discussions within the management team, it was agreed that, during the next stage of the project, the focus would be on safety, carpooling and passenger experiences.

Interviews with the development team acknowledged that the next step in the development process needed to be taken. Safety had been identified as essential to the continued service development process. Safety was defined as social proximity which is a relationship-based approach of assigning travel opportunities. Essentially, social proximity is a way of measuring travel needs and user profiles (e.g., by using self-reported data) then leveraging the information to match drivers and travelers. The development process during the second stage was based on the carpooling concept, the idea being that residents could sell services by offering travel in their private vehicles. The development team stated that carpooling would enable the efficient use of the existing resources in the region, while also contributing to increased public transport use. Instead of adding more buses to the system, the management of Blekingetrafiken could allow private vehicles to act in collaboration with public transport in a new mobility service. Management discussions then focused on ideas regarding how to integrate a carpooling service into a mutual Mobility as a Service, together with existing public transport.

In addition to safety, how the service was experienced was also an important ingredient of a Mobility as a Service prototype. The reasons mentioned by the development team included satisfied users, a willingness to reuse the service, and recommending the service to others. Based on these ideas, the development group put more resources and energy into developing a Mobility as a Service digital application prototype. However, the development team faced a problem with its business model. In Sweden, it is illegal for private and public organizations to charge for travel in private cars if a taxi license is not held (except for completely private agreements between two parties whereby only the actual cost of the journey is split between the driver and the passengers). Thus, management could not financially compensate drivers who were willing to offer their services within the mobility application.

During the next stage of the development process, the development team needed to seek a solution whereby residents without a financial interest could be integrated into the new service. The solution was to make the platform work without any formal incentives that were regulated by the authority. The incentive structure was left to the individuals using the system to either share costs or offer free rides to others, with these (potential) transactions taking place outside the borders of the mobility service platform, unbeknownst to the service provider (authority).

From the management perspective, a number of insights had been gained into the different development stages. It was articulated that it was surprising to learn that no public transport provider in Sweden had previously developed a Mobility as a Service that integrated carpooling with public transport; as a consequence, there were no previous experiences to build on. There was a need to offer a safe mobility service where the experience needed to be at the center. From a management perspective, it was a challenge to develop a service which had safety as a basic concept. At the same time, management expressed this as an opportunity to deliver social benefits to a greater extent than is customary in traditional public transport. Another insight was that, given the development of new mobility services, the legislation for carpooling in Sweden needed to be revised. In order to succeed in developing and implementing Mobility as a Service, which includes the integration of public transport with travel in private cars, the obstacles for driving others need to be reduced. In Sweden, as in many other countries, a special license is required in addition to a driver's license when it comes to driving others around and charging the traveler for the ride. In 2019, this kind of license costs about SEK 6,000 (approximately €600). In addition, a taxi operating license is also required, which requires a capital adequacy of SEK 100,000. The conclusion here is that there is a need to investigate whether or not people with valid drivers' licenses and roadworthy cars could be compensated when providing Mobility as a Service in the future.

Management acknowledged that the implementation process required a great physical presence and visibility in the geographical area to inform people about the new service, to encourage potential users to download the digital application, and to encourage private car owners to offer their services. A lesson learned during this work was how difficult it was to bring the new service to people's attention and to get them to use the app. The developers said that they believed that this type of mobility service would become more popular in the coming years; however, in order for that to become a reality, people would need to change their travel habits. The notions of getting others to want to carpool and being willing to carpool need to become behaviors that are as widely accepted as the notion of using public transportation.

Figure 1. The Hämta application. Adding a trip as a driver (to the left) and looking for a trip as a passenger (to the right)

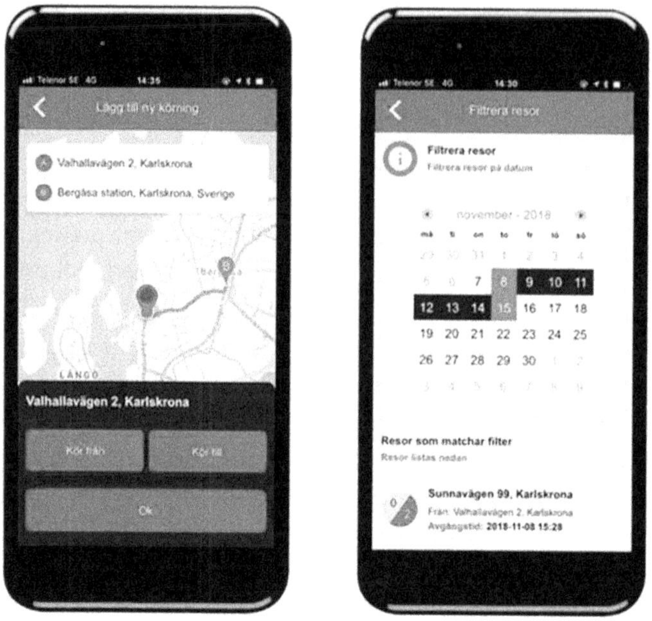

User Perspective

Hämta was introduced by Blekingetrafiken in a predefined rural test area (part of Blekinge County) with a population of approximately 600, and with a population density close to 230 per square kilometer. As described earlier, the basic function of the Hämta platform was to match people offering and requesting trips via a web application. As shown to the left of Figure 1, the driver adds upcoming trips by marking the start and finish of the trip on a map (or by typing in the addresses), in addition to the date and time of the trip. Passengers (to the right of Figure 1) search through the registered trips by selecting specific dates or origins/destinations, and then receive information if there are any potential matches. The passenger will be notified as soon as the driver accepts or rejects the trip. When the driver subsequently starts out on a trip after agreement is reached, he or she will activate the trip in the application which then sends a signal to the passenger.

As can be seen in Figure 2 (to the left), the application also merges the carpooling service with the public transport service by providing maps of all the public transport stops in the region. The user can thus also choose to carpool for only a part of his/her trip, to a bus stop or train station and then switching to public transport, instead of carpooling to his/her final destination, leading to more potential opportunities for

Figure 2. Connecting with the public transport network using the application

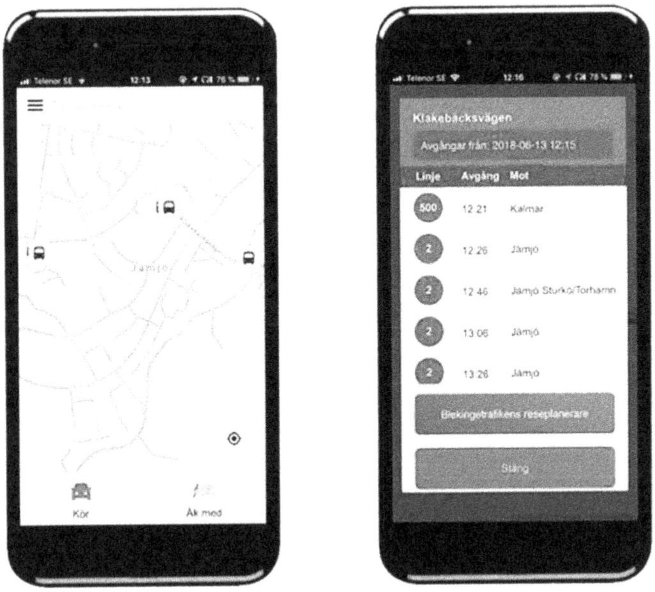

finding a match with a driver. Another feature of the application is the possibility of being updated with real-time public transport information (to the right in Figure 2), just by clicking on a bus stop icon on the map. The user can then choose to connect to a Travel Planner with the possibility of buying tickets for public transport.

A map of the potential stops along the way is shown to the left and a timetable for the selected stop is shown to the right.

Of the 600 people living in the test area, ten percent (N=61) downloaded the Hämta application. During the test period, 147 trips were registered via the platform; however, far too few of these finally led to any actual carpooling trips. Thus, no analyses of the trip characteristics or user demographics were possible. Interviews with registered users on the subject of their experiences and thoughts gave some valuable insights. Many agreed that the idea underpinning the carpooling feature is very good and should be used and developed. Environmental reasons were put forward as a very important aspect, but the social dimension was also mentioned as a positive. Commuting to work was mentioned as the usual purpose of carpooling. Although there was a very positive attitude toward it, the general opinion regarding the application was that it was too difficult to use. Interviews with both the users and the developers of the service pointed to the app being "too complicated" and having "too many steps", and thus not being user-friendly. Users also commented on the fact that there was no way of requesting a trip and setting a certain time and date without being a driver, in turn making it less appealing and with less sense of control.

Although it is understood that the reason for this feature is current legislation, users still see this as a major obstacle. It is also pointed out that there are other, nationwide, carpooling platforms run by non-profit organizations that have these features up and running. Although the Hämta application does not compete for the same trips (that are of a more long-distance nature), the regulations need to be updated to make it possible for public organizations to provide new services like this.

The keyword initially used when starting the process of developing the service was "safety". Indeed, this was an important feature as expressed by the users of the service. One comment made during an interview with a person who also knew others who had used the service, over the course of a few months early on, was that both they and the users they were in contact with generally only carpooled with people that they either knew, or knew of, and that they never carpooled with anyone unfamiliar to them. Although the application gave the impression, and the sense, of providing safe journeys, knowing a bit more about the people using the app would have been appreciated; in other words, a more community-based application. It was also expressed that making a successful trip via the application would not necessarily lead to the continued use of the platform. Although some of the participants occasionally continued to carpool, this was instead organized via SMS or Facebook, bringing control over who could join the trip. Again, this is in line with the notion of a more community-based solution.

Flexibility was also mentioned as a barrier to use. Acting as a driver hampers your possibility of being flexible regarding trips, as you have agreed to drive from A to B at a certain time. Being a passenger, and relying completely on someone else, is even less flexible as you may additionally have to adjust your own schedule in order to get the ride you want. It was also mentioned that the carpooling application, even if perfectly refined, does not have the possibility of meeting all transport needs. This was expressed without giving an answer to the question of which types of trips and activities would be most suitable. Discussions also involved who and which type of people the service was "meant" for: Was it for commuting to school or work, or was it for enhancing the accessibility of the elderly? It was expressed that, since the application was perceived as complicated, difficult and time-consuming to use, the elderly, especially those without drivers' licenses and in need of such a service, may in fact be excluded.

Users also commented on the virtual non-presence of the app in the media and promotions, saying that it might have been used more if it had been notably present. Although they were aware that it had been introduced in a limited area of 600 inhabitants, they would still have appreciated some way of being reminded of its existence. The perceived lack of communication on the part of the authority may also explain the uncertainty users had expressed with respect to whom the application was meant for.

The interviews concluded with discussions on the potential development of the new mobility service. The first thing that came up was the need to make it easier to use, with fewer stages to go through before finding matches. The possibility, as a passenger, of adding requests for trips was seen as a necessary feature. Another suggestion was to include more social cues in the app in the form of 'Who is there?'. It is interesting to note that, during the discussion with the users, no explicit indication was given as regards whether or not such changes and improvements would make the users change their behavior and use the service more than they did during the test period.

DISCUSSION

The aim of this chapter is to discuss lessons learned from developing Mobility as a Service in rural areas. The main question to reflect on is the regional public transport authority's possibility of acting as the provider of such a service. Blekingetrafiken initiated the project in order to learn more about future trends and residents' travel needs, and to explore the possibility of developing the Mobility as a Service concept in order to offer, especially to residents in rural areas with a limited supply of public transport, an expanded service without expanding the number of buses in service. From interviews with management, we gain insight into the development process that has taken place over several stages. These stages have in turn provided valuable insights regarding the development of the MaaS concept. From the management perspective, the development process initially dealt with capturing travel needs. One working tool involved arranging design sessions with potential travelers. Design sessions are appropriate when management wants to create a solution based on user needs, and when management wants to create something new that has not been attempted before. Based on what management learned from the sessions, it was possible to develop a prototype focusing on both safety and experiences, from a user perspective. The user was defined as the traveler (end user), but the solution also included the integration of public transport with carpooling whereby the user could also be a provider (driver) within the system. The integration of these different actors was modified during the development process.

First, it was assumed that residents would be able to offer travel in their private cars, with some compensation as a benefit and motivator to be included in the new mobility concept. This approach was assumed to reduce the barriers by helping people to earn some money by selling trips to neighbors and members of the local community. This idea was in line with the sharing economy. However, management discovered a legal obstacle to this approach since Sweden prohibits driving others in return for payment without a taxi license.

Another solution was then developed which instead excluded the financial compensation part, moving the financial arrangement outside of the Mobility as a Service concept. Although not in line with what was first intended, from the management perspective, it was important to find a legal solution. Based upon this design, it can be debated, of course, whether or not the application should be seen in terms of Mobility as a Service. In line with Jittrapirom et al. (2017), we would argue, however, that it is not the financial incentives per se that constitute MaaS, but the actual service/offering of combined mobility. Blekingetrafiken's initiative can be seen as an experiment whereby public transport, jointly with travelers, is trying to explore, initiate, implement and evaluate Mobility as a Service under the management of public transport. As has been shown in this chapter, the project was not an immediate success. As one of the members of the development group pointed out "…with so few participants, it's like being alone on Facebook". This does not mean that the project lacks merit, it is more a learning process. Several stages and actors are involved: i.e. the public transport provider (management), private actors who can offer their services (trips in their private cars), and the end users (passengers). In this chapter, management has pointed to several insights regarding travelers' needs, opportunities/obstacles for residents as regards offering travel in their private cars, and legal obstacles to offering the intended service. The users have indicated that, from an environmental and social perspective, they are generally positively inclined toward the mobility service, but that it needs substantial improvement, and requires guidance regarding for whom and what it is most suitable.

In the title of the chapter, we ask if it is an (im)possible mission to have public transport as a provider of MaaS in rural areas. The results of this case study indicate that, yes, it is a possible mission; however, more effort is required to put all features in place, legally and experientially. However, a new and equally important question then emerges: would it be an (un)successful mission? The actual use of the provided mobility service has, however, been very limited, making it hard to draw any conclusions regarding the potential success of rural MaaS. A recent meta-analysis of studies between 2014 and 2018 highlighted a number of factors of importance to carpooling (Olsson, Maier & Friman, 2019), which may also be important to MaaS. None of those studies were from rural contexts; nevertheless the meta-analysis found similar important motives to those highlighted by the users in this MaaS case, e.g. social and environmental concerns, but also financial costs. Thus, will MaaS solutions, in the form of carpooling integrated with public transport, be sufficient, or even necessary, for rural areas? For now, we cannot give a conclusive answer. We need a better understanding of the potential of MaaS schemes provided by private entities, public agencies (e.g., public transport provider), and a combination of both. For rural areas, other solutions than integrating carpooling may have greater success, and it is therefore important to follow the demonstrations that are currently being conducted (see, for example, Schweiger 2017 on alternative rural MaaS demonstrations in the US).

From a user perspective, environmental concerns are raised as an important value, or motive, speaking in favor of MaaS. By pooling people's journeys, it is possible to contribute toward a better environment. This positive effect will only be achieved, however, if the mobility service does not create more journeys on an aggregated level. Several users also highlight the social motive for using MaaS. If the system functions well, it may be of great value to the elderly or to those without a car in terms of reducing social exclusion. The notion of age-related use and the potential problem of the elderly using (complicated) application-based services may not, however, be easily solved just by making things easier. The meta-analysis found that carpooling generally declines with age. To attract the elderly, we need to better understand the requirements and experiences of this segment of the population. There is a huge potential here, as the proportion of the elderly in society is increasing, with a strong need for an inclusive and socially-sustainable society. Also related to MaaS are some financial motives in that people voluntarily refrain in acquiring private cars or are financially compensated when they pick up and drive others.

In summary, a range of different motives and values may be important when using MaaS in rural areas. Given that there are conflicting motives in using MaaS, the management of these opposing motives merits greater research attention. For example, does acknowledgment of the different user motives make communicating the service confusing and less attractive, or might it be able to increase its appeal to include a wider group of users and thus increase its effectiveness? New services will be interpreted by people through the lenses of their motives and values. Services which are consonant with salient motives are more likely to lead to behavior change.

From a management point of view, it may be more effective to target multiple values and motives rather than making a tradeoff between environmental, social or economic sustainability. This reasoning is supported by a quote from one of the participants in the development group, who stated that "...mobility as a service should not be a barrier limiting people's travel, it should support them in daily life...".

One important aspect not raised during the interviews, but identified in the meta-analysis by Olsson, Maier and Friman (2019), is density, which is the critical mass of people with the possibility of using the service within the area or system. In other words, before polishing the quality aspects, there needs to be enough people who will potentially use the application and provide trips. In the rural context, this is difficult, and a major obstacle. Here, we see two potential solutions: (i) a committed and respected person (or group) in the community could be given the role of ambassador in order to constantly promote the service and keep it alive. By using the service themselves, and acting like 'social glue', the members of the community may more easily see the potential of the system; (ii) instead of having a small rural area as a starting point, initiating the mobility solution at larger workplaces close to the rural area would provide a greater density and more possibilities of interacting and making agreements. Another possibility is using mobility services other than carpooling in the MaaS scheme. Based on this case, we cannot rule out whether providing other services would have made this demonstration more successful or not. Very few mobility services are available in rural areas, which is one of the biggest challenges, but it would be worth investigating potentially other mobility services that can be included in a rural MaaS scheme.

Another issue worth discussing is the notion of safety and security, raised by both management and user. This was, indeed, the very foundation of the project. It also seems that the views expressed, after using the service, confirm this issue since the users said that they generally wanted to know who they were riding with, being of the opinion that the users connected to the application should be made visible. Interestingly, in the meta-analysis by Olsson, Maier and Friman (2019), safety was found to be of very little importance to actual carpooling behavior. Thus, the question is whether safety is more important to people in rural areas, or whether it is only *believed* to be so? This is something that needs to be addressed in future research.

SOLUTIONS AND RECOMMENDATIONS

Mobility as a Service creates travel opportunities in rural areas. This chapter presents important lessons and future challenges, and also describes related solutions and recommendations. User involvement throughout the different phases of the development of MaaS is valuable but cannot end with a prototype. Discussions and user input need to form an ongoing process, including continuous evaluations in order to ensure a proper match with user needs. New technical discoveries and solutions are constantly emerging; since such solutions are an essential part of mobility services, an awareness of and openness to completely new MaaS-like solutions is needed, both from a management and a user perspective. Travel is a complicated service that involves people's daily behavior featuring multiple activities. Thus, one recommendation when introducing new mobility solutions is implementing one or more soft policy measures in parallel (Richter et al. 2010) for maximum effect when it comes to using MaaS. Another recommendation involves matching the new service with people's degrees of motivation relating to travel behavior change (Friman, Huck & Olsson, 2017; Friman, Maier & Olsson, 2019); a strategy shown to be effective during several behavioral interventions relating to unhealthy and unsustainable behavior. Last, but not least, from previous research (Olsson, Maier & Friman, 2019), it is known that density is a crucial factor, leading to the recommendation to introduce the new service in larger school districts and/ or at workplaces.

REFERENCES

Baxter, P., & Jack, S. (2008). Qualitative case study methodology: Study design and implementation for novice researchers. *Qualitative Report, 13*, 544–559.

Brake, J., & Nelson, J. D. (2007). A case study of flexible solutions to transport demand in a deregulated environment. *Journal of Transport Geography, 15*(4), 262–273. doi:10.1016/j.jtrangeo.2006.08.006

Davison, L., Enoch, M., Ryley, T., Quddus, M., & Wang, C. (2012). Identifying potential market niches for Demand Responsive Transport. *Research in Transportation Business & Management, 3*, 50–61. doi:10.1016/j.rtbm.2012.04.007

Friman, M., Huck, J., & Olsson, L. E. (2017). Transtheoretical model of change in travel behavior interventions: An integrative review. *International Journal of Environmental Research and Public Health, 14*(6), 581. doi:10.3390/ijerph14060581 PMID:28556810

Friman, M., Maier, R., & Olsson, L. E. (2019). Applying a Motivational Stage-Based Approach in order to Study a Temporary Free Public Transport Intervention. *Transport Policy, 81*, 173–183. doi:10.1016/j.tranpol.2019.06.012

Heikkilä, S. (2014). *Mobility as a Service-A proposal for action for the public administration, Case Helsinki* (Unpublished Master's thesis). Aalto University, Finland.

Jittrapirom, P., Caiati, V., Feneri, A. M., Ebrahimigharehbaghi, S., Alonso González, M. J., & Narayan, J. (2017). Mobility as a service: A critical review of definitions, assessments of schemes, and key challenges. *Urban Planning*, 2(2), 13–25. doi:10.17645/up.v2i2.931

Nourbakhsh, S. M., & Ouyangref, Y. (2012). A structured flexible transit system for low demand areas. *Transportation Research Part B: Methodological, 46*(1), 204–216. doi:10.1016/j.trb.2011.07.014

Olsson, L. E., Maier, R., & Friman, M. (2019). Why do they ride with others? Meta-analysis of factors influencing travelers to carpool. *Sustainability, 11*(8), 2414. doi:10.3390u11082414

Parker, C., Walker, C., & Johnson, R. (2011). What can we learn from car sharing experiences in the UK? *Proceedings of the Institution of Civil Engineers – Transport, 164*(3), 181-188. 10.1680/tran.2011.164.3.181

Richter, J., Friman, M., & Gärling, T. (2010). Review of Implementations of Soft Transport Policy Measures. *Transportation: Theory & Application, 2*, 5–18.

Schweiger, C. (2017). *Bringing Mobility as a Service to the United States: Accessibility Opportunities and Challenges*. Conference paper presented at ITS World Congress 2017, Montreal, Canada.

Sochor, J., Karlsson, M. & Strömberg, H. (2016). Trying Out Mobility as a Service. Experiences from a Field Trial and Implications for Understanding Demand. *Transportation Research Record: Journal of the Transportation Research Board, 2542*.

Velaga, N. R., Beecroft, M., Nelson, J. D., Corsar, D., & Edwards, P. (2012). Transport poverty meets the digital divide: Accessibility and connectivity in rural communities. *Journal of Transport Geography, 21*, 102–112. doi:10.1016/j.jtrangeo.2011.12.005

Wang, C., Quddus, M., Enoch, M., Ryley, T., & Davison, L. (2015). Exploring the propensity to travel by demand responsive transport in the rural area of Lincolnshire in England. *Case Studies on Transport Policy, 3*(2), 129–136. doi:10.1016/j.cstp.2014.12.006

Chapter 6
Mobility on Demand:
Evolving and Growing Shared Mobility in the Suburbs of Northern Virginia

Susan A. Shaheen
https://orcid.org/0000-0002-3350-856X
University of California, Berkeley, USA

Adam Cohen
University of California, Berkeley, USA

Emily Farrar
University of California, Berkeley, USA

ABSTRACT

For as long as there have been cities, there have been suburbs. Shared mobility—the shared use of a vehicle, bicycle, or other travel mode—is an innovative transportation strategy that enables users to have short-term access to a transportation mode on an as-needed basis. Shared mobility can enhance access and reduce social exclusion in lower-density environments and provide transportation options to carless and public transit-dependent households, particularly in areas without high-quality, fixed-route public transportation service. This chapter discusses the design and evolution of suburbs and how this impacts the transportation network. Additionally, this chapter reviews suburban applications of shared mobility services and provides a case study of shared mobility service deployments in Northern Virginia. The chapter concludes with key takeaways and a discussion of the potential future of shared mobility services in lower-density built environments.

DOI: 10.4018/978-1-7998-1614-0.ch006

INTRODUCTION

Lower-density built environments have been linked to higher vehicle miles traveled (VMT) and vehicle trips, which in turn are often associated with travel behavior, air quality, and climate change challenges. Shared mobility, Mobility on Demand (MOD), and Mobility as a Service (MaaS) present alternatives to auto-centric transportation. These emerging services can encourage multimodal trips, provide affordable transportation options, increase the accessibility of public transit, and encourage use of active modes. While these mobility services are most often associated with dense urban areas, there are many possible applications to serve both suburbs and edge cities (Shaheen, Cohen, Yelchuru, & Sarkhili, 2017).

This book chapter presents potential shared mobility use cases for lower-density environments. These applications are contextualized in a case study of Northern Virginia, a geographical region with a mixture of suburban and edge-city environments as part of the greater Washington D.C. metropolitan area. While these innovative services can offer many benefits (as noted above), there are also limitations to deploying shared and on-demand mobility particularly in low-density environments. This chapter is intended to introduce key applications and challenges for shared mobility in suburbs and edge cities.

This chapter is organized into six sections. First, the book chapter explains the methodology. Next, the chapter describes the evolution of suburbs and edge cities and how low-density environments affect travel behavior. The third section provides definitions of shared mobility services, MOD, and MaaS as well as a framework for applying these transportation services to suburban and edge-city built environments. In the fourth and fifth sections, the chapter provides a case study of shared mobility services in Northern Virginia and a broader discussion of challenges facing shared mobility services in lower-density environments. The final section concludes with future considerations for shared mobility, MOD, and MaaS.

METHODOLOGY

This book chapter employs a multi-method qualitative approach to research shared mobility, MOD, and MaaS; the evolution of transportation networks in suburbs and edge cities; and existing case studies on on-demand mobility in suburban and edge city settings. First, the authors conducted a literature review to document existing definitions of shared mobility, MOD, and MaaS, the influence of the built environment on travel behavior, and case studies of these services in low-density built environments. The book chapter supplements the literature review with expert interviews and an Internet-based review for the North Virginia case study. The expert interviews included a variety of policymakers and practitioners representing the private sector; local, state, and federal public agencies; and academia. The purpose of the interviews was to ask about land-use and built environment classifications, opportunities and challenges of on-demand mobility in low-density built environments, and best practices for employing shared mobility in suburban and edge-city contexts. Finally, the chapter applies a framework from Shaheen et al., 2017 to describe typologies of the built environment. MOD and MaaS are quickly evolving concepts thus, it is possible that recent literature and case studies may have been inadvertently omitted.

DESIGN AND EVOLUTION OF SUBURBS

The built environment can be categorized into five common typologies, shown in Figure 1. The U.S. Department of Transportation has defined these typologies as follows (Shaheen et al., 2017):

- **City centers** comprised of central business districts (CBDs) and surrounding neighborhoods. City centers have the highest concentration of jobs;
- **Suburban** environments characterized by high-levels of low-density residential uses with fewer jobs than residences;
- **Edge cities** that present some features of city center employment mixed with suburban form. They tend to have large concentrations of office and retail space, often paired with multi-family residences;
- **Exurban** environments with low-density residential development within the commute shed of a larger and denser urbanized area; and
- **Rural** environments characterized by low-density light industrial, agricultural, and other resource-based employment.

For the purposes of this chapter, the authors focus primarily on shared mobility in the context of classic "suburbs" (predominantly low-density residential) and "edge cities" (mixed-use employment centers located outside of urban centers). These two environments are discussed in greater detail in the following sections.

Defining "Suburbs"

For as long as there have been cities, there have been suburbs. Suburbs trace their origins to the Sixth Century BCE in Babylon. Cicero used the term "suburbani" to describe the large estates of wealthy Romans on the city's periphery (Columbia University Press, 2012). In North America, early streetcar suburbs were built across the continent along horsecars and later cable and electric streetcar lines. In the

Figure 1. Five Common built environments
Source: Shaheen et al. 2017.

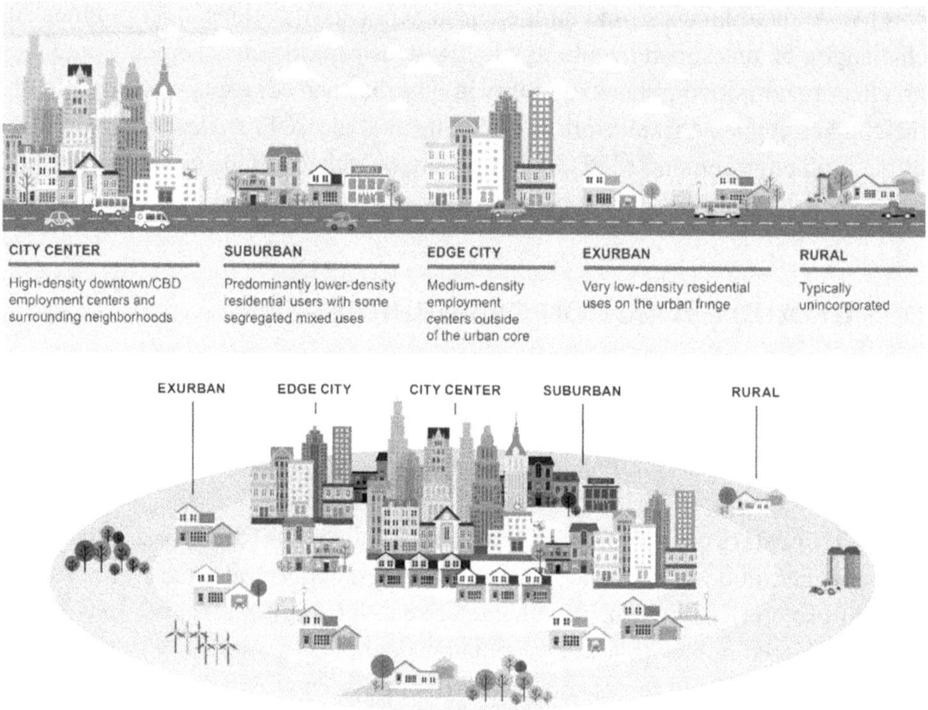

post-war years, North American suburbs were re-imagined around auto mobility with the growing popularity of private vehicles, the Interstate Highway Act, and the conversion of streetcars into bus lines.

There is no consensus on what constitutes or how to precisely define a suburb (Forsyth, 2012). Suburbs have been characterized, defined, and categorized across numerous dimensions ranging from location and transportation modes to culture and physical appearance (Forsyth, 2012). A number of early North American writers suggested that the suburb reflected the character, behavior, and culture of middle-class society (Lansbury, 1970), while others defined suburbs primarily by their location, land-use, density, and governmental structure (Kurtz & Eicher, 1958). Airgood-Obrycki and Rieger (2019) examine suburban definitions in the literature and categorize three common definitions: 1) *census-convenient*, which defines suburbs as any place that falls outside of Office of Management and Budget (OMB) defined cities but within metropolitan boundary areas; 2) *suburbanisms*, which proposes a continuum of suburban ways of life that highlight a range of key characteristics (e.g., single-family dwelling occupancy, homeownership, and automobile commuting); and

3) *typology*, which seeks to categorize specific suburban types, providing additional detail about the built form, location within the metro, demographics, and/or the history of a suburb. In this book chapter, the authors employ the USDOT definition for suburbs, which falls within the definition typology.

A close examination of suburbanization reveals subtle yet remarkable differences in urban and environmental design, often driven by their age, location, transportation modes, and density. Pre-war suburbs built around railroad and streetcar lines tend to be more walkable than their post-war automobile-centric counterparts. But a closer look reminds planners and policymakers that urban form and density also matter. Anecdotally, most people portray suburbs as high levels of low-density residential (typically between 4 to 10 dwelling units per acre) uses with fewer jobs than residences (commonly referred to as bedroom communities). However, this singular view of suburbs fails to recognize the wide diversity and opportunity of innovative mobility solutions to serve a wider array of suburban densities, land-use contexts, and trip purposes.

Defining "Edge Cities"

Between 1800 and 2000, the percentage of Americans living in urban areas increased from less than 5 percent to nearly 80 percent (Bouston, Bunten, & Heary, 2013). In spite of this shift from rural to urban areas, most of the post-World War II growth has occurred in suburbs outside of central cities. Over the past 30 years, a number of these suburbs have urbanized into "edge cities" with employment centers and densities more emblematic of city centers and street patterns similar to suburbs (Garreau, 1992).

The term "edge city" was coined in the early 1990s by Joel Garreau, a journalist, to describe the increasing densification and mixed-use nature of suburbs, recognizing diversity in suburban form (Garreau, 1992). As Garreau (1992) notes, edge cities tend to have large concentrations of office and retail space often paired with multi-family residences, resulting in work trips toward the edge city in the morning and away from it in the evening. Edge cities do not exist in isolation, but they compete directly with existing city centers within their metropolitan areas. According to Garreau (1992), edge cities have at least 5 million square feet of office; 600,000 square feet of leasable retail (a benchmark that may need to be re-defined in an era of online commerce and reductions in brick and mortar retail square footage); more jobs than bedrooms; and higher residential densities than suburbs (typically 10 to 50 dwelling units per acre) (Garreau, 1992). Many edge cities have developed around suburban transportation nodes, often highway interchanges, rail lines, or both.

In a review of existing edge-city definitions, Scheer and Petkov (1998) note the importance of the core commercial area, rather than the residential surroundings, in defining and classifying an edge city. The authors also differentiate edge cities from traditional cities, describing edge city centers as highly developed, independent nodes located within a less-dense development. The surrounding development and residential areas may not adhere to traditional forms such as: radial street hierarchies, a name and identity, economic dependency, or a degree of political control by its residents. The edge-city concept also has received criticism in the literature. Lang (2000) critiques Garreau for conflating all non-downtown office space with office space that is located specifically in an edge city. Lang (2000) notes the existence of "edgeless cities" that account for two-thirds of U.S. office space outside of downtown areas. Readers can perhaps gain a better understanding of edge cities by how Lang describes their counterpart: edgeless cities lack the density or cohesiveness of edge cities, are not mixed-use, pedestrian-friendly, and are not easily accessed by public transit (Lang, 2000).

The State of Mobility in Suburban and Edge City Environments

Existing research indicates that residents of suburban developments tend to have higher vehicle miles traveled (VMT) and walk less than those of higher-density developments (reviewed in Saelens, Sallis, & Frank, 2003; Bauman & Bull, 2007; and Martin et al. 2016). The effects of the built environment on travel behavior have been extensively studied, with many studies finding that the built environment has a statistically significant impact on travel behavior (Ewing & Cervero, 2010). However, self-selection for certain types of built environments complicates this relationship, obscuring the extent to which encouraging multimodality and mixed-use development can reduce VMT and encourage active transportation. While the built environment appears to impact travel behavior, so do attitudes and residential preferences (Ewing & Cervero, 2010; Handy, Cao, & Mokhtarian, 2005).

Suburban land-use and density, coupled with urban form (e.g., the physical characteristics that make up built-up areas such as: shape, size, and configuration of the built environment) are typically not well suited for fixed-route public transportation service. Limited access and lack of walkable urban form can create social isolation for carless and carlite households in suburban settings. Given the difficulty of changing land-use patterns and transportation infrastructure in the near- to medium-term, these challenges are likely to persist.

SHARED MOBILITY SERVICES

Today, shared mobility has the potential to impact suburban mobility in some contexts. Shared mobility - the shared use of a vehicle, bicycle, scooter, or other travel mode—is an innovative transportation strategy that enables users to have short-term access to a transportation mode on an as-needed basis. Shared mobility includes various passenger modes and courier services to meet the diverse needs of users. The most advanced passenger services incorporate trip planning and booking; real-time information; and fare payment into a single user interface. Passenger modes facilitated by shared providers can include carsharing, bikesharing and scooter sharing (sometimes collectively referred to as shared micromobility), ridesharing (carpooling and vanpooling), transportation network companies (also known as TNCs, ridesourcing, and ridehailing), microtransit, shuttle services, public transportation, and other innovative and emerging transportation solutions. Shared mobility courier services can include app-based courier network services (CNS), robotic delivery vehicles, and aerial delivery services (e.g., drones) (Shaheen et al., 2017). Shared mobility also includes emerging mobility technologies, such robotic delivery urban air mobility (inclusive of passenger mobility and goods delivery), which could have implications for suburban mobility and goods access in the future. Please refer to Table 1 below for a description of common shared mobility passenger services that are expanding into suburban settings outside of the urban core.

Emerging Concepts in Shared Mobility

In cities around the world, innovative and emerging shared modes are offering residents, businesses, travelers, and other users more options to access mobility, goods, and services. On both sides of the Atlantic, two parallel approaches to multimodal access to public and private transportation services are emerging. In North America, consumers are assigning economic values to transportation services and making mobility decisions (including the decision not to travel and instead have a good or service delivered) based on cost, travel and wait time, number of connections, convenience, and other attributes – a concept commonly referred to as Mobility on Demand (MOD) (Shaheen et al., 2017). On the other side of the Atlantic in Europe, services that allow travelers to enroll for mobility services in one bundled service are gaining popularity – a concept known as Mobility as a Service (MaaS) (Sochor, Arby, Karlsson, & Sarasini, 2018).

Table 1. Common shared mobility services

Service	Definition
Bikesharing (also known as shared micromobility)	Provides users with on-demand access to bicycles at a variety of pick-up and drop-off locations for one-way (point-to-point) or roundtrip travel. Bikesharing users access bicycles using one of three bikesharing models: 1) station-based bikesharing (users access bicycles via unattended stations); 2) dockless (users may access (unlock) a bicycle and park it at any location within a predefined geographic region); and 3) hybrid bikesharing systems (users may check out and return bicycles either through a station or non-station location). Bikesharing fleets are commonly deployed in a network within a metropolitan region, city, neighborhood, employment center, and/or university campus.
Carsharing	Individuals gain the benefits of private-vehicle use without the costs and responsibilities of ownership. Individuals typically access vehicles by joining an organization that maintains a fleet of cars and light trucks deployed in lots located within neighborhoods and at public transit stations, employment centers, and colleges and universities. Typically, the carsharing operator provides gasoline, parking, and maintenance. Generally, participants pay a fee each time they use a vehicle.
Courier Network Services (CNS)	Provides for-hire delivery services for monetary compensation via an online application or platform (such as a website or smartphone app) to connect couriers using their personal vehicles; bicycles; or scooters with freight (e.g., packages, food).
Drones	A short-range unmanned aerial vehicle (or UAV) that can transport small packages, food, or other goods.
Microtransit	Privately or publicly operated, technology-enabled transit services that typically use multi-passenger/pooled shuttles or vans to provide on-demand or fixed-schedule services with either dynamic or fixed routing.
Ridesharing (also known as carpooling and vanpooling)	Facilitates formal or informal shared rides between drivers and passengers with similar origin-destination pairings.
Robotic Delivery	Offer short-range unmanned ground-based delivery of packages, food, or other goods using a small conveyance robot.
Scooter Sharing (also known as shared micromobility)	Users gain the benefits of a private scooter without the costs and responsibilities of ownership. Individuals typically access scooters by joining an organization that maintains a fleet at various locations. The scooter operator usually provides gasoline, parking, and maintenance. Generally, participants pay a fee each time they use a scooter. Scooters can be accessed via unattended stations or accessed (unlocked) and returned (parked) to any location within a predefined geographic region. Scooter sharing includes two types of services: Standing electric scooter sharing using shared scooters with a standing design with a handlebar, deck and wheels that is propelled by an electric motor. The most common scooters today are made of aluminum, titanium, and steel. Moped-style scooter sharing using shared scooters with a seated-design, either electric or gas powered, that generally having a less stringent licensing requirement than motorcycles designed to travel on public roads
Taxis	Provide prearranged and on-demand vehicle services for compensation through a negotiated price, zone pricing, or a taximeter. Trips can be made by advance reservations (booked through a phone, website, or smartphone application), street hail (by raising a hand or standing at a taxi stand or specified loading zone), or e-Hail (dispatching a taxi driver using a smartphone application).
Transportation Network Companies (TNCs) (also known as ridesourcing and ridehailing)	Provides prearranged and on-demand transportation services for compensation, which connect drivers of personal vehicles with passengers. Smartphone mobile applications facilitate booking, ratings (for both drivers and passengers), and electronic payment. TNCs also includes "ridesplitting," in which customers can choose to split a ride and fare in a TNC vehicle (where available).
Urban Air Mobility	The safe and efficient system for air passenger and cargo transportation within an urban area, inclusive of small package delivery and other urban Unmanned Aerial Systems (UAS) services, which supports a mix of onboard/ground-piloted and increasingly autonomous operations.

Adapted from Cohen & Shaheen, 2016; Shaheen, Cohen, & Zohdy, 2016; Shaheen et al., 2017

Mobility on Demand (MOD)

The US Department of Transportation (USDOT) defines MOD as an innovative, user-focused approach that leverages emerging mobility services, integrated public transit networks and operations, real-time data, connected travelers, and cooperative Intelligent Transportation Systems (ITS) to allow for a more traveler-centric, transportation system-of-systems approach, providing improved mobility options to all travelers and users of the system in an efficient and safe manner (Sheehan & Torng, 2016). MOD is an innovative concept based on the principle that transportation is a commodity where modes have economic values that are distinguishable in terms of cost, journey time, wait time, number of connections, convenience, and other attributes. MOD enables consumers to access mobility, goods, and services on demand by dispatching or using shared mobility, delivery services, and public transportation strategies through an integrated and connected multimodal network (Shaheen et al. 2017). MOD promotes choice in personal mobility, leverages emerging and existing technologies and big data capabilities, encourages multimodal connectivity and system interoperability, and promotes innovative business models that enhance traveler experience. MOD has three major guiding principles: 1) traveler centric and consumer driven, 2) data connected and platform independent, and 3) multimodal and mode agnostic. Technology enables an interoperable and multimodal transportation MOD ecosystem. MOD, as envisioned by the US Department of Transportation, culminates in the management of supply and demand across mobility services through an integrated transportation systems management and operations approach that is coordinated among the public and private sectors and the traveling public. MOD also encompasses decision-support systems to: 1) aggregate real-time, historic, and predicted system condition information; 2) analyze alternative response strategies to address current or predicted problems; 3) assess the tradeoffs associated with strategies that support a number of operational objectives that vary dynamically; and 4) produce recommended strategies for implementation by system operators to guide and influence consumer choice (Shaheen et al. 2017). In summary, MOD consists of how people make mobility decisions, how they move, how they consume goods and services, and the stakeholders that make these actions possible (Shaheen et al., 2017).

A subset of literature on routing and automated vehicle systems define MOD as one-way vehicle sharing using small, electric cars (e.g., Mitchell, Borroni-Bird, and Burns, 2010; Pavone, 2015). It is important to note that this definition is inconsistent with the definition used by the USDOT.

Mobility as a Service (MaaS)

In Europe, another evolving concept known as MaaS is gaining popularity. MaaS represents a shift from personally owned modes of transportation toward traditional or innovative services (e.g., shared mobility) aggregated into a service offering. Fundamentally, MaaS restructures the mobility distribution chain by integrating the products and services of mobility providers and supplying them to users as a single service. Typically, a digital platform creates and manages trips that users can pay for via a single account. A distinguishing feature of MaaS is giving users the option to purchase MaaS products, such as monthly subscription plans that best fit a user's or household's needs. These subscriptions can include a certain amount of each transportation service (e.g., public transportation, bikesharing, carsharing, taxis, etc.) and are similar to other service bundles, such as mobile phone plans where the user pays one price for the combination of a multiple-service elements (e.g., talk, text, data, roaming, long distance, etc.).

Brokering travel with suppliers, repackaging, and reselling it as a bundled package is another distinguishing characteristic of MaaS (Matyas & Kamargianni, 2018; Durand, Harms, Hoogendoorn-Lanser, & Zijlstra, 2018; Hietanen, 2014). For example, in Gothenburg, Sweden the first MaaS deployment known as UbiGo operated as a pilot between November 2013 to April 2014. UbiGo repackaged existing transportation services (e.g., public transit, taxi, bikesharing, and carsharing) into a one-stop, monthly, paid subscription service for the entire household. UbiGo subscriptions started at approximately €135 or 185 USD per month at the time of the trial, although the average subscription was approximately €200 or 280 USD per month. The pilot program contributed to a reduction in household vehicle ownership and increased use of bikesharing, carsharing, public transportation, and taxis. More recently, UbiGo relaunched another pilot in Stockholm in March 2018.

Sochor et al. (2018) establishes a MaaS framework that describes four levels of varying integration:

- Level 0 (No integration);
- Level 1 (Information Integration) – The MaaS service is primarily a travel planning tool funded through advertising or taxpayer funds. Level 1 service providers aggregate and display data, but they do not have a fiduciary responsibility to ensure data fidelity;
- Level 2 (Booking and Payment Integration) – Service providers integrate trip booking and payment to enhance customer convenience and encourage multimodal travel. For service providers, level 2 grows the potential customer base, but it also increases potential competition by offering transportation

services alongside other service providers. Because level 2 integrates ticketing and payment, data fidelity becomes key;

- Level 3 (Service Offer Integration) – MaaS is intended to serve as a comprehensive alternative to private-vehicle ownership by bundling transportation services together and offering subscription packages. Level 3 emphasizes meeting a household's complete mobility needs rather than a single trip between an origin and destination;
- Level 4 (Integration of Societal Goals) – Adds value by employing incentives, gamification, and other policies to impact traveler choices to influence societal and environmental outcomes.

In a literature review of MaaS that identified 16 peer-reviewed journal articles and conference papers, Utriainen and Pöllänen (2017) identified the following as key characteristics of MaaS:

- The integration of traditional and innovative transportation modes (i.e., shared mobility) (Melis, et al., 2018; Melis, Prandini, Sartori, & Callegati, 2016);
- The option for pay-as-you-go and subscription pricing (Pangbourne, Mladenovic, Stead, & Milakis, 2018);
- A single platform, where users can plan, book, pay, and get tickets for their trips (Ambrosino, Nelson, Boero, & Pettinelli, 2016; Hensher, 2017; Hietanen, 2014; Kamargianni, Li, Matyas, & Schäfer, 2016);
- Multiple stakeholders (customers, service providers, apps, public agencies, etc.) (Kamargianni et al., 2016; Kamargianni & Matyas, 2017; Melis A., 2017; Ozaki, 2018);
- The use of information communications technology (i.e., smartphone apps) (Hilgert, Kagerbauer, Schuster, & Becker, 2016; Jittrapirom et al., 2017; Melis A., 2017); and
- A customized mobility experience allowing users to modify available trips based on traveler preferences (Hensher, 2017).

Similarities and Differences Between MOD and MaaS

Based on these definitions, MaaS differs from MOD in a few key ways. First, MOD focuses on the commodification of passenger mobility and goods delivery and transportation systems management, whereas MaaS primarily focuses on passenger mobility aggregation and service bundling. Second, MaaS integrates existing and innovative mobility services into one single digital platform where customers purchase mobility service packages tailored to their individual needs (ranging from per trip fares to bundled subscription mobility services). In contrast, MOD leverages passenger mobility and goods delivery services to enhance accessibility, while simultaneously focusing on balancing supply and demand to match changing conditions across the transportation system. There are also similarities between MOD and MaaS. Both are focused on providing travelers with more seamless travel options (i.e., routing, booking, and payment) for all trip segments, including shared mobility and public transportation, to improve the user experience and enable more informed transportation choices.

Suburban Applications of Shared Mobility

In North America, the first shared mobility initiatives (carsharing and bikesharing) launched in 1994. Initial deployments of shared mobility emphasized walkable, high-density, mixed-use urban locations. Over the past decade, shared mobility has expanded to an increasing array of locations and use cases. Broadly, these can be categorized into different trip types/use cases: 1) first-last-mile connections to public transportation; 2) public transit replacement; 3) late-night transportation; 4) paratransit; 5) point-to-point mobility; and 6) closed-door applications. Table 2 summarizes these examples and summarizes opportunities and challenges of applying shared mobility services in low-density environments. Shared mobility represents an alternative to personal-vehicle travel; however, there can be challenges to implementing these services in auto-centric built environments due to the lack of density and an urban form that reinforces auto ownership, presents challenges for public transportation, and often lacks adequate infrastructure for pedestrians and micromobility.

Increasingly, the public and private sectors are identifying opportunities for public-private partnerships and integrated mobility solutions, such as MOD and MaaS, to augment public transportation in lower-density environments. Achieving multimodal integration typically includes three components: 1) fare integration; 2) information (or digital) integration; and 3) infrastructure (or physical) integration. Table 3 presents descriptions of each component, as well as an existing case study that implements the component. However, it is important to remember that case studies of multimodal integration often apply to a regional context and are not necessarily specific to suburban applications (e.g., regional transportation apps and integrated fare payment solutions). Continued efforts toward fare, information, and infrastructure integration can support MOD and MaaS deployments in lower-density built environments, such as suburban, exurban, and rural areas.

Table 2. Suburban applications of shared mobility and opportunities to leverage MOD and MaaS

Suburban Application	Example	Opportunities and Challenges	Applicable Modes
First-and-Last Mile Connections to Public Transportation Access to and from public transit stations can be a significant barrier to public transportation use in suburban and edge city neighborhoods. Most people are comfortable walking less than ¼ mile to and from public transit stops. This can pose a particular challenge in suburbs and edge cities with non-grid street layouts, lower densities, and automobile-oriented urban forms that can frequently increase the distance to public transit stops or increase the perception of distance (e.g., a person may feel that walking is farther than it actually may be). Shared mobility, MOD, and MaaS can help overcome "first-and-last mile" challenges.	Summit, NJ partnered with TNCs to increase rail ridership, while delaying or foregoing the construction of additional parking capacity. As part of the program, participants with existing parking permits can be eligible to have their ride costs waived (freeing additional parking capacity), and participants without a parking permit pay 2 USD each way, equal to the cost of daily parking (increasing public transit capacity above parking limitations).	**Opportunities** • Reduces VMT and congestion, if personal vehicle trips are replaced • Can increase walking and use of active modes • Serves as a feeder to public transit **Challenges** • Parking garages at public transit stations may incentivize people to drive the first and last mile • Lack of population density may minimize profitability of sharing schemes • Shared micromobility requires existing infrastructure (i.e., sidewalks, bike lanes)	carsharing, shared micromobility (bikesharing and scooter sharing), microtransit, ridesharing, taxis, and TNCs
Public Transit Replacement Underperforming public transit services in lower- density environments with lower ridership, limited coverage, or infrequent schedules may be replaced with shared mobility, MOD, and MaaS services.	Arlington, TX has contracted with the microtransit service provider Via to replace its fixed-route transit service with demand-responsive microtransit. Via offers on-demand rides within a defined service area in the city for a 3 USD flat fee between the hours of 6am and 9pm, Monday through Saturday.	**Opportunities** • May result in cost savings for public transit agency • May improve coverage of public transit network • Dynamic routing may reduce wait times **Challenges** • May have higher cost per customer ride • Will likely need to be subsidized by a public agency to maintain affordability • Unbanked users and those without smartphones may have difficulty accessing on-demand services	microtransit
Late-Night Transportation In suburban and rural areas, late-night transit services can be difficult and costly to provide. Shared mobility, MOD, and MaaS services may be able to help fill gaps in the transportation system when fixed-route rail or bus services may not be available or cost prohibitive.	In Pinellas County, Florida, the Pinellas Suncoast Transit Authority launched "TD (Transportation Disadvantaged) Late Shift," a program that allows economically disadvantaged riders to request up to 25 for-hire rides (e.g., taxis and TNCs) per month for their work commute in-between the hours of 10pm and 6am, as long as they are within the county lines.	**Opportunities** • Supplements public transit for times that may be costly to provide service • Provides safe, affordable transport for late-night workers **Challenges** • Will likely need to be subsidized by a public agency to maintain affordability • Shared services that rely on contract labor may not have consistent coverage spatially or temporally	microtransit, taxis, and TNCs
Closed-Campus Applications (Roundtrip and One-Way Trips) A fleet of automobiles, bicycles, or other modes to facilitate access within a campus or closed-network applications (e.g., office parks, colleges/universities, and planned unit developments).	Rossmoor, a senior adult community in Walnut Creek, CA, deployed a survey to assess interest in an electric vehicle (EV) carsharing program A total of 443 Rossmoor residents participated in a survey that ran from December 2009 through May 2011. The surveys found a high level of interest in EV carsharing in a master-planned community for older adults, although it has not yet been implemented at this location (Shaheen, Cano, & Camel, 2013).	**Opportunities** • Minimizes vehicle traffic and need for parking within closed communities • Shuttles and microtransit can provide transport for those unable to drive • Opportunities for fixed route within closed campus • Dedicated riders always present **Challenges** • Limited population may have difficulties supporting program costs • Vehicles or devices may be underused • Inequitable strategy for general population; may draw users from public transit	carsharing, microtransit, shuttles, and shared micromobility

Table 3. Components of multimodal integration

Type of Integration	Description	Case Study
Fare Integration	Fare integration allows travelers to use a single card or account to pay for different travel modes (i.e., a fare card that can be used for commuter rail as well as to unlock a carsharing vehicle). Fare integration enhances traveler convenience and can reduce barriers to using shared mobility and public transportation (e.g., providing options to store cash payments on a fare card for unbanked households).	In the San Francisco Bay Area, the Clipper Card allows users to access a variety of public transit networks in the region, as well as the region's Bay Wheels system with semi-integrated fare payment.
Information Integration	Information integration refers to the ease with which a traveler has access to information concerning: 1) different transportation modes, 2) wayfinding, 3) trip planning, 4) fares, and 5) connection points.	In Denver, Colorado, the Regional Transportation District in Denver shares real-time public transit information via an application programming interface (or API), with multimodal vendors and the public through the GoDenver app (Aguilar, 2016; Centennial Innovation Team & Fehr and Peers, 2017).
Infrastructure Integration	Infrastructure integration refers to the physical co-location of public transportation with shared modes (sometimes referred to as multimodal integration or mobility hubs). Mobility hubs are locations that contain multiple transportation modes and include mixed land uses such as employment, housing, retail, and public transit. These hubs are designed to facilitate transit-oriented development and intermodal connections between different transportation modes--both public and private.	The San Diego Association of Governments (SANDAG) has developed eight prototypes to show how infrastructure integration could be implemented in different contexts. SANDAG's prototypes includes service facilities and infrastructure for carsharing, micromobility (scooter sharing and bikesharing), TNCs, microtransit, flexible curb space, electric vehicle charging, package delivery, mobile retail services, loading zones, and other shared modes and infrastructure.

CASE STUDY OF SHARED MOBILITY NORTHERN VIRGINIA

The Washington metropolitan area includes the federal district (Washington D.C.) and parts of Maryland, Virginia, and West Virginia. According to the U.S. Census Bureau, the Washington D.C. metro is the sixth largest metropolitan area in the country with an estimated population of 6.2 million (Clabaugh, 2018). According to the 2013-2017 American Community Survey (ACS) five-year estimate, the mean travel time to work for residents is 30 minutes. Public transportation accounts for the greatest modal share for commuting (35.4 percent), followed by single occupant vehicles (34 percent), walking (13.2 percent), carpooling (5.4 percent), and cycling (4.6 percent) (see Figure 2). Six percent of the region telecommutes. The remaining

Figure 2. Commuting modal share in Washington, D.C.

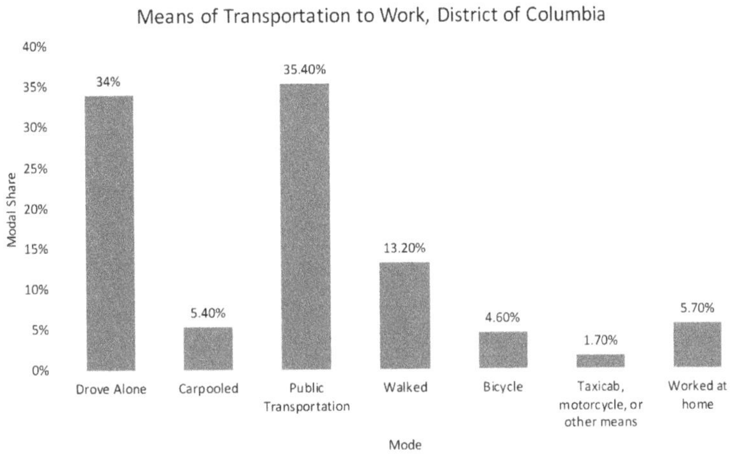

trips are made by other alternative modes (US Census Bureau, 2017). Some trends affecting travel behavior in Northern Virginia include a relatively higher income and cost of living and a greater percentage of individuals with a college or other advanced degree. Additionally, there is a high concentration of employment outside of the city center of Washington D.C.

Northern Virginia includes numerous edge cities with high concentrations of office and retail employment, such as Herndon/Reston, Tysons Corner, the Rosslyn/Ballston corridor, Crystal City/Pentagon City, and Alexandria. As of Quarter 1 2017, the Edge Cities of Northern Virginia comprise an estimated 130 million square feet of office space and 1.4 million non-farm jobs (Cushman and Wakefield, 2017). In comparison, Washington D.C. had an estimated 108 million square feet of office and 780,000 non-farm jobs. The surrounding areas in Fairfax and Arlington Counties are comprised of predominantly low-density residential communities (Cushman and Wakefield, 2017). See Figure 3 for a map of cities and communities in Northern Virginia.

Carsharing in Northern Virginia

Northern Virginia demonstrates numerous applications and use cases for shared mobility in a suburban and edge city land-use context. Both Enterprise and Zipcar each offer approximately 80 roundtrip carsharing vehicles in Arlington, Ballston, and Rosslyn. Car2go has maintained a fleet of free-floating one-way carsharing vehicles since 2015. Please note that car2go (now Share Now, a Daimler and BMW carsharing service) will cease operations in North America in February 2020 but

Figure 3. Map of Northern Virginia

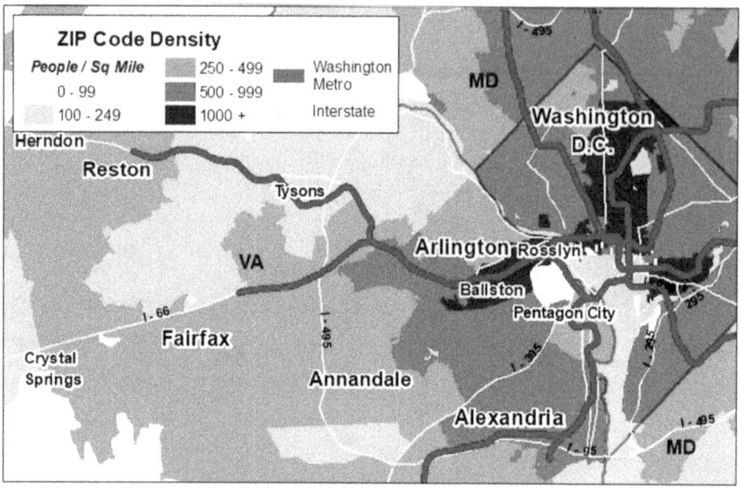

continue operations in Europe. In the early years of carsharing in Northern Virginia, Arlington County offered a pilot risk-sharing partnership with both Zipcar and Flexcar (later acquired by Zipcar). This partnership was based on a "subtraction model" in which Flexcar and Zipcar needed approximately $1,200 per vehicle per month to break-even. Arlington County, Flexcar, and Zipcar deducted the revenue generated from the total needed per vehicle to determine the per month subsidy that was needed, if applicable (Shaheen, Cohen, & Roberts, 2006). At present, the Washington Metropolitan Area Transit Authority (WMATA) issues a request for proposal (RFP) to encourage a carsharing operator to locate vehicles at its metro rail stations. WMATA's RFP process can serve as a method for initiating a joint contract, lease, or real estate use agreement between their public transit agency and the carsharing operator. Enterprise CarShare won the most recent competitive bid (Zauzmer, 2015) and currently has vehicles located at 45 of the system's 91 Metrorail stations (WMATA, 2017). The Arlington County risk-sharing partnership and the WMATA RFP process provide models that local governments and public agencies can use to encourage shared modes in lower-density areas.

Bikesharing in Northern Virginia

In addition to carsharing, Northern Virginia has had multiple shared micromobility deployments across a variety of use cases. ViaCycle briefly operated the Patriot Bikeshare program, a 20-bicycle campus bikesharing system at George Mason University in Fairfax, VA between 2012 to 2013. The program was available to

Table 4. Corporate headquarters located in Tyson's Corner

Booz Allen Hamilton	MITRE Corporation	Capital One	Freddie Mac
Gannett Company	Hilton Worldwide	Exelis	MicroStrategy
Octagon	Logistics Management Institute	Primus Telecom	SAIC
Space Adventures	Spacenet	Sunrise Senior Living	USA Today

faculty, staff, students, and guests for cycling on- and off-campus (Pullias & King, 2012). In 2018, the university relaunched the former Patriot Bikeshare bikes through a Patriot Green Grant. As of August 2019, students can check out one of 10 bikes from recreational centers on the university campus as part of a campus bikesharing pilot program (George Mason University Parking and Transportation, 2019).

In 2010, Capital Bikeshare launched a public bikesharing program in Washington D.C. with 400 bicycles (Shaheen, Martin, Cohen, & Finson, 2012). Since launching, the program has expanded its fleet 825 percent with over 3,700 bicycles across Washington D.C. and Northern Virginia (Goldchain, 2017). Edge city deployments across Northern Virginia include: Alexandria, Rosslyn/Ballston Corridor, Crystal City/Pentagon City, Tysons Corner, and Reston.

Tysons Corner, in particular, offers a model for shared micromobility in an edge city/office park setting. As noted previously, Tysons Corner is comprised of approximately 22M square feet of office space (based on conservative estimates) and includes two regional shopping malls. The corporate headquarters of 16 major corporations are located in Tysons Corner (See Table 4 below).

As of 2010, Tysons Corner was a census-designated place with a population of 19,627 with a total land area of 4.27 miles (a density of 4,600 people per square mile). The area is bordered by two freeways designed for high speed, uninterrupted traffic flow to the East and North and the WMATA silver line to the South and West (with three metro stops in the vicinity). These characteristics result in large areas of Tysons Corner that are not readily walkable and present first-and-last mile challenges connecting to Metrorail (see Figure 4 below for an aerial photograph of Tysons Corner).

Capital Bikeshare, a station-based system, has located ten kiosk stations (totaling 114 docks), which are roughly evenly distributed across the Tysons Corner area. Tysons Corner offers a prime example of how bikesharing can serve different types of use cases in an edge city and office park-type setting. Bikesharing can serve as a first-and-last mile connection to rail transit. Additionally, bikesharing can serve mid-day users for both work-related and lunch trips. Even if employees drive to

Figure 4. Photo of Tyson's Corner

work, bikesharing can help minimize mid-day trips and VMT, while providing an active transportation alternative throughout the workday.

In addition to these higher density shared micromobility deployments, Capital Bikeshare has also sited bikesharing kiosks in more suburban settings. For example, the program has an 11-dock kiosk at Crescent Apartments in Reston. Crescent Apartments features 181 garden-style affordable apartments on 16.5 acres (approximately 11 dwelling units/per acre). The property is owned by the Fairfax County Board of Supervisors and managed by the Fairfax County Redevelopment and Housing Authority (FCRHA). These stations represent an opportunity for residents to travel around the relatively low-density complex without using a personal vehicle, as well as a transportation option that residents can use to access nearby commercial areas.

Ridesharing in Northern Virginia

A number of studies have also documented the use of casual carpooling (also known as "slugging") between Northern Virginia and Washington D.C. A 2006 study counted 6,459 riders and 3,229 drivers (9,688 total participants) using casual carpooling during the morning commute on a typical weekday between Virginia and the District (Vanasse Hangen Brustlin, 2006). A separate online survey of slugging users in Northern Virginia found that the majority (60 percent) participated as passengers, while 12 percent were drivers and 28 percent were both passengers and drivers (Oliphant, 2008). Drivers reported departure flexibility as the primary reason for driving instead of riding. The top reason for choosing to be a rider was the desire to save on the cost of gasoline, followed by a preference to do other things during the drive. The study also found that 85 percent of respondents slugged roundtrip and a large percentage of respondents had used slugging for extended periods (e.g., 40 percent of female and 45 percent of male respondents had been slugging for more than five years).

Miscellaneous Shared Mobility Services in Northern Virginia

For-hire and demand-responsive services, such as TNCs, taxis, and microtransit, can also provide first-and-last mile and point-to-point connectivity in suburbs and edge cities. TNCs provide prearranged and on-demand transportation services for compensation, which connect drivers of personal vehicles with passengers. Smartphone applications are used for booking, ratings (for both drivers and passengers), and electronic payment (Rayle, Dai, Chan, Cervero, & Shaheen, 2016). Taxi services provide prearranged and on-demand vehicle services for compensation through a negotiated price, zone pricing, or a taximeter. Trips can be made by advance reservations (booked through a phone, website, or smartphone application); street hail (by raising a hand or standing at a taxi stand or specified loading zone); or e-Hail (dispatching a driver using a smartphone application). With microtransit, private-sector transportation providers may offer either fixed-route or flexible-route services, as well as scheduled or on-demand/dispatch services alongside public transportation or in lieu of public transit where fixed-route service lacks density or ridership to efficiently support a high level of service (e.g., public transit headways 20 minutes or less). In May 2019, microtransit service provider Via expanded to Alexandria, Virginia. Using the Via app, passengers select their pick-up and drop-off location and confirm their ride. The app's algorithm pairs riders into pooled rides and directs passengers to a nearby corner – a virtual bus stop – for pick up and drop off (Via, 2019).

Othcr prospective shared mobility services that could be deployed in suburban and edge city areas include scooter sharing and microtransit. With scooter sharing, users gain the benefits of a private scooter without the costs and responsibilities of ownership. Individuals typically access scooters by joining an organization that maintains a fleet of scooters at various locations. The scooter operator usually provides gasoline, parking, and maintenance. Generally, participants pay a fee each time they use a scooter. Trips can be roundtrip or one way. Currently, eight private dockless micromobility providers, including JUMP electric bikesharing and numerous scooter sharing providers, are licensed to operate throughout the District (District Department of Transportation, 2019).

Table 5. Summary of shared mobility services from the Northern Virginia case study

Suburban Application	Services Offered (Private Operators)
First-and-Last Mile Connections to Public Transportation	• One-way carsharing *(Share Now)* • Roundtrip carsharing located at public transit stations *(Enterprise CarShare)* • Station-based bikesharing *(Capital Bikeshare)* • Dockless bikesharing *(JUMP)* • Ridesharing/carpooling*(CarpoolNow, Waze)*
Public Transit Replacement	• Roundtrip carsharing *(Zipcar and Enterprise CarShare)* • Ridesharing/carpooling • Microtransit *(Via)*
Late-Night Transportation	• TNCs *(Lyft, Uber)*
Closed-Campus Applications (Roundtrip and One-Way Trips)	• Campus bikesharing *(Patriot Bikeshare, the second phase pilot program is ongoing)*

DISCUSSION

Table 5 below provides a summary of the shared mobility services offered throughout Northern Virginia, classified by the suburban application framework discussed earlier in the book chapter. As illustrated by the North Virginia case study, shared mobility is arriving in suburbs and edge cities – in the form of pilot programs, public-private partnerships, and commercial services.

To maximize the potential benefits of many shared modes (i.e., reducing single occupancy vehicle trips), system integration is typically needed to ensure that travelers can seamlessly connect between modes. The need for multimodal integration is of particular concern in the suburbs and edge cities, where spatial factors and auto-centered landscapes may preclude public agencies from offering extensive, frequent public transit services. For example, the lack of a continuous network of sidewalks or bike lanes could severely inhibit the expansion of shared micromobility services in suburbs and edge cities. Similarly, higher road speeds and the prevalence of fast-moving vehicles creates barriers to use and safety concerns. Table 6 below, adapted from Shaheen et al. (2017), describes barriers to shared mobility uptake in suburban and edge-city environments according to the STEPS framework. Each of these barriers represent unique challenges that providers and cities will need to address to ensure equitable, sustainable shared mobility options. The STEPS framework is defined as follows (Shaheen et al., 2017):

Table 6. Challenges for MOD in suburban and edge-city environments

Constraint	Suburbs	Edge Cities
Spatial	• First- and last-mile public transit connection (e.g., connections from public transit to large employment centers) • Lack of existing infrastructure to support active transportation (i.e., bike lanes, pedestrian-friendly streetscapes) • Limited school drop-off space	• First- and last-mile public transit connection (e.g., connections from public transit to large employment centers)
Temporal	• Infrequent public transit • Limited auto alternatives for off-peak hours	• Commute hour congestion • Limited auto alternatives for off-peak hours
Economic	• Lack of affordable alternatives to auto ownership	• Lack of affordable alternatives to auto ownership
Physical	• Lack of mobility options that can contribute to social isolation for people with disabilities, older adults, and others with limited mobility options	• Lack of accessible public transit services
Social	• Social stigma for suburban bus users	• Negative perceptions of public transit

Adapted from Shaheen et al., 2017

- **Spatial** factors that compromise daily travel needs (e.g., excessively long distances between destinations, lack of public transit within walking distance). Spatial factors can also include distance from community resources such as: grocery stores, retail centers, educational institutions, parks, and others;
- **Temporal** barriers that inhibit a user from completing time-sensitive trips, such as arriving to work (e.g., public transit reliability issues, limited operating hours, traffic congestion);
- **Economic** barriers are direct costs (e.g., fares, tolls, vehicle ownership, and delivery costs) and indirect costs (e.g., smartphone, Internet, credit card access) that create economic hardship or preclude users from completing basic travel or receiving goods and services;
- **Physical** and cognitive limitations that make using standard transportation modes difficult or impossible to use (e.g., infants, older adults, and people with disabilities); and
- **Social** barriers (includes social, cultural, safety, and language barriers) that inhibit a user's comfort with using transportation (e.g., neighborhood crime, poorly targeted marketing, and lack of multi-language information).

Improving mobility in these environments will likely require communities to increase investments in active transportation infrastructure, enhance accessibility, bridge spatial and temporal accessibility, and increase access to personal vehicle

Table 7. Key concepts of complete streets

	Elements
Pedestrians	• Adequate and unobstructed walking space • Adequate lighting, benches, trees, and shading • Roadway separation and on-street parking • Easy access to walkable destinations • Safe and frequent crossings
Micromobility	• Dedicate clearly marked lanes (or other micromobility infrastructure) for bicycles, scooters, and other low-speed modes • Adequate micromobility parking • Destinations accessible by micromobility
Transit	• Connectivity to micromobility and pedestrian networks • Micromobility parking • Walkable and bikeable distances between stops and stations
Rights-of-Way Allocation	• Dedicate rights-of-way for shared micromobility pick-up, drop-off, and charging
Digital Infrastructure	• Digital technology is the link between travelers (mobility consumers) and mobility service providers. Enabling seamless digital connections and reduce traveler inconvenience and create a digital network effect, where a network of connected shared services supports greater shared mobility use.

Adapted from Active Transportation Alliance, n.d.

travel alternatives. Complete streets - a transportation policy and design approach that requires streets to be planned, designed, operated, and maintained to enable safe, convenient, and comfortable travel and access for users of all ages and abilities regardless of their transportation mode – is one policy strategy communities can employ to enhance active transportation and shared micromobility use in suburban and edge-city built environments. Table 7 lists selected elements of complete streets. In addition to providing a physical landscape for shared mobility, communities and office/retail campuses can encourage shared mobility use by partnering with operators to provide mobility strategies. For example, communities can work with station-based bikesharing operators to site stations, develop protected infrastructure, and offset deployment costs. Communities can also partner with shared mobility operators to provide first- and last-mile connections to public transit, such as including discounts on TNC or microtransit trips that end or begin at transit stations.

CONCLUSION

Shared mobility, MOD, and MaaS may be able to help overcome a number of equity issues commonly associated with suburbs, such as affordability, job access, and social inclusion. It can also provide choice to travelers who typically must choose between public transit or a private vehicle for trip making. After housing, transportation is the second largest expense for American households, taking up 19 percent of the average American family's income. For auto-dependent suburbs, this proportion climbs up to 25 percent (Federal Highway Administration, 2017). By enhancing accessibility through affordable alternatives to private-vehicle ownership, shared mobility may be able to bridge service gaps within the existing transportation network particularly where public transit service is unavailable, geographically limited, or infrequent.

As cities and technologies have evolved, societies have moved on from wheeled carts and horses to horseless carriages and modern cars. Today, this evolution continues. Technology is changing the way we move, allowing us to reimagine how we use and interact with cars. The integration of transportation modes, real-time information, and instant communication and dispatch – all possible with the click of a mouse or a smartphone app – is redefining auto mobility. Rather than rendering cars obsolete, the convergence of on-demand shared travel ensures that vehicles retain their fundamental importance. It is a recognition that we cannot undo eight decades of urban form. A private automobile may be appropriate for many trips, but by offering travelers options, mobility consumers can exercise choice. Just because one lives and works in the suburbs does not mean one has to rely on a private vehicle for every trip. Shared mobility can help to bridge the first-and-last mile connection so a suburban traveler can take public transit to work. Or if a traveler does drive, shared mobility may allow that traveler to use shared modes for mid-day trips or facilitate carpooling.

Some service modes, such as shared micromobility (e.g., bikesharing and scooter sharing), can support active lifestyles. Constructing grade separated bicycle, scooter, and pedestrian facilities can serve a dual purpose of providing active transportation and recreational facilities. More broadly, shared mobility may offer innovative mobility options to improve access and mobility, while reducing social exclusion and isolation. Finally, shared mobility may be able to enhance services and reduce the costs of public transportation by providing flexible on-demand service options where fixed-route (or frequent service) is not efficient or practical due to lower density.

In the future, the convergence of automation, electrification, and sharing holds the potential to reshape mobility. Many forecast automated and connected mobility strategies — offering an array of transportation services to urban residents — will expand from cities to the suburbs. These changes are contributing to the commodification and aggregation and bundling of transportation services. While the impacts of these changes are not fully known, what is clear is that these innovations will likely have a disruptive impact on urban and rural areas, and society. Ultimately, this could lead to the replacement of many privately-owned vehicles with mobility services in the suburbs. Given the potential benefits of shared mobility, more understanding is needed to explore its role in suburban locations today and in the future.

ACKNOWLEDGMENT

This research received no specific grant from any funding agency in the public, commercial, or not-for-profit sectors. The authors would like to thank the American Planning Association, the California Department of Transportation (Caltrans), the Mineta Transportation Institute, the U.S. Department of Transportation for their generous support of shared mobility research. Special thanks goes to the shared mobility operators, industry experts, and policy makers, who make this research possible. The contents of the chapter reflect the views of the authors and do not necessarily indicate sponsor acceptance.

REFERENCES

Active Transportation Alliance. (n.d.). *Complete Streets, Complete Networks. Chapter 1: Basics*. Retrieved from: http://atpolicy.org/resources/design-guides/complete-streets-complete-networks-design-guide/

Aguilar, J. (2016, August 15). Centennial teams up with Lyft for free rides to light rail station. *The Denver Post*. Retrieved from: http://www.denverpost.com/2016/08/15/lyft-centennial-team-up-for-free-rides-light-rail-station/

Airgood-Obrycki, W., & Rieger, S. (n.d.). *Defining Suburbs: How Definitions Shape the Suburban Landscape*. Academic Press.

Ambrosino, G., Nelson, J., Boero, M., & Pettinelli, I. (2016). Enabling intermodal urban transport through complementary services: From Flexible Mobility Services to the Shared Use Mobility Agency: Workshop 4. Developing inter-modal transport systems. *Research in Transportation Economics*, *59*, 179–184. doi:10.1016/j.retrec.2016.07.015

Bauman, A. E., & Bull, F. C. (2007). *Environmental Correlates of Physical Activity And Walking in Adults and Children: A Review of Reviews*. London: National Institute of Health and Clinical Excellence.

Blue, L. A. (n.d.). Retrieved August 27, 2019, from https://www.bluela.com/

Boustan, L. P., Bunten, D., & Hearey, O. (2013). *Urbanization in the United States, 1800-2000 (No. 19041)*. Cambridge, MA: National Bureau of Economic Research. doi:10.3386/w19041

Centennial Innovation Team, & Fehr and Peers. (2017). *goCentennial Final Report*. Centennial, CO: goCentennial. Available at: http://www.centennialco.gov/uploads/files/Government/Iteam/Go%20Centennial%20Final%20Report_for%20web.pdf

Clabaugh, J. (2018). Washington metro population climbs to 6.2 million. *WTOP*. Available at: https://wtop.com/business-finance/2018/03/washington-metro-population-climbs-to-6-2-million/

Cohen, A., & Shaheen, S. (2016). *Planning for Shared Mobility*. Chicago: American Planning Association. Retrieved from https://www.planning.org/publications/report/9107556/

Columbia University Press. (2012). *The Columbia Electronic Encyclopedia* (6th ed.). New York: Columbia University.

Cushman and Wakefield. (2017). *Northern Virginia Market Report*. Washington, DC: Cushman and Wakefield. Retrieved from: http://www.cushmanwakefield.com/~/media/reports/unitedstates/CW_VA_Survey_2Q17.pdf

District Department of Transportation. (2019). *Dockless Vehicle Permits and Operators*. Retrieved from: https://ddot.dc.gov/page/dockless-vehicles-district

Durand, A., Harms, L., Hoogendoorn-Lanser, S., & Zijlstra, T. (2018). *Mobility-as-a-Service and changes in travel preferences and travel behaviour: a literature review*. KiM Netherlands Institute for Transport Policy Analysis.

Ewing, R., & Cervero, R. (2010). Travel and the built environment: A meta-analysis. *Journal of the American Planning Association*, *76*(3), 265–294. doi:10.1080/01944361003766766

Federal Highway Administration. (2017). *Transportation and Housing Costs Factsheet*. Washington, DC: U.S. Department of Transportation. Retrieved from: https://www.fhwa.dot.gov/livability/fact_sheets/transandhousing.pdf

Forsyth, A. (2012). Defining suburbs. *Journal of Planning Literature*, *27*(3), 270–281. doi:10.1177/0885412212448101

Garreau, J. (1992). *Edge City: Life on the New Frontier*. New York: Anchor Books.

George Mason University Parking and Transportation. (2019). Patriot Bike Check Out Program. *George Mason University*. Retrieved from: https://transportation.gmu.edu/bicycle-programs/check-out/

Goldchain, M. (2017, March 9). Since launch, Capital Bikeshare has expanded by 825 percent. *Curbed*. Retrieved from: https://dc.curbed.com/2017/3/9/14871330/capital-bikeshare-growth-data

Handy, S., Cao, X., & Mokhtarian, P. (2005). Correlation or causality between the built environment and travel behavior? Evidence from Northern California. *Transportation Research Part D, Transport and Environment*, *10*(6), 427–444. doi:10.1016/j.trd.2005.05.002

Hensher, D. (2017). Future bus transport contracts under a mobility as a service (MaaS) regime in the digital age: Are they likely to change? *Transportation Research Part A, Policy and Practice*, *98*, 86–96. doi:10.1016/j.tra.2017.02.006

Hietanen, S. (2014). *Mobility as a service The New Transport Paradigm*. ITS & TransporManagement Supplement. Eurotransport, 2-4.

Hilgert, T., Kagerbauer, M., Schuster, T., & Becker, C. (2016). Optimization of Individual Travel Behavior through Customized Mobility Services and their Effects on Travel Demand and Transportation Systems. *Transportation Research Procedia*, *19*, 58–69. doi:10.1016/j.trpro.2016.12.068

Jittrapirom, P., Caiati, V., Feneri, A.-M., Ebrahimigharehbaghi, S., Alonso González, M., & Narayan, J. (2017). Mobility as a Service: A Critical Review of Definitions, Assessments of Schemes, and Key Challenges. *Urban Planning*, *2*(2), 13–25. doi:10.17645/up.v2i2.931

Kamargianni, M., Li, W., Matyas, M., & Schäfer, A. (2016). A Critical Review of New Mobility Services for Urban Transport. *Transportation Research Procedia*, *14*, 3294–3303. doi:10.1016/j.trpro.2016.05.277

Kamargianni, M., & Matyas, M. (2017). The Business Ecosystem of Mobility-as-a-Service. In J.M. Crites (Ed.), *96th Transportation Research Board Annual Meeting* (pp. 8-12). Washington, DC: Transportation Research Board.

Kurtz, R. A., & Eicher, J. B. (1958). Fringe and Suburb: A Confusion of Concepts. *Social Forces*, *37*(1), 32–37. doi:10.2307/2573776

Lang, R. E. (2003). *Edgeless Cities: Exploring the Elusive Metropolis*. Brookings Institution Press.

Lansbury, R. (1970). The Suburban Community. *The Australian and New Zealand Journal of Sociology*, *6*(2), 131–138. doi:10.1177/144078337000600205

Martin, E., Shaheen, S., Zohdy, I., Chan, N., Bansal, A., Bhattacharyya, A., ... Yeung Yam Wah, C. (2016). *Understanding Travel Behavior Resarch Scan. (No. FHWA-PL-17-025)*. Washington, DC: U.S. Department of Transportation.

Matyas, M., & Kamargianni, M. (2018). The potential of mobility as a service bundles as a mobility management tool. *Transportation*, 1–18.

Melis, A. (2017). Responding to Climate Change: Lessons from an Australian Hotspot. *Urban Policy and Research*, *36*(1), 114–115. doi:10.1080/08111146.2017.1329803

Melis, A., Mirri, S., Prandi, C., Prandini, M., Salomoni, P., & Callegati, F. (2018). Integrating personalized and accessible itineraries in MaaS ecosystems through microservices. *Mobile Networks and Applications*, *23*(1), 167–176. doi:10.100711036-017-0831-z

Melis, A., Prandini, M., Sartori, L., & Callegati, F. (2016). Public transportation, IoT, trust and urban habits. In *International Conference on Internet Science* (pp. 318-325). Springer. 10.1007/978-3-319-45982-0_27

Mitchell, W. J., Borroni-Bird, C. E., & Burns, L. D. (2010). *Reinventing the automobile: Personal urban mobility for the 21st century*. MIT Press. doi:10.7551/mitpress/8490.001.0001

Oliphant, M. (2008). *The native slugs of northern Virginia: A profile of slugging in the Washington, D.C. region (Major Paper for Master of Sciences in Urban and Regional Planning)*. Blacksburg, VA: Virginia Polytechnic Institute.

Ozaki, H. (2018). Technical standardization of ITS and Asian initiatives for intelligent mobility. *IATSS Research*, *42*(2), 72–75. doi:10.1016/j.iatssr.2018.05.002

Pangbourne, K., Mladenovic, M., Stead, D., & Milakis, D. (2018). The case of Mobility as a Service: a critical reflection on challenges for urban transport and mobility governance. In G. Marsden & L. Reardon (Eds.), *Governance of the Smart Mobility Transition* (pp. 33–48). Bingley, UK: Emerald Group Publishing. doi:10.1108/978-1-78754-317-120181003

Pavone, M. (2015). Autonomous mobility-on-demand systems for future urban mobility. In Autonomes Fahren (pp. 399-416). Springer Vieweg. doi:10.1007/978-3-662-45854-9_19

Pullias, B., & King, P. (2012, October 1). Patriot Bikeshare Program Comes to Mason. *George Mason University News*. Retrieved from: https://www2.gmu.edu/news/1140

Rayle, L., Dai, D., Chan, N., Cervero, R., & Shaheen, S. (2016). Just a better taxi? A survey-based comparison of taxis, transit, and ridesourcing services in San Francisco. *Transport Policy*, *45*, 168–178. doi:10.1016/j.tranpol.2015.10.004

Saelens, B. E., Sallis, J. F., & Frank, L. D. (2003). Environmental correlates of walking and cycling: Findings from the transportation, urban design, and planning literatures. *Annals of Behavioral Medicine*, *25*(2), 80–91. doi:10.1207/S15324796ABM2502_03 PMID:12704009

Scheer, B. C., & Petkov, M. (1998). Edge City Morphology: A Comparison of Commercial Centers. *Journal of the American Planning Association*, *64*(3), 298–310. doi:10.1080/01944369808975987

Shaheen, S., Cohen, A., Yelchuru, B., & Sarkhili, S. (2017). *Mobility on Demand Operational Concept Report* (No. FHWA-JPO-18-611). United States Department of Transportation. Intelligent Transportation Systems Joint Program Office. Retrieved from: https://rosap.ntl.bts.gov/view/dot/34258

Shaheen, S., Cohen, A., & Zohdy, I. (2016). *Shared Mobility: Current Practices and Guiding Principles (No. FHWA-HOP-16-022)*. Washington, DC: U.S. Department of Transportation.

Shaheen, S. A., Cano, L. A., & Camel, M. L. (2013). Electric Vehicle Carsharing in a Senior Adult Community in the San Francisco Bay Area. In *92nd Annual Meeting of the Transportation Research Board* (pp. 1-13). Washington, DC: Transportation Research Board.

Shaheen, S. A., Cohen, A. P., & Roberts, J. D. (2006). Carsharing in North America: Market growth, current developments, and future potential. *Transportation Research Record: Journal of the Transportation Research Board*, *1986*(1), 116–124. doi:10.1177/0361198106198600115

Shaheen, S. A., Martin, E. W., Cohen, A. P., & Finson, R. S. (2012). Public bikesharing. In *North America: early operator and user understanding (No. CA-MTI-12-1029)*. San Jose, CA: Mineta Transportation Institute.

Sheehan, R., & Torng, G. W. (2016). *MOD Fact Sheet #1: Overview*. Retrieved from: https://www.transit.dot.gov/research-innovation/mobility-demand-mod-sandbox-program

Sochor, J., Arby, H., Karlsson, I. M., & Sarasini, S. (2018). A topological approach to Mobility as a Service: A proposed tool for understanding requirements and effects, and for aiding the integration of societal goals. *Research in Transportation Business & Management*, *27*, 3–14. doi:10.1016/j.rtbm.2018.12.003

U.S. Census Bureau. (2017). *Commuting characteristics by sex, 2017* (S0801). Retrieved from: https://factfinder.census.gov/

Utriainen, R., & Pöllänen, M. (2018). Review on mobility as a service in scientific publications. *Research in Transportation Business & Management, 27*, 15–23. doi:10.1016/j.rtbm.2018.10.005

Vanasse Hangen Brustlin, Inc. (2006). *Dynamic Ridesharing (Slugging) Data*. Richmond, VA: Virginia Department of Transportation.

Via. (2019). *Via Announces Expansion to Alexandria, Virginia*. Via. Retrieved from: https://ridewithvia.com/2019/05/via-announces-expansion-to-alexandria-virginia/

WMATA. (2017). *Car Sharing*. Retrieved from: https://www.wmata.com/service/car-sharing.cfm

Zauzmer, J. (2015, May 15). Zipcar loses out to Enterprise on contract to park at Metro stations. *The Washington Post*. Retrieved from: https://www.washingtonpost.com/news/dr-gridlock/wp/2015/05/15/zipcar-loses-out-to-enterprise-on-contract-to-park-at-metro-stations/?utm_term=.c703f38cf4b1

ADDITIONAL READING

Cohen, A., & Shaheen, S. (2016). *Planning for Shared Mobility*. Chicago: American Planning Association; Available at https://www.planning.org/publications/report/9107556/

Dai, D., & Weinzimmer, D. (2014). *Riding First Class: Impacts of Silicon Valley Shuttles on Commute & Residential Location Choice*. UC Berkeley: Institute of Transportation Studies. Available at: https://escholarship.org/uc/item/2jr7z01q

International, S. A. E. (2018). *Taxonomy and Definitions for Terms Related to Shared Mobility and Enabling Technologies* (J3163). Available at: https://www.sae.org/standards/content/j3163_201809/

Shaheen, S., & Cohen, A. (2019). *Shared Micromobility Policy Toolkit: Docked and Dockless Bike and Scooter Sharing*. UC Berkeley: Transportation Sustainability Research Center. Available at: https://escholarship.org/uc/item/00k897b5

Shaheen, S., Cohen, A., Dowd, M., & Davis, R. (in press). *A Framework for Integrating Transportation into Smart Cities: State of the Practice in 20 US Cities*. San Jose, CA: Mineta Transportation Institute. More information available at: https://transweb.sjsu.edu/mctm/research/utc/Theoretical-Framework-and-Index-Transportation-Innovation-Smart-Cities-Early-Understanding-and-Best-Practices

Shaheen, S., Cohen, A., & Farrar, E. (2018). *The Potential Societal Barriers of Urban Air Mobility UAM*. Washington, D.C.: National Aeronautics and Space Administration; Available at https://escholarship.org/uc/item/7p69d2bg

Shaheen, S., Cohen, A., Yelchuru, B., & Sarkhili, S. (2017). *Mobility on Demand Operational Concept Report* (No. FHWA-JPO-18-611). United States. Department of Transportation. Intelligent Transportation Systems Joint Program Office. Available at: https://rosap.ntl.bts.gov/view/dot/34258

Shaheen, S., Cohen, A., & Zohdy, I. (2016). *Shared Mobility: Current Practices and Guiding Principles (No. FHWA-HOP-16-022)*. Washington, D.C.: U.S. Department of Transportation; Available at https://ops.fhwa.dot.gov/publications/fhwahop16022/fhwahop16022.pdf

Shaheen, S., Cohen, A., Zohdy, I., & Kock, B. (2016). *Smartphone Applications to Influence Travel Choices: Practices and Policies* (No. FHWA-HOP-16-023). Available at: https://ops.fhwa.dot.gov/publications/fhwahop16023/fhwahop16023.pdf

Sochor, J., Arby, H., Karlsson, I. M., & Sarasini, S. (2018). A topological approach to Mobility as a Service: A proposed tool for understanding requirements and effects, and for aiding the integration of societal goals. *Research in Transportation Business & Management*, *27*, 3–14. doi:10.1016/j.rtbm.2018.12.003

KEY TERMS AND DEFINITIONS

Edge City: Medium-density employment centers outside of the urban core. Edge cities present some features of city-center employment mixed with suburban form. They tend to have large concentrations of office and retail space, often paired with multi-family residences.

Mobility as a Service (MaaS): Integrates existing and innovative mobility services into one single digital platform where customers purchase mobility service packages tailored to their individual needs (ranging from per trip fares to bundled subscription mobility services).

Mobility on Demand (MOD): MOD enables consumers to access mobility, goods, and services on demand by dispatching or using shared mobility, delivery services, and public transportation strategies through an integrated and connected multimodal network. Advanced MOD services incorporate trip planning and booking, real-time information, and fare payment into a single user interface (Shaheen, Cohen, Yelchuru, & Sarkhili, 2017).

Shared Micromobility: The shared use of a bicycle, scooter, or other low-speed mode enabling users to have short-term access to an active or low-speed motorized transportation mode on an as-needed basis.

Shared Mobility: The shared use of a vehicle, bicycle, or other travel mode enabling users to have short-term access to a transportation mode on an as-needed basis. Passenger modes facilitated by shared providers can include carsharing; bikesharing; and scooter sharing (sometimes collectively referred to as shared micromobility); ridesharing (carpooling and vanpooling); transportation network companies (also known as TNCs, ridesourcing, and ridehailing); microtransit; shuttle services; public transportation; and other innovative and emerging transportation solutions. Courier services can include app-based courier network services (CNS), robotic delivery vehicles, and aerial delivery services (e.g., drones).

Suburban: Predominantly lower-density residential uses with some segregated mixed-uses. Sometimes described as containing fewer jobs than residences.

Chapter 7
Innovative Mobility Solutions in Rural Areas With Case Studies

Marian Gogola
University of Žilina, Slovakia

Dana Sitányiová
University of Žilina, Slovakia

ABSTRACT

Rural areas still have some disadvantages in comparison to urban areas. Access to jobs, education, or health services provided for inhabitants in rural areas are limited. Another important topic is mobility, which is crucial to meet the travel expectations of inhabitants in rural areas. Finding new forms of mobility capable of covering all areas and meeting the requirements of rural inhabitants is the most important challenge. This chapter deals with mobility in rural areas from the perspective of demographics, social change, and accessibility to transport services. Moreover, this chapter offers various case studies dealing with innovative solutions for mobility in rural areas resulting from the RUMOBIL project.

INTRODUCTION

Living in urban areas has important advantages such as the close availability of various services ranging from health service, shopping, and education, to cultural and sports events. One of the consequences of the lack of such opportunities in rural areas is urban drift. In particular young people decide to move to bigger cities in the

DOI: 10.4018/978-1-7998-1614-0.ch007

hope of greater opportunities for employment or better education. As a consequence, many cities become increasingly overcrowded putting pressure on different kinds of infrastructure and services. Moreover, the exodus to cities together with the negative natural growth has led to a population decline in many EU rural territories, generating substantial gaps between regions of the same country and Europe as a whole. The territorial dimension of economic weakness and social exclusion has been widely recognized in European regions by O'Donnell, O'Donovan, and Elmusharaf (2018). There are also considerable differences in regional demographic patterns across the European Union (EU) from overcrowded big cities with relatively young populations to more remote, peripheral rural regions that may have declining population numbers. Demographic changes in the EU including rural regions are likely to be of considerable importance in the coming years. Above all, the EU population will continue to age as a result of a consistently low level of fertility and extended life expectancy. The social and economic consequences associated with population ageing are likely to have profound implications across the EU, both nationally and regionally. Among other issues, there is the challenge of developing the rural areas to make them more attractive to young people in terms of employment prospects, education and training opportunities, health services, social life, etc.

As many European rural regions suffer from an ongoing population decline and dynamic ageing, so services of public interest retreat from these areas. This, however, brings enormous challenges for the provision of public services, such as public transport (PT). Transport plays a key role in responding to the problems of rural social exclusion and out-migration of the young generation from these areas. On the one hand, population decrease and limited funds to finance PT result in reduced PT offers. On the other hand, mobility needs in regions where people are aging quicker than elsewhere change and even increase. The large distances between services and rural population centers often cause difficulties for people without access to private transport. In the light of these challenges, it is important, both at EU and national level, to review and adapt existing transport policies, because changes in demography also have an important impact on the future demand on all forms of transport.

Many scientific works deal with this problem. Among others, the study of Leibert and Golinski, (2017) focuses on problems of demographical changes such as depopulation and ageing in German rural regions, leading to a discussion about the future provision of services of general interest and equal living conditions in the context of the future spatial development of Germany. In their paper, they argue that the peripheralization approach (peripheralization is understood as the process of producing and reproducing peripheries) is a helpful tool to better understand how the interaction of out-migration, dependence, disconnection, and stigmatization shape the future of rural regions and lead to their gradual weakening.

The importance of mobility in rural areas was studied by Milbourne, and Kitchen, (2014). Their study highlights mobility as an important shaper of rural lifestyles and rural places. They draw on empirical materials from a recent community study in rural Wales to reveal the difficulties associated with practicing everyday mobility in rural settings. Some studies (Rogers, Dufty-Jones, Steele, 2015) examined the problems and role of housing in rural areas.

In recent years, in several countries (including Slovakia, Czech Republic, Hungary, and Poland) the increase of road transport has also resulted in the significant growth of negative impacts on the environment, increasing congestion in urban areas or increases in road accidents. In parallel, in passenger traffic, the growth of individual automobile transport is reflected by a significant decline in the performance of public passenger transport (rail, bus and public urban transport).

As said before, the significant demographic changes which mostly all the European countries have been experiencing in the last decade are affecting travel behavior. Moreover, this trend is not homogenous across different areas of Europe. For instance, the current demographic development in East European countries is a process similar to the one that took place in the developed Western and Northern Europe from the mid-60s to late 70s. There are also significant demographic changes such as age, gender, household composition or income of its members. This matter is emphasized in rural areas, where accessibility is not properly ensured. Accessibility can be defined as the proximity or facility for spatial interaction and so, more generally, as access to health care, education, work and other services (e.g., shopping centers). Hence, the importance of public transport supply (mainly for rural areas which highly depend on provided transport services or transport infrastructures) is increasing.

BACKGROUND – NEW DEMAND PATTERNS AND CHALLENGES FOR PUBLIC TRANSPORT

Bearing in mind that rural areas are characterized by low population density, large distance from city centers, spatial scarcity of facilities and very often undeveloped public transport network, residents of rural areas are transport-deprived and often dependent on passenger cars (Ranković Plazinić, Jović, 2013). The concerns of policymakers and transport planners now lie with the problems of how to maintain public transport, on which the mobility of many people depends, and how to increase its attractiveness to car users. A significant shift from car to public transport would reduce congestion and at the same time improve the efficiency of all road-based transport operations, as well bring important environmental benefits. A complex understanding of the determinants of public transport demand will help transport planners to ensure sustainable transport in the future.

Below, the most important socio-economic aspects concerning demand patterns for PT in the future have been identified.

- **Age**: There are big differences in the travel behavior of children, young people, adults, and older people, due to the different types of activities they are interested in. Children are primarily interested in educational and play activities, young people mainly in educational and social activities, adults in work-related activities and pensioners are primarily interested in social and leisure activities. These activities also influence the distance of their travel. This tends to be shorter for children and old people, while it can be considerably longer for young people and adults. The ageing of the population will change household structures (smaller family units) so social institutions will be required more and more to replace family care. More resources will be needed for structures and services dedicated to aged people: their specific needs will have to be addressed in urban planning, infrastructures, and services design.
- **Income and car Ownership:** People with a higher income, are not only *able* to travel more and longer but often they are *required* to travel (e.g. to work). This is very often linked to car ownership. Those who can count on a higher income often prefer to travel by a private car, while those who cannot afford it are forced to use public transport services, whose presence and quality becomes even more essential in rural areas (where services like car-sharing are very uncommon). On the one hand, the possibility of moving by car positively affects the development of the economy, employment and influences the way people live, increasing social inclusion, but on the other hand it has a very negative impact on the environment and traffic congestion, matters already very serious in urban or suburban areas.

Traffic congestion is something that the population of cities and commuters around the world have to face every day, costing enormous amounts of both time and money. The processes of urbanization, disurbanisation, and suburbanization play a central role. Urbanization is defined as the movement of the population from rural to urban settlements and from smaller settlements to larger ones, while disurbanisation is the movement from inner cities to their suburbs or surrounding communities. Both these processes should be seen not only as a change in population distribution and spatial structure but also as a change in the way of life.

What has emerged from the study conducted in RUMOBIL is that there is a difference between trends in Western European countries (Germany and Italy) and Eastern European countries (Slovakia, Czech Republic, Hungary, Poland). While in Eastern countries, apart from a few cases of urbanization, the overall trend is suburbanization due to economic reasons, in Western ones urbanization prevails, although to a lesser degree. It can be said that the Eastern countries are experiencing a process that Western ones have already lived in the last thirty years, after a massive urbanization process. It must also be said that the big cities of the Western countries already have enough developed suburban areas so that any change in the trend is slower and lighter.

Approximately half of the population of Germany lives in the 30 major urban German agglomerations, such as Berlin, Hamburg, Munich, and Cologne, but also in smaller cities including Münster, Freiburg, Leipzig, and Dresden. The research work by Henning (2018) helps to understand the demographic processes and development that occurred there in recent years, revealing some distinct trends of re-urbanization in some of the most densely populated areas. Examination of the average annual change in the populations of the selected urban agglomerations highlights the regional differences in these trends. Contrary to the national population decline, which is predicted to continue, 26 out of the 30 major agglomerations show population growth between 2008 and 2013. In 25 of these areas, population growth in the city center is even higher than in its suburban area. This can be seen as evidence of a very recent trend of re-urbanization in these places.

Suburbanization is generally seen as a negative phenomenon, although it brings some positive aspects. On the one hand, it is true that if part of the population moves to the metropolitan areas, it means that these sites will require investment, infrastructure, facilities that increase social inclusion. On the other hand, it means longer and more expensive trips for commuters and high effects on traffic congestion, air pollution, and noise (if there is no efficient alternative to a private car). Furthermore, there are negative impacts on the protection of natural habitats and on quality of life (social issues, such as emptying the inner city, unequal distribution of jobs and people's homes).

Providing public transport in low-density rural areas is usually cost-inefficient, although necessary in a lot of cases, and it has been a challenge for many governments all over the world. Nowadays, several cities have multiple modal choices at their disposal, most often offered in a combination (conventional or automated metro, light rail or tramway, bus or BRT, or waterborne) plus other less conventional modes (cable car, tram-train or monorail). Cities with long-established public transport systems have problems particularly with ageing rail infrastructure and rolling stock and now need rehabilitation and automation to improve their operational efficiency and capacity. Besides challenges linked to quantitative growth, public transport must make significant qualitative improvements in order to become more attractive. The notion of quality in public transport should also be created through operational excellence, which includes enhanced frequency, punctuality and reliability of the service thanks to optimized network design and service performance. Likewise, smart ticketing and integrated travel information contribute to making public transport customer/user-friendly, while facilitating accessibility for all citizens. As far as rural areas are concerned, we can see some different needs and user types that lead to different demand patterns, like the following:

- Inhabitants in rural areas without ownership of individual means of transport (i.e. cars)
- Inhabitants with specific mobility needs (disabled, etc.)
- Inhabitants in low densely populated areas which need to commute daily (school children students, commuters, etc.)
- Inhabitants who travel irregularly (pensioners, etc.)
- Tourists who consider particular rural areas interesting for any reason.

These groups refer to low populated areas where people are not able to provide adequate demand for regular public transport operation. To solve this, we can see various approaches. The first one consists of the development of regular operations based on short and regular intervals supported by integrated public transport organization. This is mainly for rural areas where the population is increasing. The population in these rural areas can take advantage of the proximity of economically strong cities.

The second approach is suitable for less populated areas and is based on the creation of good conditions for development of either public transport service on demand (so-called Demand Responsive Transport - DRT) or a combination of individual transport (cars, bikes) and PT.

Apart from new innovative forms of demand-based public transport service or combination of individual transport and public transport, the lack of efficient passenger information systems should not be underestimated. This may strongly reduce the use of public transport (where it is present) and increase car ownership. The provision of accurate, current information on arrival and departure times and stations can pose a challenge. Passengers from suburban, rural and remote areas need more reliable and sophisticated travel information compared to the urban area, because, unlike in urban areas, passengers in rural areas are provided with very limited transport facilities and alternatives, and generally make longer journeys.

Mobility and its quality are key elements of assessment standards in the countries of the European Union (EU). In general, transport is one of the basic sectors that significantly affect socio-economic development and growth. In passenger transport, quality depends largely on satisfying the everyday needs of citizens, including the level of access to work, schools, shops, accessibility to social care and leisure activities. The main objective of this chapter is to point out and discuss various approaches to enhancing public transport services in rural areas based on case study experiences.

Based on regional EU statistics findings, in 2016 in inland Europe the majority of trips were made by passenger car (83%)[1]. The study of Wolny (2019) confirmed this and explains car dependency in suburban regions in Poland. In general, transport services in rural areas have their own specification: the accessibility to services or budget. Smith, Hirsch, and Davis (2012) proposed minimum transport needs with an emphasis on household costs in rural areas. Operation and provision of g public transport in rural areas, usually on a cost-inefficient basis, has been a challenge for many governmental authorities all over the world. For some groups of passengers public transport is crucial because of lack of car ownership. It is necessary to find the right approach for authorities to combine many aspects within a transport system. Each low-density area can have its own optimal combination, depending on the wishes of the stakeholders and the circumstances of the system. Therefore, three ingredients are crucial for success: the presence of financial means from the perspective of the authority providing public transport, cooperation between stakeholders and flexible supply of scheduled and on-demand transports.

The study of Velaga et. al (2012) discovered that rural communities face a range of challenges associated with accessibility and connectivity which apply in both the physical and virtual sphere. Constraints in rural transport infrastructure and services are often caused by limitations in the development and resilience of technological infrastructures. In this context, there is a significant disparity between urban and rural communities. McDonagh, (2006) explains the transport policy instruments and transport-related social exclusion in rural areas of Ireland.

Certainly, there is a bigger difference between urban and rural or suburban areas. However, we can see there is still a lack of sufficient statistical data on public transport in such areas. The use of public transport in the EU reached its highest level since 2000, with a total of 57.9 billion journeys made in 2014 according to UITP's new Statistics Brief, 'Local public transport in the European Union' (2016). But we can say it is mainly in and around the urban areas. 2014 was the first year of distinct growth in demand for public transport after years of stable demand following the start of the economic crisis in 2008. The highest total demand in 2014 for bus, tram, metro and suburban rail was recorded in Germany (10.9bn journeys), UK (7.7bn) and France (7.6bn). Between 2013 and 2014, 'growth leaders' France, Italy, Poland, and the UK had a combined increase of 600 million journeys, driving up the total EU figure, see Figure 1.

Figure 1. EU-28 Public transport journeys by bus, tram and metro per urban inhabitant in 2014 and evolution trend in the previous 5 year
Source: UITP, 2016

Of the 57.9 billion public transport journeys made in 2014, 55.8% were by bus, 16.1% by metro, 14.5% by tram and 13.6% by suburban rail. The developments mask significant national variations, however, which are quite closely linked to national employment figures. 17 EU countries saw higher ridership in 2014 compared to 2010 but only seven had sustained growth: Austria, France, Germany, Lithuania, Malta, Sweden, and the UK. Bulgaria was the only country where ridership dropped every year since 2000. Encouragingly, countries such as Spain, Ireland, and Italy that have been impacted by the crisis, saw a return to growth in 2014. In EU capital cities, the average annual percentage growth in demand (2010-2014) was highest in Brussels; demand per capita is approximately 2.5 times higher in capital cities than the national average.

In general, there are various approaches to public transport organization. In many countries, local or regional public transport is organized by municipalities or their companies, and by regional or county government which subsidizes the service. The advanced system of public transport organization is represented by integrated public transport systems where all regional and urban transport systems are organized by one so-called integrator (e.g. in Germany or Austria there are the public transport unions, so-called Verkehrsverbunds).

Most regions in the Czech Republic have a similar organization . In Slovakia, meanwhile, there are only 2, although others are in the preparation phase. However, what is important from the legislative framework is the recommendation to have a Public Transport Service Plan according to Act Nr.56/2012 in Collection of acts on road transport in Slovakia[2]. In Slovakia, it is obligatory for each town, city or region. This plan determines the requirements of public transport supply based on transport demand respecting changing socio-demographic characteristics.

There also is a similar approach in Germany. For instance, in the Public Transport Plan for Saxony-Anhalt for 2010 -2025 (2018)[3] the level of and target for public transport service considering the future forecast is also stated. The public transport timetable is planned on the headway interval (so-called Taktplan). The plan also states the goals in updating the transport infrastructure and maximum travel time to reach important towns and cities to get to work or to schools and points out different approaches needed in various regions.

In economically strong countries like Germany, Switzerland, Austria or Netherlands, high-frequency public transport can also be provided in suburban or rural areas which are profiting from the proximity of big urban areas, or high population density. The problem occurs in economically weak regions or rural areas which do not have enough finance to support high-frequency public transport. In these areas, the population without a car is dependent on PT service.

New Challenges of Public Transport

Public transport is facing many challenges and opportunities that result from changes occurring within the sector itself, as well as from external trends affecting its wider socio-economic environment. Growing urbanization leads to an increasing demand for transport, which requires a corresponding increase in mass transit supply to absorb it. The study by Wollner (2015) already described some new concepts of mobility in rural areas, in the CASI project they investigated a large collection of sustainable innovations from all over Europe illustrating well the future of sustainable mobility in rural areas.

Moreover, new mobility services are offered to solve problems with public transport demand and supply. Besides challenges linked to quantitative growth, public transport must make significant qualitative improvements to become more attractive. Passengers expect the same level of quality services and connectivity from public transport vehicles and stations as they already have in their environment and living space. Such services build on basic requirements, which include comfort, security, and cleanliness. The notion of quality in public transport should also be created through operational excellence, which includes enhanced frequency, punctuality and reliability of the service thanks to optimized network design and service performance. Likewise, smart ticketing and integrated travel information contribute to making public transport customer/user-friendly, while facilitating accessibility for all citizens. All these tools offer a huge potential to attract new public transport customers and consequently increase revenues.

The world is changing fast and urban areas are springing up everywhere, driven by the growth of cities of all sizes. Combined with rapid economic growth and the emergence of a large middle class in emerging countries, these trends are shifting the world's center of gravity to the South-East. Globally, some 1,000 cities of more than 500,000 inhabitants are already facing major mobility problems, due to the near impossibility of providing adequate infrastructure to keep pace with the ever-increasing popularity of the private car. In the meantime, in Western countries, car use seems to have reached a ceiling. Young people according to the UITP Public Transport Trends Report (2015) are now more interested in all the latest mobility solutions than in car ownership. In urban and rural areas, this is leading to new mobility behavior. Concerning rural areas, the statistics show different trends there, so we would like to present some good examples that can be taken for inspiration.

South-Moravian Region

Interesting results are provided by the study by Stastna et al. (2015) focused on integrated public transport in South-Moravia ensured by the Integrated Transport System of the South Moravian Region. This joint venture company involves 21 state-owned, communal and private companies in unified tariffs and schedules, operated from one place with mutual interconnection of various public transport systems. It covers 672 of 673 communes in the region (both urban and rural) and an additional 55 communes in neighboring regions including small overlaps to Slovakia and Austria. It operates over an area of 8,117 km^2 with 1,221,212 inhabitants (2010), including villages with small populations and out-migration trends. The system is under permanent monitoring and the routes, schedules, and other characteristics are adapted according to the needs of inhabitants. The system consists of 331 public transport lines, including 25 railway lines, 11 tramway lines, 13 trolleybus lines, 79

Figure 2. The modal split according to the settlement type
Source: The Association of German Transport Companies, 2016

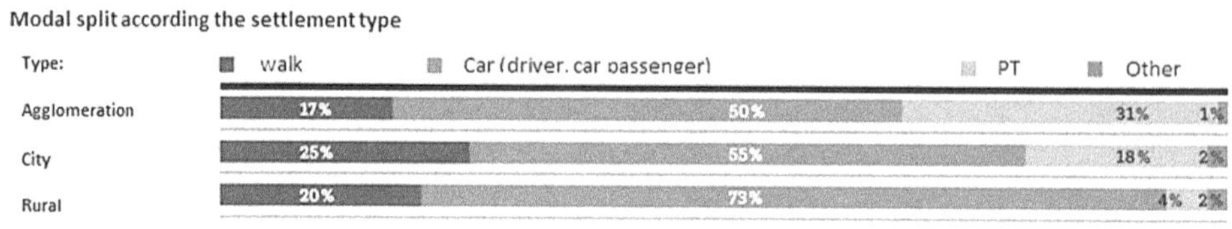

bus lines of urban transport in 10 cities and towns (including Brno) and 193 regional bus lines altogether with about 3500 stops. Railway and tramway lines outbalance in radial directions, whereas buses mostly on the tangential ones. The system generates more than 20,000 daily connections. It operates with 1300 vehicles. The average number of public transport connections according to the study is 27. The reachability analysis (analysis of total travel time required to get from the origin village or small town to the destination urban center in minutes) showed that all villages or towns are a maximum of 40 minutes from the closest urban center. The output from this case study shows that rural areas can profit from a PT offer which is based on the integration of PT services or other services related to mobility.

Mittelthüringen

Another interesting example is related to Public Transport Union Mittelthüringen. The Verkehrsverbund Mittelthüringen - VMT (2018) is a consortium of 13 transport companies which offer regional local transport as a simple, environmentally friendly and inexpensive mobility solution. The territory of the VMT includes the cities of Erfurt, Weimar, Jena, and Gera as well as the counties of Gotha, Weimarer Land, and the Saale-Holzland-Kreis. In this case, passengers from rural areas also benefit from a single public transport system. The system covers urban as well as rural areas with frequent public transport service.

But based on the evidence from UITP, rural inhabitants are more likely to use and need cars. This is mainly caused by the lower accessibility of other forms of transport mode. The statistic of the Association of German Transport Companies (2016) shows and confirms the data about modal split in various types of settlements, see Figure 2. It provides interesting information about the relationship between urban-rural typology and changes in demand for transport.

Based on UITP, there is a correlation between the modal split in well-served urban areas and public transport service provided (Figure 2). Demand for PT in rural areas is very low. Due to this fact, low density rural areas cannot be fully supplied by common public transport services, and new approaches are sought.

New Approaches and Forms of Public Transport

Rural areas are facing problems with everyday mobility. Analysis of the current status in particular countries clearly indicates that rural areas are facing changes, especially in travel demand. However, various mobility solutions have already been testing new ways of supplying public transport services in many countries. These methods respond better to uncertain travel demand. The role that taxis or other demand-responsive transport services could have, alongside conventional public transport, in meeting the transport needs of rural dwellers were examined.

Two principal approaches prevail. The first one is the development of regular operation based on short and regular intervals supported by integrated public transport organization. This is suitable mainly for rural areas where the population is increasing, or in the case when these rural areas can benefit from the proximity of economically strong cities. But what if the rural area is represented by low-density population and small settlements distributed in the area? In this case, it is recommended to establish the minimum public transport frequency service which will guarantee the minimum number of connections to or from cities to the rural area. This can be ensured by the implementation of standard quality criteria in the contracts between operators and governments or municipalities.

The second approach is suitable for less populated areas. It is based on the creation of good conditions for the development of a combination of an on-demand and conventional public transport, promising improved mobility and increased service coverage. Some studies, for instance (Wang et. al, 2015) already tested the propensity of such transport in a rural area. There are many examples of such a service as Dial a bus in Belgium[4], Germany[5] or the UK[6]. Even if DRT services are not an innovation, new recent technologies enable their large scale involvement. Nowadays, new solutions based on the concept of Mobility as a Service (MaaS) are emerging. This includes the development of smartphone apps which help to manage, monitor or order transport services for customers. Various forms of mobility combining different modes of transport including non-motorized supporting the multimodality including bike-sharing, carsharing, carpooling, Bike+Ride, Park+Ride, e-scooter-sharing, etc. are also suitable for rural areas with low transport demand.

Figure 3 presents the basic concept for an innovative solution in rural mobility. Conventional modes such as trains or buses are still represented, but it is necessary to also add individual modes and apply the new MaaS services. This can be converted in the new framework called the SMART RURAL MOBILITY, Figure 3.

Figure 3. The right mix for solving mobility in rural areas

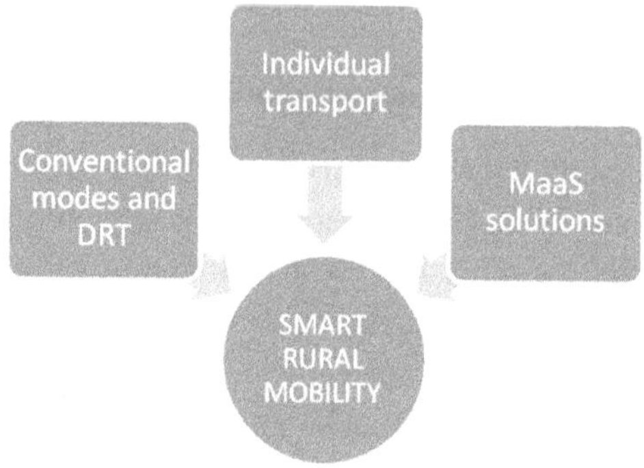

Mobility as a Service in Rural Areas

As mentioned above, rural areas have specific characteristics that need to be properly addressed towards guaranteeing and adjusted mobility to its population transport demand. According to the study provided by Barreto, Amaral and Baltazar (2018), MaaS - a key component of any future mobility system - will play an important role in enhancing social inclusion in rural areas. Creating the opportunity to integrate various transport modes and functionalities, this system will allow access to information, in a personalized way. As they presented in the pilot cases, MaaS can integrate various transport modes, allowing the rural population to increase their quality of life.

Eckhardta, et al. (2018) presented a paper providing characteristics of rural areas for MaaS development in Finland based on a project co-funded by the Ministry of Agriculture and Forestry of Finland. The paper presents rural mobility SWOT analysis and challenges, as well as goals and vision. Solutions for MaaS services in terms of collaboration, services & markets, planning & decision-making, and technology & information are proposed. The next steps to be taken in rural MaaS development in Finland are examined there too.

In our chapter we would like to describe some of the case studies with the pilots of new mobility services in various regions of Europe during the RUMOBIL project. The RUMOBIL consortium demonstrated a roadmap – RUMOBIL strategy towards smart rural mobility including MaaS concept elements divided into three fields of concern:

1. New approaches and transport services to link rural areas to national and EU transport;
2. Improvement of access points to public transport networks to render offered services more attractive;
3. Enhancement of passenger information to promote the use of public transport in rural areas.

CASE STUDIES: RUMOBIL PROJECT

Alternative transport services and new innovative solutions may play a key role in keeping rural people engaged in mainstream society. The main objective of the RUMOBIL project (supported by the Interreg Central Europe program in the period April 2016 – May 2019) was to improve public transport in nine rural areas by better connecting those areas to the national and European transport networks. The project provided answers to the core-periphery dichotomy in the CENTRAL EUROPE area. Thanks to the project, the gap between peripheral and less accessible regions and the area's well-connected centers could be reduced.

Quicker and more comfortable access to public transport positively contributes to the quality of life in rural areas and supports business activities there. To meet that aim, a set of evidence-based policy recommendations was elaborated at the end of the project to improve the capacities of local and regional decision-makers responsible for the design and coordination of public passenger transport in Central European regions affected by demographic change. These recommendations were presented in the form of the RUMOBIL Model, a strategy which provides insights into already proven good practices and novel solutions. This means that new public transport solutions, new transport planning knowledge, and new strategic recommendations were transferred to decision-makers, public officials and transport professionals.

Eight pilot actions, performed in rural regions of some Central European countries (Germany, Italy, Croatia, Czech Republic, Poland, Hungary and Slovakia), allowed the testing of a number of innovative applications during the period 2017-2019 in order to determine how sparsely populated peripheral areas can be better linked to a primary, secondary or tertiary transport node (access to European and national passenger transport networks). These pilots were experimental trials, small-scale, short-term projects, which helped partners learn how a large-scale project might work in practice in the future in the region. The pilot projects provided a platform for the partner organizations to test measure, prove value and reveal deficiencies before spending a significant amount of time, energy or money on a larger-scale project. All pilot projects began with a proposal listing the objectives of the action.

Figure 4. Typology of settlements in partners' regions
Source: RUMOBIL, D.T2.6.2 Quantitative evaluation, 2019

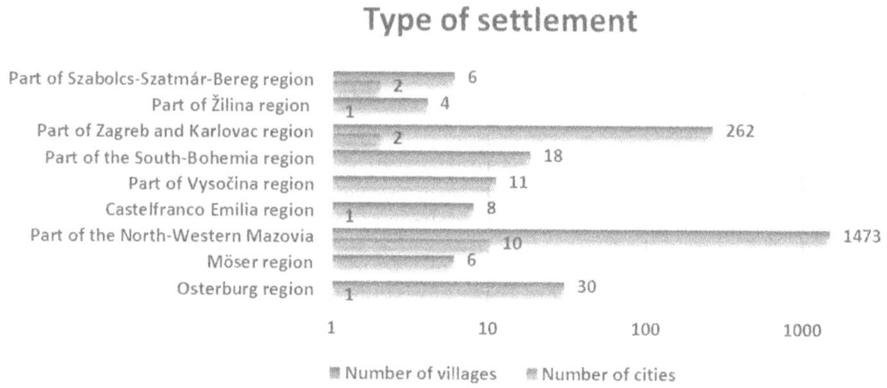

A set of monitoring indicators was defined to measure outputs against the goals of the pilots and RUMOBIL project as well. In total 51 indicators were examined. All partners gathered data from their pilot regions with the aim of giving an overview of the functional features of the existing transportation system and baseline situation in the region before and after pilot implementation. Data were used to prepare a transport demand prognosis for the future in mid-term and long-term horizon, measure user satisfaction with the pilot measure, and measure project impact in the future. Partners selected their set of indicators, choosing a mixture of different types of indicator suitable for individual pilot projects and providing data for all purposes mentioned above. Because there were various types of pilots, the common comparable set of indicators was defined. While some indicators were unique or tailored to a specific pilot project context, some common indicators were required, too, with the aim of comparing results among partners. That data was used for evaluation of pilot performance (ex-post evaluation), assessing the value for money of the pilot, valuation of user acceptance and satisfaction, and for transport demand prognosis. Indicators were quantitative and qualitative and divided into five groups: Geographical, Socio-demographical, Operational, Economical and Qualitative. Some result indicators are reviewed in this chapter.

The Base Situation in Regions

Spatial and Socio-Demographic Characteristics

All partner regions are rural areas, however with different spatial or socio-demographic characteristics. The size of area covered by pilot measures (area where the PT service is accessible) in partner regions ranges from the biggest one 5,339 km^2 – region in North-Western Mazovia (PL) located around the two railway lines Nasielsk – Sierpc, and Sierpc – Kutno and the smallest one covering area 71 km^2 – Microregion of Kaplice (CZ), a part of South Bohemia region.

Population density varies between 44 inh/km^2 in Osterburg region (D) and 321 inh/km^2 in Castelfranco Emilia (I). Most partner regions are predominantly rural; regions, where the rural population live in settlements under 5,000 inhabitants account for 50% or more of the total population and villages represent the main type of settlements, Figure 4.

However, some of the regions are very close to big cities (Zagreb, Bologna, Modena), which explains why they were assigned as urban/rural or intermediate regions. The information about the spatial distribution of settlements was useful to understand spatial implications and relationship between urban and rural environment. Concerning demographic patterns, the biggest population lives in the Mazovia region – 365,657 inh. (2017), which represents 60% of the RUMOBIL target group, while the least populated area is a part of South-Bohemia region – 3,263 inh. (2017), representing 1% of the RUMOBIL target group.

South-Bohemia, Vysočina (CZ) or Szatmár-Bereg region (HU) represent low-density rural areas with poor public transport service. For example, Vysočina Region situated in the center of the Czech Republic differs from neighboring regions by its dissected territory, higher altitude, and low residential density. Disintegrated residential structures very often cause people, especially young and qualified inhabitants to move away from smaller municipalities. Economic performance of this Region in comparison with other regions of the Czech Republic is below the average. Then there are rural areas which are near smaller cities as the Mazovia region, Žilina region and Ozalj (Karlovac) region (HR) benefiting from their position. Another case is represented by CastelFranco Emilia region, which in fact lies between the suburban area of Modena and Bologna.

Most of the partner regions suffer from population decline. An exception is Castelfranco, where the population growth rate was +1.11%/year (in period 2011-2014). The town Castelfranco Emilia in Modena Province, the site of the pilot project, is located about 13km east of Modena; it has a population of 32,174 inhabitants (2014). The town is situated on the main regional and national rail network that allows easy connections to the main cities of the region, and from there, particularly

Figure 5. Populations by age groups in partners' regions
Source: RUMOBIL, D.T2.6.2 – Quantitative evaluation, 2019

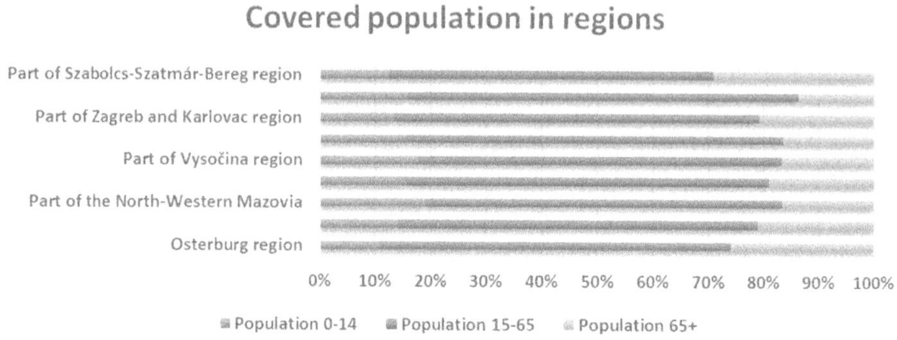

Figure 6. Ageing index in partners' regions and countries
Source: RUMOBIL, D.T2.6.2 – Quantitative evaluation, 2019

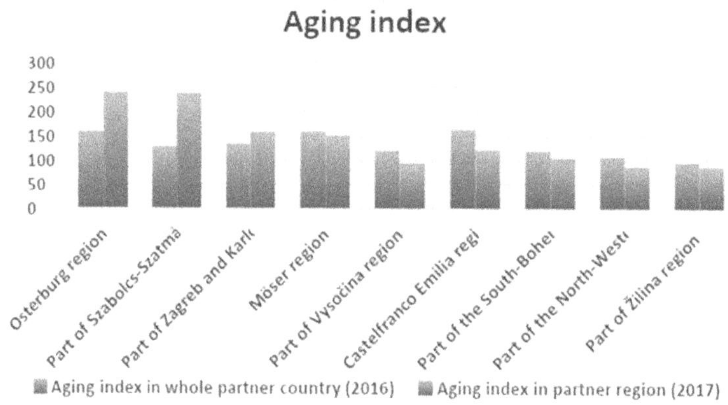

from Bologna (25 km distant from Castelfranco Emilia), it is possible to enter trains to many Italian and European destinations. The territory of the municipality is quite extensive and includes eight minor villages that consist of predominantly rural areas and no relevant production or industrial facilities are present in the territory. The proximity to two important towns such as Modena and Bologna, the lower cost of housing than in the main cities, the presence of public services and in particular the good train service, have fostered a strong demographic growth in the last 25 years accompanied by significant growth in commuting. The increase of population was initially in the small villages around Castelfranco Emilia and this, some years ago, has created a kind of repopulation of rural-agricultural areas that were more attractive than the main city. This attractiveness has created new needs, principally for people that are getting older, to access the main services located in the central town and cities.

Ageing is a problem in all the RUMOBIL partner regions. Especially in Szabolcs-Szatmár-Bereg County, where people in the group 65+ represent 29% of the total population, Figure 5.

One of the indicators of population ageing is the ageing index, or the elder child ratio, defined as the ratio of elders (individuals 65+) to children (youths under the age of 15). Populations with ageing index scores of 15 or less are described as young, whereas populations with aging index scores of 30 or greater are considered old. Populations in all partners' countries are old. When we compare the ageing indices in the partner regions with the ageing indices in the relevant countries we see that the worst situation is in Osterburg and Szatmár-Bereg region and the best situation in Castelfranco Emilia region, Figure 6.

Consistently low birth rates and higher life expectancy will transform the shape of the age pyramid in partners' regions; probably the most important change will be the transition towards a much older population structure, a development which is already apparent in the many EU Member States.

Demographic and social factors are very influential in public transport demand. Population growth, as a leading factor, and population characteristics (e.g., age structure, the declining proportion of young people, ageing) have a big impact on the use of public transport in rural regions. Without understanding the transport system dynamics and more importantly, the behavior of public transport travelers, it would be difficult to make accurate transport demand forecasts, which are necessary for marketing, service planning and fare policy purposes for regions.

Operational or performance indicators are widely used in transport planning and evaluation purposes. Performance measurement can support public transport planning in rural areas in many ways. Respecting the aims of the RUMOBIL project, selected operational indicators allowed the determination of whether resources were used efficiently and equitably, identification of potential problems and wins, and verification of whether a particular pilot achieved its predicted targets. They were also used for comparison among partners, even if there were different types of pilots. Some pilots intended to launch new services (Saxony-Anhalt, South-Bohemia), others focused on the improvement of existing services (Vysočina, HŽPP) or interchange hubs or stops (Žilina, Szatmár-Bereg), and yet others implemented new technology in the operation of an existing service (Mazovia, Castelfranco). In all cases partners needed to collect information about the public transport service in the target area before and after pilot implementation focusing on the particular mode of transport, which was related to the pilot, however some indicators were dealing with data on

Figure 7. Spatial accessibility of PT in partners' regions
Source: RUMOBIL, D.T2.6.2 – Quantitative evaluation, 2019

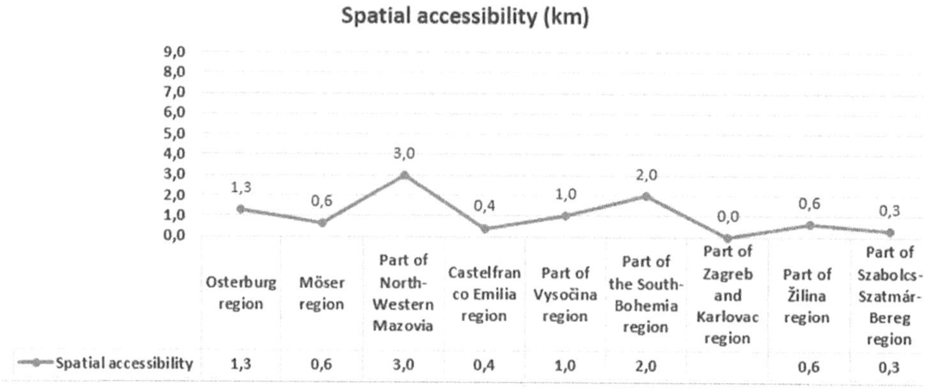

Figure 8. Correlation between population density and spatial accessibility in partners' regions
Source: RUMOBIL, D.T2.6.2 – Quantitative evaluation, 2019

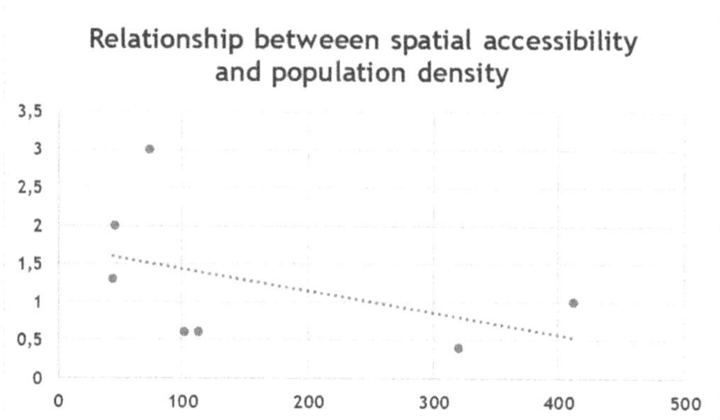

the whole public transport service in the target area focusing on transit of passengers from target area to central town or city.

Most of the partners' areas are served by bus and rail transport except the Vysočina region which is served only by bus transport. German and Italian regions are also served by DRT service. Accessibility of public transport is one of the key attributes of its quality, which indirectly affects its competitiveness in relation to car transport because the advantage of car transport mainly lies in its door-to-door concept. Spatial accessibility was defined as a distance to overcome to get to the public transport stop or station (km). The best situation is in Szatmár-Bereg with average distance 0.3 km to the nearest stop, Figure 7.

Regarding the result of the correlation analysis (Figure 8), there is a mild correlation between population density and spatial accessibility (correlation coefficient -0.53) in partners' regions. HŽPP did not provide data on spatial accessibility.

There are different characteristics of target areas in partner countries in terms of size, population density and transport mode evaluated. The most stops are in the Castelfranco Emilia region, where the population density is very high. There is one more important **intermodal point** in each target area of partners' countries, except Mazovia with two of them.

In rural areas, the journey very often starts with driving a car or riding a bike and ends with using public transit. In this case, it is very important to have an appropriate parking lot or parking places or facilities close to public transit stations or stops. The situation is very different in partner countries. There is a lack of parking places, especially in the Mazovia and Ozalj region.

Reachability was defined as total travel time required to get from origin (average distant village) to the destination (in minutes) connecting the rural environment with the central (bigger) city. The biggest distance is between Ozalj and Zagreb (around 60 km), the average travel time by train from Zagreb to Ozalj is 82 min. The best result is in the case of Castelfranco served by DRT service, however, the biggest distance between the most distant village and the central city is only 12 km (Figure 9). If we compare the ratio between travel time and distance for the Italian region (1.5) and Croatian region (1.36) the reachability is better in Croatian region using the train.

Optimal price, time values and service frequency of public transport are very important factors for passengers' behavior. Higher frequency means less waiting, probably the worst part of using public transport. At the same time, increasing frequency may be the best investment that can be made to get people out of their cars. However, it is not easy to ensure the required frequency in rural areas and it is necessary to solve the frequency optimization problem. The solution should satisfy a given origin-destination demand and a constraint on the available fleet of buses or other transport modes. Frequency varies from 9 connections per day in Vysočina (bus connections) and Ozalj region (rail connections) to 144 in the Casterfranco Emilia region including regular bus + on-demand service + train connections per working day (average – 6 rides per hour).

All partners provide a regular operation regime, except the German and Italian partners. Flexible transport service can cover a range of different mobility offers, where services are flexible in one or more of the dimensions of time, route, vehicle allocation, vehicle operator, type of payment or passenger category. Flexible transport services are usually used as a part of the public transport mix in areas where the demand is too low to support conventional public transport. MLV introduced a

Figure 9. Reachability in Castelfranco region
Source: RUMOBIL, D.T2.6.2 – Quantitative evaluation, 2019

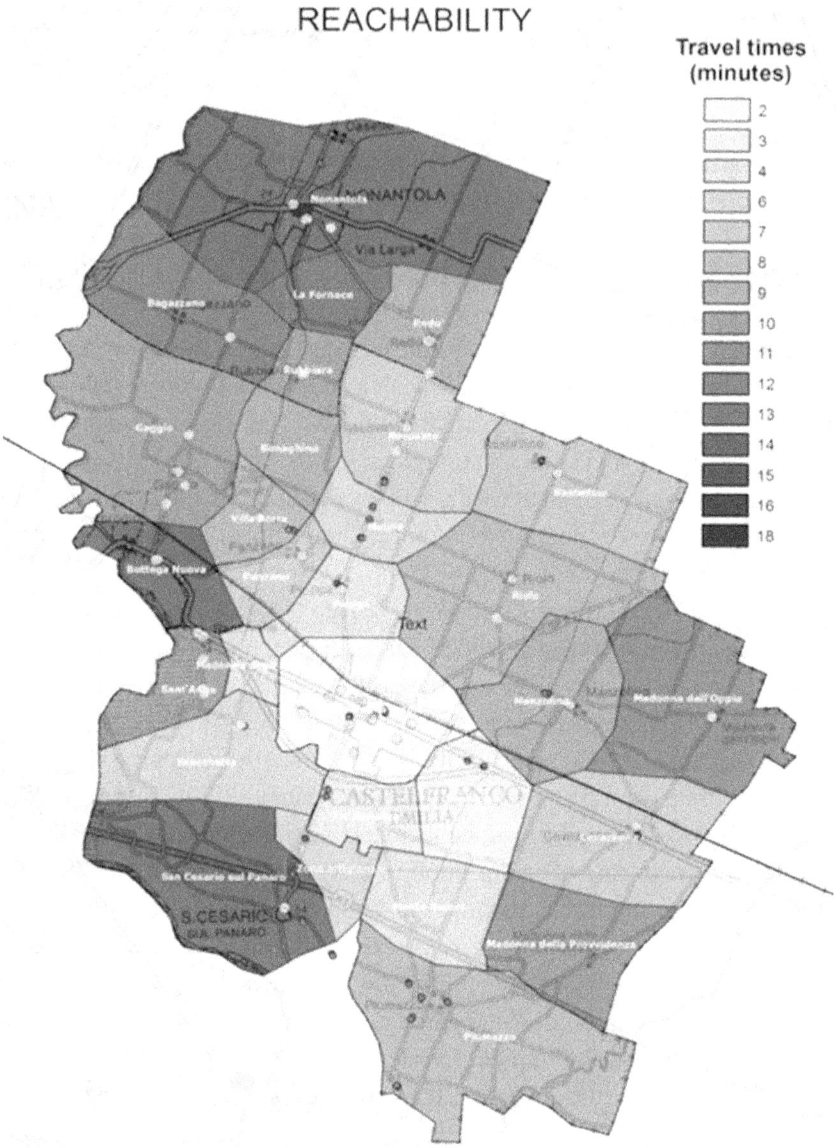

Figure 10. Capacity offered by PT services in partners' regions
Source: RUMOBIL, D.T2.6.2 – Quantitative evaluation, 2019

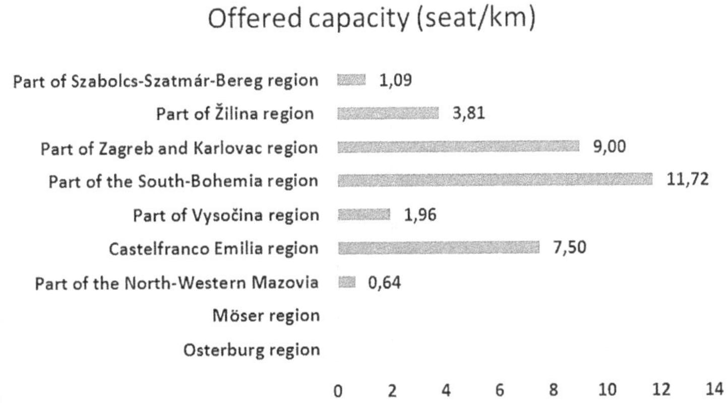

Figure 11. Parking fee in the central city in partners' regions
Source: RUMOBIL, D.T2.6.2 – Quantitative evaluation, 2019

new type of service – Burgerbus, characterized by the fact that bus drivers are local volunteers. In this way, service costs are significantly reduced compared to a traditional service. DRT service in Castelfranco is provided by aMo as a combination of regular and flexible, possibly on-demand service.

The offered capacity is equal to the number of seats available multiplied by the number of kilometers ridden. Even if there are different types of public transport service available in partners' area, we can compare the offered capacity regardless of the capacity of vehicles (from low-capacity minibuses to medium or high capacity trains). This indicator is very important to provide information for capacity

Figure 12. Travel costs to reach the central city in partners' regions
Source: RUMOBIL, D.T2.6.2 – Quantitative evaluation, 2019

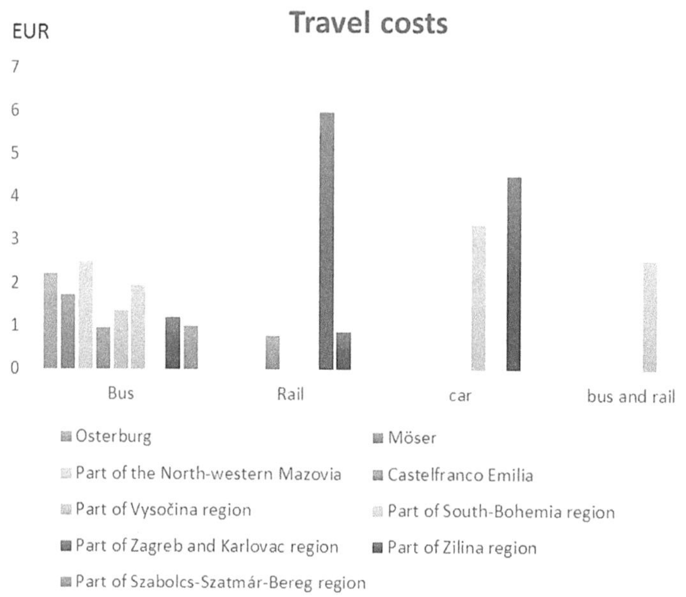

normalization, comparing the offered capacity with other indicators such as the number of passengers, Figure 10.

In most of the partners' regions, there were no substantial changes in capacity offered in the target area after pilot project implementation, Figure 10. The same capacity was offered in Ozalj, Mazovia and Vysočina region; there was a decrease of 5% in Castelfranco, an increase of 25% in Mazovia and an increase of 12% in South-Bohemia (where a new line was introduced). MLV did not provide information on this indicator for Möser and Osterburg region. Travel costs include all travel costs necessary for a trip to the central city. They were calculated especially for a single or intermodal trip by public transport. Some partners also provided costs for individual modes of transport (bus, car, rail) for a standard trip on working day. The highest price was calculated for the trip by train from Ozalj to Zagreb, however, it is also the longest trip (approx. 60 km), Figure 12.

Parking fees were calculated as a standard parking fee for all-day parking in the target city. The highest parking fee is in Zagreb, on the contrary parking is free of charge in Castelfranco, Mazovia, Szabolcs-Szatmár-Bereg County, Möser and Ostreburg, where the central city usually represents small towns, Figure 11.

Lessons Learned from Pilot Projects

Lessons learned from the pilot measures were summarized in the RUMOBIL Strategy – a decision-support model that proposed a set of general recommendations and verified solutions to reduce the critical issues of mobility in rural and peripheral areas and outline the relative implementation progress. The Strategy was contextualized with consideration given to the very different characteristics of each territory in terms of needs, limits, and potential. The objective of the RUMOBIL strategy (2018) is to contrast the population decrease, guaranteeing the transferability of the strategy outside the partner territories and beyond the end of the project. In order to pursue this objective, the RUMOBIL Strategy mainly takes into account three key aspects, i.e. the target users, to whom the solutions have been addressed; the users' needs, which should be satisfied; and transferability. The general target user of the solutions identified is the wide group of inhabitants of rural areas, focusing on more specific groups depending on the particular needs to be satisfied in each area (e.g. disabled, elderly people and/or students). In rural and peripheral areas and regarding mobility questions, the need for an effective and useful public transport system is surely the most urgent request to be satisfied. Indeed, in many partners' territories the improvement of PT has been highlighted, in terms of enhancement of the supply (frequency, no. of daily departures, etc.), creation of more appealing services for users, more accessibility to PT systems for all users in rural areas in general, and disabled and elderly people in particular, and promotion of the systems in order to raise citizens' awareness of the services. The RUMOBIL Strategy can be considered "transferable", as it is a unique model that allows decision-makers and stakeholders to identify the most appropriate solutions to be adopted in every rural area.

Considering all the inputs collected and analyzed during the project, the Strategy was built on the identification of solutions, in terms of concrete actions, and recommendations emerged from:

- Good practices in Central Europe,
- The current technological state and the correlations with economic and demographic issues,
- Analysis of the main requirements of a good "service strategy" able to satisfy the users' needs.

Solutions and recommendations have systematized consistently with the actions to be taken as follows:

Transportation Planning

The importance of a rational system of transportation planning and decision making has been recognized in the RUMOBIL partnership, consisting of:

- Inventorying existing travel and activity patterns in the region, respecting relationships with other regions, and the national and international context,
- Developing models of local transportation supply and demand relationships,
- Formulating transportation options,
- forecasting the effects of each option on the travel of different target groups,
- Evaluating each option on the basis of economic and other criteria.

Carrying out these steps requires an optimal regional institutional arrangement for transportation planning and decision making. The future is uncertain, especially now in the face of globalization and the high tempo of social, demographic and technological changes. It is difficult but very important to understand uncertainties and objectively estimate transport demand in rural areas. A future-oriented public transport for rural areas should consider demographic trends, socio-economic changes and the diversity of users and provide tailored offers. It is necessary to assess customer expectations and align them with service capabilities. In this way, PT supports the provision of equal living conditions in rural areas. Information analyzed during the project was examined together with respective stakeholders to prepare the decision-making processes leading to the implementation of the RUMOBIL Strategy through improved transport strategies and plans.

PT Offer and Local Development

A strong public transport is also an important factor for the prospects of rural areas. Well-designed and organized PT supports general development along PT corridors and a more homogeneous countryside. Improving public transport between urban and rural areas can increase connectivity between urban and rural areas, help develop local economies, contribute to the growth of local business and integrate the rural environment into a mainstream economy. Good public transport service offered by operators can, with the help of local government, create satisfaction by employing local people and different stakeholders, promoting heritage, sights and natural beauties. People are interested in visiting rural areas with specific tourist attractions and routes, and if combined with natural adventure, sport or other activities, these kind of routes are excellent for attracting visitors, and thus PT passengers. With people coming, the local population has a reason to stay in a rural area and provide services. For tourist destination areas, encouraging a transfer from cars to public

Figure 13. Thematic ride and promotional material
Source: RUMOBIL, DELIVERABLE D.T1.6.1 – Work paper summarizing the learning from RUMOBIL pilots, 2019

transport can reduce congestion and parking or environmental problems in areas often visited for their tranquility, natural or historic landscapes. PT also presents the potential for engaging with visitors, enhancing the visitor experience and attracting new markets. The elements to which attention should be given are personnel, cooperation and coordination of all stakeholders promoting existing sights (natural, cultural, infrastructural, etc.) to expand tourism in the region. Good examples of this approach are presented by pilots from the Ozalj region and South Bohemia region.

Ozalj Region

Ozalj is a small town in central Croatia (Karlovac County), located 65 km far from Zagreb, the Croatian capital. The official name of the pilot project implemented there was "Tourist trains for improving rural mobility and economy". Pilot project activities included the organizing of special tourist trains to the Ozalj area. Each of them was linked to a special event in this region. A total of 39 trains were organized. Provision of pilot project train lines and pilot project communication activities was the responsibility of HZPP (HŽ Passenger Transport Limited Liability Company, the national rail operator). Representatives of the City of Ozalj were responsible for creating the content and the events at the destination – the administrative area of the City of Ozalj. When organizing each special train and its content, HZPP also cooperated with local stakeholders, like the municipality of Ozalj, and 20 other stakeholders representing local organizations, and businesses. HZPP aimed at increasing the number of train passengers by introducing new train services between isolated rural areas in Karlovac County and the secondary transport hub in Karlovac (Zagreb-Rijeka railway – access to the Mediterranean TEN) to the primary hub – Zagreb. The new service was designed to suit both the needs of commuters and the itineraries of tourists and to generate additional users. A marketing campaign, promoting the pilot, was conducted by external experts. They created a Facebook page, arranged radio commercials and TV appearances as well as prepared leaflets so that the public was informed about the new rail service, Figure 13. Most trains were planned to operate and provide services on the following Zagreb – Jastrebarsko - Karlovac – Ozalj – Kamanje route (with possible stops in some stations along the way to allow passengers to board, depending on each train theme). The pilot project helped to enhance the local economy, especially the tourist industry, and supported local producers, shopkeepers, and cultural organizations by attracting more people to visit the Ozalj region. Good lessons were learned around the coordination of the transport offering, cooperation with different local stakeholders and creation of interesting events. Special tourist trains were organized from early spring to late autumn, including for example the "Slava Raškaj tour through the city of Ozalj", "Spring awakening", "Honey tour through Zaluka plant", "Days of wine", "The old villages games", "City beach swimming", "Štrudlafest", etc. As a positive result, the City of Ozalj recorded an increase in overnight stays of guests from Zagreb and the surrounding area and received very positive comments from people who took a train trip several times.

Riding thematic trains in the Ozalj region is not just about getting people from place to place. The pilot aimed to stimulate the development of sustainable rural tourism, providing unique tourist product relating to local natural or cultural heritage and to benefit other economic sectors. It answered a question of how the local community can be involved and what the role of local entrepreneurs could be.

Kaplice – South-Bohemia Region

The area of this pilot project was a rural area with a low population density close to the town of Kaplice, situated very close to the main railway line České Budějovice – Linz, part of the TEN-T network. The local transport node is a railway station located 5 kilometers from the town of Kaplice. The station is not very well used for traveling to the region compared to the direct bus lines.

The main railway service includes two-hour interval international trains to Linz, of which half are Prague-Linz express trains and the other half are semi-fast regional trains České Budějovice – Linz, as complementary connections to regional trains from České Budějovice to the border crossing, which are in the summer season reinforced by other connections designed especially for tourists visiting Šumava National Park. All villages and towns have their own bus lines, but they are often slow, especially connections to the regional capital, the City of České Budějovice. For example, the bus ride from Besednice (central village) to České Budějovice

Figure 14. Promotional material for a new bus line
Source: RUMOBIL, DELIVERABLE D.T1.6.1 – Work paper summarizing the learning from RUMOBIL pilots, 2019

takes about 1 hour, some passengers have to change bus or train connections one or two times, and the interchange time is usually very long.

Another problem is the service on weekends, and public holidays, when the transport service is very bad: some villages have only two or three connections per day, some none at all. In addition, bus time schedules are not well synchronised with trains in most cases. The pilot bus line aimed to change the current situation of public transport in the area. Passengers were provided not only the new pilot bus line, but they also received access to the regional and national rail transport network. The second objective was to learn how to use rail transport more to reach the pilot region and to make the region more visible and attractive. The ultimate goal, perhaps the most important, was to improve the quality of life in the pilot area and to keep the population here, to increase the number of people in the region. The new pilot bus line led from the railway station Kaplice to operate in neighboring municipalities and settlements/solitudes with the destination city of Benešov nad Černou. The line helped to improve the availability of public transport in the area (including weekends and public holidays, especially in places where access was not available), but the interest of the local populations did not meet the level of demand that would lead to the inclusion of this pilot line to the public transport obligation in future. To change this unfavorable situation, local governments should encourage and promote public transportation more.

Bearing in mind the low demand of local people for new bus lines it is necessary to better understand their requirements and prepare a more tailored approach to marketing than what's currently offered. It will require a willingness to take personalization to the next level, evolving to offer exactly the right product, to the right customer, at the right time. A concept of MaaS is also to provide people with easy access to travel information so they can be better informed as to the different ways to undertake their trips, or perhaps a service allowing access to multiple transport providers via a monthly user fee. Personalized travel marketing, accompanied by other hard and soft measures, and the possibility for participants to give feedback on the existing public transport services in their region, could be an efficient way to increase awareness and encourage people to use public transport more often in rural areas. The secondary objective, to allow tourists access to the area, was fulfilled successfully, see Figure 14. Despite the short test period of the pilot project, which covered the tourist season for only 3 months, there was considerable interest from both individual tourists and organized groups (senior clubs, tourist clubs, scouts, as well as schools that used the bus for school trips).

Integration Coordination and Intermodality

The significantly inadequate mobility options in rural areas very often result from the lack of spatial coverage of the area by public bus transportation services, with railway transport available only to the inhabitants residing in the vicinity of the railway lines. The availability of public transportation, and especially, the better management of routes and the establishment of integrated systems in public transportation contribute to the change of many negative trends in rural areas. Integrated transport systems aim to provide residents with high quality, easily accessible transport network across a whole region. Travelers should be able to move using different modes of transport - such as rail, bus service, cars, on-demand service, bicycles, and even feet. The network should have a unified tariff and payment system. Connections should be easy to reach, resulting in fewer transfers, reasonable costs, a reliable service, and convenient travel times. Consistent with pilot targets, actions facilitating integration have been implemented in several pilot regions (Saxony-Anhalt, Mazovia, Vysočina, Žilina, and South Bohemia), which also included new services implementation, optimization of interchange nodes and intermodal options.

Vysočina Region

The Vysočina Region aimed at improving the offer of public transport connections as a key condition for the development of mobility of the population, in terms of linking rural areas to major settlements and, through them, to transport networks of national and EU importance. The pilot strived to test and verify the demand for new public bus transport connections in a territory with fragmented settlement structure and in the context of the preparation of the "Public Transport of Vysočina" concept. The tested measures combined "standard" existing solutions, designed by the current regional public transport system, with ongoing preparations to launch the future integrated system "Public Transport of Vysočina". These connections were ensured by 4 public bus transport operators based on long-term public service commitment contracts with the Vysočina Region. To sum up the whole trial period (March 2017 – June 2018), a total of 34 new public bus transport connections were introduced, operated on 4 lines, both on working days and at weekends. During this period, buses on pilot connections covered a total distance of almost 204 000 kilometers and transported more than 75 000 passengers. The pilot connections contributed to the improvement of accessibility of rural municipalities and their connection to nodes of regional or subregional importance, that are significant centers of education, employment, and various public services for the inhabitants (regional capital Jihlava, cities of Třebíč and Havlíčkův Brod, the town of Velké Meziříčí). These centers also serve as traffic junctions and interchange points between transport modes.

Figure 15. New facilities in Rajecké Teplice railway station
Source: RUMOBIL, DELIVERABLE D.T1.6.1 – Workpaper summarizing the learning from RUMOBIL pilots, 2019

One of the pilot lines also strengthened the connection to the locality of Světlá nad Sázavou and Lipnice nad Sázavou, which is an attractive tourist destination as well as an important interchange point with the railway.

Žilina Region

The area of the pilot project was a rural area around the small spa town Rajecké Teplice. The town serves as a local transport node for the surrounding countryside area with railway and bus stations. The railway transport is operated by Železničná spoločnosť Slovensko, a. s. on the regional line Žilina – Rajec. Railway transport is partially integrated with urban transport in Žilina city. However, urban, suburban and bus transport services between Žilina and Rajec are operated by different operators. The transport schedules were not fully harmonized and connecting interchange points were missing. Railway transport did not fulfill the criteria of its capacity. Before pilot project implementation integration and intermodality were not supported properly in the Rajecké Teplice region. The main aim of the project was to restore the small rural intermodal connection hub in the town of Rajecké Teplice to better serve the local people and to promote intermodal traveling options in this rural territory. The improved station acts now as an interface between the various modes of transport (rail, bus, car, cycling, walking) and will potentially also increase the ratio of people using public transport connections to the central city Žilina and consolidate the entire public transport system in the pilot area. Another aim of the pilot project was to examine how the higher level of multimodal hub services would influence the behavior of commuting passengers. With the newly rebuilt multimodal node, the Žilina self-governing region expected the stabilization of the decline in the number of passengers despite the aging of the population. They expected better coordination of different types of transport and at the same time a gradual increase in the number of passengers in the period of two or three years.

The project aimed to upgrade the space of the bus station, install monitoring panels providing information about departures and arrivals of public transport lines, ensure barrier-free access from bus to the railway station and accommodate this connecting point with an adequate number of safe bike kickstands, Figure 15. Another important aspect was security. Among other infrastructural improvement works, CCTV (Closed-circuit television) was installed at the intermodal station in Rajecké Teplice to ensure safety at the station, to make the intermodal node more attractive to passengers by increasing their feeling of safety and preventing damage to parked vehicles, or bicycles.

The term integrated mobility is used when different mobility service providers collaborate and market their options to travel in integrated way, by supplying information, reservations, bookings or payments. The aim is to offer a product where all these services are integrated. It is time for transport companies to transform their mobility portfolio and use new solutions & technology for multimodal MaaS platforms.

Flexible Transport Systems

Flexible transport systems bring a promising approach to improving the efficiency and performance of passenger transportation services in rural areas. They provide passengers with flexibility in choosing routes, times, modes of transport, service provider or payment systems. The flexibility can be based on the integration of different modes of transport, or possibly spanning multiple service providers. Service is more sophisticated, comfortable and cost-effective. The concept of flexible transport is not new. There are many existing systems, like DRT services, shared taxicabs, or car-clubs. This concept is especially suitable within rural areas, which usually suffer from a lack of service availability and demand uncertainties. Flexible transport systems could be integrated into existing PT services and aim to support it reaching low-density areas. They can also cover specific time frames. In some cases, they can represent the only PT alternative.

Möser and Osterburg

Partners in Saxony-Anhalt tested a new bus service supplementing the existing regular bus service on selected days of the week, on specific routes, connecting rural settlements with higher-level transportation node (e.g. railway station). Whereas Saxony-Anhalt provides a good main transport network (rail service and interregional bus services), many rural/peripheral areas are not adequately connected to the main transport network. Residents of those areas cannot benefit from Saxony-Anhalt's main transport network despite higher mobility demands. New approaches are

Figure 16. Burgerbus service vehicle in Osterburg
Source: RUMOBIL, DELIVERABLE D.T1.6.1 – Work paper summarizing the learning from RUMOBIL pilots, 2019

therefore necessary to better connect peripheral areas affected by the demographic change to the transport network. The MLV together with its in-house transport agency NASA aimed to introduce a new bus service operating according to the demand of residents. A "Citizens' Bus", in German "Bürgerbus" operated by local communities and volunteers represents a highly innovative and cost-effective approach for connecting rural/ peripheral areas. The approach was tested in two municipalities Möser and Osterburg, with the buses running between remote villages and bigger towns/secondary or tertiary transport hubs. The Citizens' Bus service in Osterburg and Möser is arranged by the local community and its residents. A minibus with 8 passenger seats is driven by volunteers. Both Osterburg and Möser attracted enough voluntary drivers to launch the service. The administrations of Osterburg and Möser organized the operation of the bus. Buses with fixed routes and running times connect smaller settlements to both the city centers and train station.

Accessibility for people with reduced mobility was ensured by using a low floor bus equipped with a ramp for wheelchairs and mobility scooters, Figure 16. GPS transmitters and devices for ticket sales were installed in Citizens' Buses and software was developed to arrange ticket sales and GPS tracking in the bus. This will later allow integration of the Citizens' Buses into real-time information systems. Furthermore, software to arrange the operation of the bus (planning of voluntary drivers, etc.) was purchased.

Spatial Densification of Stops/Stations

Accessibility of the public transport system is an essential indicator that guarantees equal traveling opportunities for all people. Accessibility to PT mostly affects low-income people, the elderly and people with disabilities. Access to public transport routes is among the public transport indicators that are important for these social groups of people. The proximity of homes or target destinations to public transport networks is a key factor supporting the usage of public transport. One of the most

Figure 17. Number of bus stops in pilot areas in Saxony-Anhalt
Source: RUMOBIL, DELIVERABLE D.T1.6.1 – Work paper summarizing the learning from RUMOBIL pilots, 2019

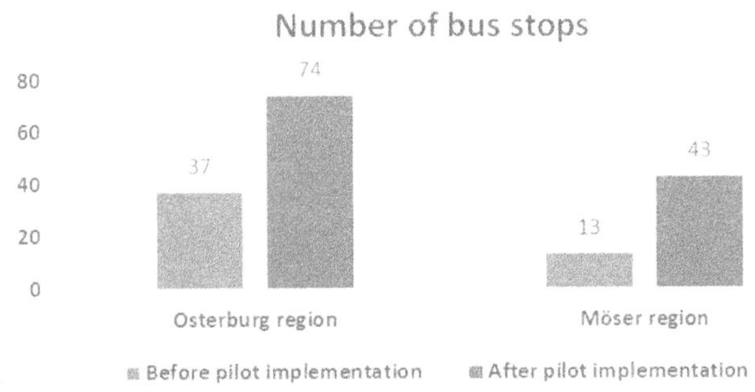

frequent strategies for increasing proximity to public transport is to densify the stops. Both the location and the spacing of bus stops significantly affect service performance and passenger satisfaction as they influence travel time. Every transit trip begins and ends with pedestrian travel. Distance to stop is considered a critical factor for accessibility of public transport, especially for vulnerable PT users. MLV tested densification of stops in the Möser and Osterburg region.

Möser and Osterburg

To make public transport in both pilot areas more attractive, the Citizens' Bus implementation was accompanied by densification of bus stops in each settlement along its routes. More precisely, in each settlement 3 to 4 bus stops were installed instead of just one, to shorten the distance between passengers' homes and bus stops. In this way, the number of bus stops substantially increased in both pilot areas Figure 17. As this approach is regarded as a trial process, the installed bus stops were only a bus stop sign and an information plate for the bus schedule. This also helped to reduce costs in the service trial period.

An Appealing Design and the Consideration of Comfort

According to the results of qualitative surveys performed among passengers in partners' regions, comfort is an important factor in residents' choice of traffic mode. Thus, improving comfort in vehicles and at bus stops and stations to attract more passengers received much attention from PT service operators and authorities in some partner regions.

Figure 18. New bus shelter and information table in Nagykálló
Source: RUMOBIL, DELIVERABLE D.T1.6.1 – Workpaper summarizing the learning from RUMOBIL pilots, 2019

Szabolcs-Szatmár-Bereg County

Hungarian partners rebuilt selected bus stops on the bus line Nyíregyháza – Nagykálló. New bus shelters were installed there. A good bus shelter – a waiting area for passengers - is an essential part of any successful PT bus service system.

From the perspective of the PT operator and infrastructure manager, a good shelter has low maintenance and energy consumption requirements and is vandal-resistant. From the passenger's point of view, an ideal shelter allows easy access to the bus, is comfortable and convenient and provides accurate information. The new design of bus stops in Nagykálló respects both these principles. Apart from the physical reconstruction of shelters, several enhancements were performed including digital passenger information boards to provide real-time information collected via GPS from buses. WiFi accessibility can also contribute to public inclusion and increasing of bus-stops' attractiveness, Figure 18.

Information and Info-Mobility Tools

New info-mobility tools are probably the ones more often considered recently. Info-mobility measures concern a lot of fields that are all important to make a service more attractive and, above all, easy to use. Many of them refer to real-time information that can improve the quality of service. Other actions focus on helping people to plan their trips, to make decisions, or to manage booking services and ticket buying. Journey planners based on information technology save time and prevent trouble during the trip. Several partners involved new info mobility tools. aMo developed a new app for users of the existing DRT service – Prontobus - to view reservations updated in real-time and with the ability to book a trip directly in the app, or through a web site. Mazowieckie Voivodeship launched the TropKM app based on the operation of GPS transmitters installed in regional trains, where the position of the train is available on mobile devices in real-time. The Žilina region, and partners from the self-government of Szabolcs-Szatmár-Bereg County, and the Municipality of Nagykálló installed info panels providing real-time information at bus stops and intermodal nodes.

Castelfranco Emilia Region

Flexible and demand-responsive transport systems have been identified as one of the promising solutions for widespread public transport in rural areas and many flexible transport services have been established. The DRT service called Prontobus was established in the Modena region to integrate with the conventional PT service (based on fixed lines and timetables). Before RUMOBIL intervention the service was available with reservations that could be made only through a call center. The call center handled reservations manually and no information about the service was provided to users. The lack of information on the DRT service was highlighted by users – through a phone survey – as one of the main difficulties in accessing it. The main aims of the project were: to significantly improve information and accessibility to the DRT service that was actually underused compared to its potential; to improve the integration of the various public transport modes existing in the territory and the quality of public transport offered; to better connect the rural areas to the main European (TEN-T) and national transport network that are available in the area; to improve the performance of the reservation center providing a tool that would simplify the activities of recording and reporting the DRT services.

The RUMOBIL pilot project tried to reach these aims by implementing a new software application connected to a web portal specifically dedicated to the information about DRT services. The system consists of:

Figure 19. Web portal for users, App for drivers on buses
Source: RUMOBIL, DELIVERABLE D.T1.6.1 – Work paper summarizing the learning from RUMOBIL pilots, 2019

Figure 20. Number of trips per month comparison
Source: RUMOBIL, DELIVERABLE D.T1.6.1 – Work paper summarizing the learning from RUMOBIL pilots, 2019

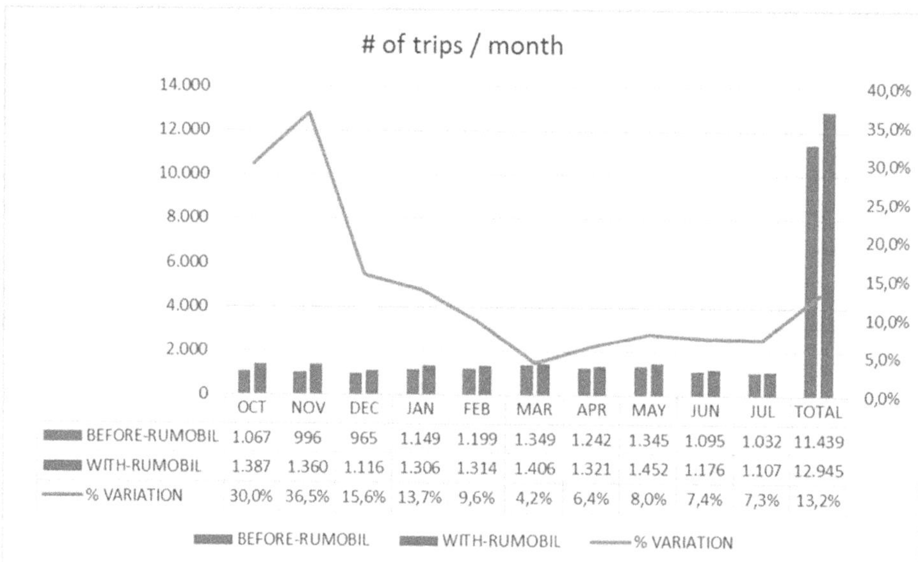

- A web portal (www.rumobil-modena.eu) for the users of the Prontobus service with information about the service and the situation of reservations updated in real-time with the possibility to register in order to receive e-mails or SMS about interesting services recognized for the following day;
- A site dedicated to the management of reservations for the call center;
- An app for users of the Prontobus service to view reservations updated in real-time and with the ability to book an existing trip directly from the app;
- An app for drivers of the Prontobus service to get information about reservation in real-time and the trips to be made (Figure 19).

Starting from the date of the press conference, the advertising activity of the RUMOBIL project began with posters posted in the main stops covered by the Prontobus service.

End-users with the mobile apps can register directly for existing on-demand services that are interesting for them, without needing to make a phone call to the reservation center. Another interesting feature of the new software is that it is possible to automatically transmit the details of the services to be performed on a tablet to the bus drivers, updated in real-time.

In the period from October to July (2017-2018), an increase of passengers was detected on the Prontobus service in Castelfranco Emilia compared to the same period of 2016-2017. The total increase in passengers was 13.3%. In the first 3 months of the implementation, the increase of passengers was higher than 25% while in the last months the increase stabilized at about 7% per month, Figure 20. This is a very good result that allows us to state that the introduction of the new info mobility service supported by a publicity campaign had a positive effect on the number of passengers, achieving the project objective to increase the number of trips.

Mobility apps with their users can be an essential source of data. The new app designed according to aMo's specific requirements was the starting point for data insights. Each transaction made in a mobility app creates valuable data for the service provider. Analysis of this data enables the determination of patterns of mobility demand from Prontobus users. As such, analysis not only provides insights into current activity at particular stops or trips, but also allow for prediction of future passenger flows, driving service improvements.

Mazovia Region

Information about rail services (e.g. timetables, position of trains) has been substantially transformed in recent years. Real-time information systems enable passengers to access that information simply. This is very useful, especially if for any reason reliability cannot be guaranteed and the situation is uncertain. If there

Figure 21. Mobile app screenshots, info kiosk
Source: RUMOBIL, DELIVERABLE D.T1.6.1 – Workpaper summarizing the learning from RUMOBIL pilots, 2019

is a long delay in the next service shown, passenger can use another mode, or do something else and come back to the stop later. However, if a user knows that the real-time information is reliable and if the train is only a few minutes away, he will probably decide to wait.

Mazowieckie voivodeship implemented a new information system in the North-western part of Mazovia at two railway lines, Nasielsk – Sierpc, and Sierpc – Kutno. The region in which the pilot project is located is about 120-150 km away from Warsaw. This is a typical farming region with low population density. There is no attractive PT offer in the target area and rural inhabitants are dependent on cars. Their usage of PT services is low. The bus network is not tailored to the needs of passengers and timetables are not cyclical. Vehicles used to transport people are old and uncomfortable. Also, they do not serve all small towns and villages, or they handle them very rarely during the day. This makes it more difficult to plan a trip or a return trip, and it takes a lot of time. In the region, there are two railway lines which run several pairs of trains during the day. Modern trains run on single-track lines with low average commercial speed. There is no cyclic timetable. At certain times of the day, the gap in the train schedule is 4 hours. Railway infrastructure and stations are neglected: they need modernization. In many cases they are distant from settlements, destroyed and under-invested, need to be repaired and possibly relocated to a better location. There is no dynamic passenger information system, and many stops have no voice prompts. In case of a train failure/delay, the passenger has no information on the possible start time of the trip.

The aim of the pilot project was to increase the number of train passengers in the target area by implementation of a new real-time passenger information system. To reach this goal, several measures were implemented that allow passengers to plan their trips more effectively, including installation of GPS transmitters on trains, an app for mobile phones, and interactive kiosks at railway stations, Figure 21. Passenger can use their mobile devices (phone/tablet), as well as desktop computers and info kiosks, to check where the train is currently located, if it is running according to the timetable, or the possible waiting times.

The mobile application "tropKM" is available on devices running Windows, iOS, and Android for all passengers of Mazowieckie Railways without limit or charge. The application makes it possible to search the appropriate train, the train location on the interactive map indicating the time of the last data update and the speed of the vehicle. The application is designed to show the delay of the train (if such occurred) a description of the DMU train type and difficulties. It is possible to check the current timetable of all trains of Mazowieckie Railways.

In the first two months of the system operation, the application was installed 2 300 times and was in use 11 500 times. After 12 months of the system operation, the application was installed 7 830 times and was in use 115 600 times. On average, the system recorded over 10 000 inquiries related to rail traffic in the RUMOBIL project target area monthly.

Participation of Stakeholders

Citizens and stakeholders' involvement is a precondition for any mobility intervention since the long-term perspective of the measure requires a high degree of support and acceptance. The main project beneficiaries were groups of residents who were awaiting the arrival of a better transportation system and services in the region. Both individuals and groups of residents, in this case, are looking forward to the benefits they reap from quicker, or more efficient transportation. Besides residents, there can be other types of beneficiaries, e.g. tourists or specific groups of residents. The stakeholder identification process is one of the most important processes in each project management because projects are undertaken to fulfill the requirements of stakeholders. Stakeholders can be a person, group or organization that has an interest or concern in mobility intervention. Stakeholders can affect or be affected by a mobility measure. In the RUMOBIL project potential stakeholders were divided into three categories:

Figure 22. Stakeholder engagment map for pilot regions in Saxony-Anhalt
Source: RUMOBIL, DELIVERABLE D.T1.6.1 – Work paper summarizing the learning from RUMOBIL pilots, 2019

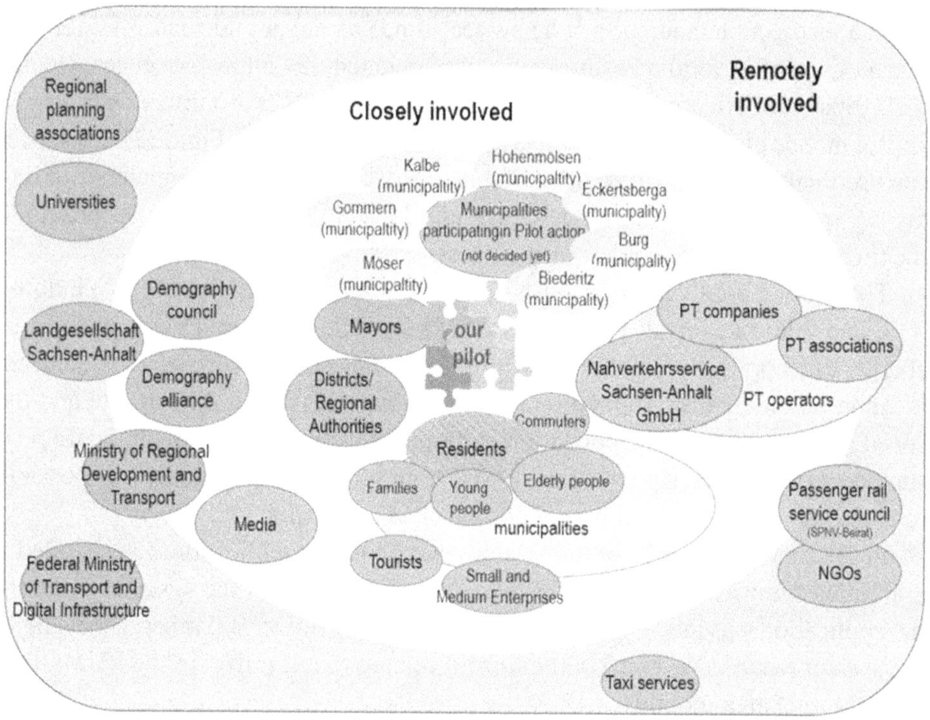

- **Primary Stakeholders (or Beneficiaries):** Those who are ultimately affected by new transport measures (e.g. different social groups — certain municipalities or village communities, elderly people, young people, employed or unemployed people, business branches, organizations, etc.).
- **Key Actors:** Those who have political responsibility (mayors, other authority levels); financial resources (public and private funds); the authority (by domain or territory); the skills and expertise (public administrations, universities, private sector) in transport and related domains (land use, environment, education, health, tourism, etc.).
- **Intermediaries:** Those who implement transport policy (PT and infrastructure operators, public administrations, police, etc.). Those who carry out major transport activities (PT operators). Those who represent pertinent interest groups (associations, chambers, cooperatives, networks, and nongovernmental organizations -NGOs). Those who inform and report on transport (authorities, operators, local media).

Involvement of stakeholders is fundamental for the start-up and the success of the measure. All partners prepared a map of stakeholders engaging (example from Germany is in Figure 22) before the project implementation and stakeholders' engagement was ensured throughout a whole pilot lifecycle. Partners took into consideration stakeholders' requirements, expectations, and perceptions. Stakeholders helped to organize a pilot and impact the outcomes that were achieved.

Saxony-Anhalt Region

The important aspect of the pilot was the involvement of local public authorities, local associations, and the local public. With their knowledge and experience of local peculiarities and possibilities, they contributed to the planning and the operation of the new bus services in the Möser and Osterburg region. Local stakeholders were involved at various levels: development of routes and bus schedule, and identification of places for additional bus stops.

They had also many proposals and ideas about the specific needs of residents. Further, the public was asked to support the project with voluntary drivers. The properly trained volunteer drivers include especially seniors. In this way, they support their local community and feel useful. This model is efficient in communities with less dense populations. Figure 22 shows the map of stakeholders who could be involved in project implementation.

Promotion Campaigns

To raise awareness of sustainable mobility among people in rural regions, various techniques may be used, such as designated educational programs, seminars, meetings, or different types of campaigns. The production of a successful sustainable mobility campaign is a complex task and requires an understanding of the general needs of those affected. The campaigns provided by partners in their regions offered an opportunity to present sustainable mobility alternatives to residents and to explain the challenges that their regions are facing. Another aim of the campaigns was to inform beneficiaries about concrete pilot measures and their advantages, as well as to test new measure and get feedback from the public. The campaigns promoted public transport as a high-quality alternative to individual car transport among all target groups and highlighted the main advantages of the new measure. The campaigns aimed to make the regional public transport more attractive for citizens and increase the share of public transport in the modal split. The campaign was done through different channels like local media, competitions, web sites, apps, social media, events, posters, printed materials, etc. Partners performed campaigns in different ways.

Figure 23. Number of trips for each user and number of trips of users per month
Source: RUMOBIL, DELIVERABLE D.T1.6.1 – Workpaper summarizing the learning from RUMOBIL pilots, 2019

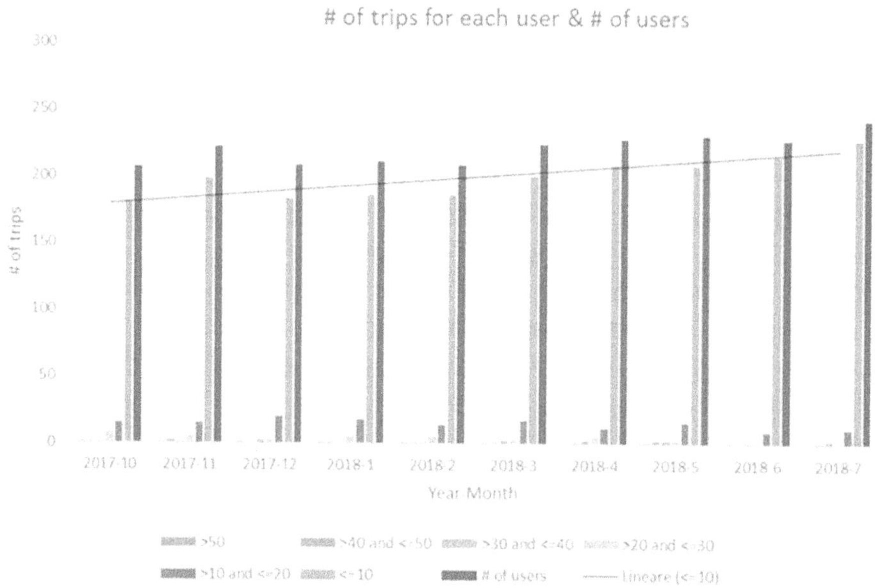

Appropriate Evaluation Methods and Tools

Assessing the impact of measures is an essential part of mobility or transport planning. Monitoring and evaluation activities deliver data about the progress of measures and the impact of measures. Monitoring activities should be carried out before, during and after implementation of intervention measures. They provide information to planners and decision-makers that allow timely identification of problems, potential successes and the need for readjustment of plans. Appropriate funding and institutional co-operation are necessary to successfully carry out monitoring and evaluation activities. A key part of the monitoring and evaluation process is the definition of indicators for which data needs to be collected before, during and after implementation. A set of indicators was defined to measure outputs against the goals of pilots and the RUMOBIL project. All partners gathered data from their pilot regions. Some measures implemented generated new knowledge on PT performance or users' travel behavior. A good example is represented by a pilot from Italy.

Figure 24. Summarization of users' survey
Source: RUMOBIL, DELIVERABLE D.T1.6.1 – Work paper summarizing the learning from RUMOBIL pilots, 2019

Castelfranco Emilia Region

aMo developed procedures to upload all data on DRT service performance recorded by the call center to a dedicated database. Using a reporting tool, it is now possible to have detailed information that was previously unknown. Before the introduction of the new RUMOBIL info tool, the only data available was the number of travelers per day, whereas now it is possible to know how the trips have been made in detail (travel time, origin-destination of the trip, users behavior, etc.). For example, an interesting analysis comparing users' trips helped to detect the trend that the number of users making few trips (less than 10 trips per month) is increasing and at the same time the number of persons making many trips each is decreasing, Figure 23.

This is a very interesting result discovering the fact that DRT Prontobus service is not designed for people doing everyday commuting (thus making many trips with the same origin-destination), but more for those making occasional trips. The RUMOBIL pilot brought a new tool for the PT operator allowing it to analyze service performance and users' travel behavior. This new knowledge can be used for research examining public transportation performance and how it may be enhanced in the future. New tools of info-mobility in the DRT services increased their attractiveness.

Figure 25. Number of passengers in partners' regions
Source: RUMOBIL, D.T2.6.2 Quantitative evaluation, 2019

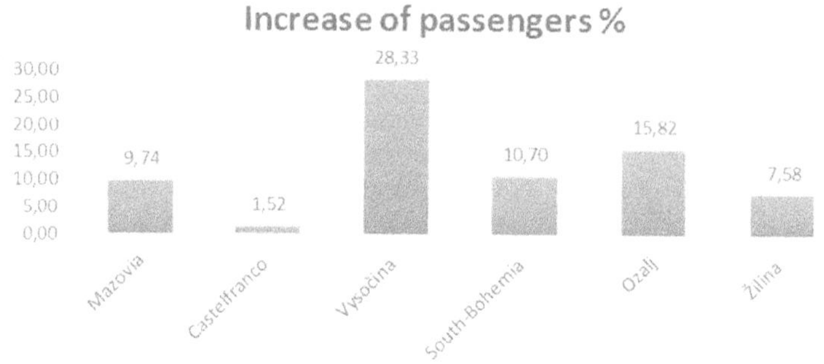

This is an important indication for future pilot replications and in fact, in Modena region, aMo began to replicate RUMOBIL on the other 5 existing DRT services.

Customer satisfaction has been considered one of the most important factors in PT service due to its direct relation to customer retention. User surveys were developed to explore the satisfaction of users with pilot measures. Qualitative indicators described the level of satisfaction with new or existing public transport service and allow passengers to evaluate the pilots. Information was gathered from the survey on the specific sample of respondents using pilot results in the area where the operation takes place. Users ranked the service on a 1-5 scale, ranging from 1 (not satisfied at all) to 5 (very satisfied). Partners included selected indicators relevant to their pilots such as infrastructure quality, accessibility, space design, comfort, access to information, availability, punctuality of service, security, safety, etc. The results of the surveys are provided in Figure 24. The users were satisfied with improved access to information, and they also appreciated accessibility and safety, while infrastructure quality or space design at bus stops are still problems in some partners' regions.

CONCLUSION

In this chapter, we discussed how regions throughout Europe were struggling to sustain mobility in sparsely populated rural areas. We can see that there is a big difference among various types of urban, suburban or rural areas that requires different approaches. Demographic changes such as the aging population and limited public finances endanger the accessibility of public transport services in many European rural areas and it is becoming increasingly difficult to ensure adequate mobility offers in these sparsely populated regions. New rural transport services require holistic, user-focused approaches. Local barriers, priorities and suitable options must be determined together with stakeholders. Increase in awareness of rural citizens requires more promotion, credit and market-related initiatives by transport operators as well as municipalities. Innovations represent the potential for exciting new transport and accessibility solutions. Geographical location and related transport accessibility and traffic services are without any doubt essential elements to create preconditions for economic development. It is well known that higher opportunities to travel mean higher work opportunities and, consequently, also more resources for leisure activities. This process is a basic principle of the economic development of an area and it is the core of the concept of accessibility.

Some outputs of the RUMOBIL project were introduced. The final RUMOBIL project deliverable is the RUMOBIL Strategy, a work paper that highlights the necessity to link rural areas to European and national transport networks and shows how this can be accomplished with the limited resources available. Furthermore, the RUMOBIL Strategy is the step thanks to which solutions that could be accepted by decision-makers for the improvement of sustainable mobility in their own rural and peripheral areas have been pointed out. The enhancement of the existing PT systems is based on the improvement of basic features (timetable, integrated tariffs, friendliness of vehicles and stations/stops, new and better interchange terminals) and in particular on deployment of new info mobility tools, that play a key role in all the layers (onboard, at the stops, on users' devices). What emerges from all pilots is that no result can be achieved without a proper and well-managed communication campaign that allows all potential users to be aware of their opportunities beyond private cars and collaboration with stakeholders. As expected, the most common objective among the pilots was to increase the use of PT in partners' rural areas, also considering the introduction of new services or more flexible ones. As shown in Figure 25, there was an increase in public transport service usage recorded in all partner regions (two partners were not able to provide accurate data).

Figure 25 represents the difference between the number of passengers before and after pilot implementation. The highest growth was in the Vysočina region, where the pilot project tested new bus lines for a new regional transport plan. Based on data from partners, we can consider all pilots as successful.

Considering the sustainability of pilots, we can see various scenarios for the tested measures in the future. All the pilots were implemented within the RUMOBIL project, funded by the Interreg Central Europe program as short-term and small scale projects. Keeping in mind the MaaS concept, information technology driven pilots (including mobile apps) were especially successful and they have been already spread to other regions in Poland and Italy. Partner testing of the concept of citizens' buses in Möser and Osterburg needs more time to evaluate the feasibility of such services in rural areas of the Saxony-Anhalt region. The Vysočina region plans to sustain pilot bus lines and include them into the new integrated transport system, which will be launched in December 2019. HZZP will dispatch thematic trains againin 2020. The Žilina region will try to advance the multimodal service in Rajecké Teplice with the help of IT solutions, and more comfort and information will be available at bus stops in Szabolcs-Szatmár-Bereg County.

Nevertheless, the RUMOBIL project allowed public authorities and entities responsible for the organisation of public passenger transport to assess which solutions could be suitable for their rural territories, and discuss and coordinate those with all relevant actors. RUMOBIL aimed to improve their planning capacities, to enable transport planners to better understand how to optimise the organisation of public transport in peripheral areas affected by demographic change, and to integrate new approaches and solutions in their strategic and operational transport planning. This approach led to the elaboration of a RUMOBIL Strategy consisting of policy recommendations regarding how public transport is to be designed to better link peripheral areas to national and European transport networks.

GENERAL FINDINGS AND RECOMMENDATIONS

Some general findings and recommendations are already mentioned in the previous section. Following facts can be emphasized for implementation of MaaS concept in rural areas. Smaller scale pilot projects usually don't cost much financially. In general, they help planners and companies to understand what works and what doesn't without the same level of risk that a bigger project would. However, they can generate success stories that can be used for vision casting. Measures proved to be effective can be implemented in other sites and reduced cost will allow mobility companies to focus their resources on enhancing and proving out their technologies.

Local conditions and circumstances should be carefully examined during implementation and supply and demand side of any region transport market should be analyzed. MaaS services were mostly presented in urban context. Some solutions for MaaS services in terms of collaboration, services & markets, planning & decision-making, and technology & information were proposed and implemented also in rural areas, especially as pilot projects funded by different EU or national programs. Before wider implementation of MaaS in rural regions we should better understand the needs of different types of rural communities and distinguish between different spatial settlement structures, demographic patterns and tendencies.

Most of public mobility services providers in rural areas still focus on collective transport based on modes such as trains or buses, usually being a fixed in many ways. These services often lack flexibility, especially regarding geographical issues.

By contrast the mix of traditional and innovative on-demand transport services can provide higher level of flexibility, since there are different conditions in rural areas compared to inner city mobility. E.g. trip distances are longer, population is dispersed and there are potentially fewer people that could share a ride. By new technological breakthroughs many of the current obstacles could be hindered and novel technologically advanced services could contribute to seamless mobility. E. g. implementing a new public transport built on autonomous shared vehicles might be the major step in the near future.

The fundamental question for policy makers is how to address different processes like urbanization or suburbanization and how to understand urban-rural relations and functions. Very often there's a need for a new intervention logic or policy instrument to help establish new partnerships or business models. Partnership should not be focused only on cooperation between administrations and transport providers. It must involve also the market actors like private firms or civil society organizations.

Some countries like Finland or Netherlands have positive experience with implementation of MaaS in rural areas, as proved in the study by Eckhardt (2018) or Geurs et al .(2018). They underlined that MaaS for rural areas are specific. In cities with many transportation options, the MaaS model provides maximum flexibility to travelers for deciding among travel modes, schedules, or price. Having centralized information give them opportunity to understand the choices and decide on the most appropriate. On the other hand, there are also benefits for service providers. They are able to better identify and fill gaps, and actively determine areas of unmet demand. By tracking searches for origins and destinations in the system and comparing them with service availability they can help especially small urban and rural communities, which lack the varied mobility options. This can be crucial also for vulnerable customers as elderly people.

The MaaS concept is relatively new and current MaaS are still operated usually at a local level. Much effort should be made into turning MaaS as a broader and integrated solution. An appropriate policy framework is necessary which allows implementation of MaaS and brings benefits of MaaS to rural travelers. Operation of such services still faces many challenges for appropriate business models. Users' perspective is important as well. Sometimes scientists or practitioners tend to be too optimistic about the possible rate at which MaaS might replace car transport or conventional public transport. Therefore, one of the most important parameters that need to be considered when implementing MaaS in regions is the perspective of users, as stated for example in the study by Lyons et al. (2019). If the users are convinced that the service is usable, they are going to choose and pay for it. This confirms also the study by Ho et al. (2018) testing various subscription plans for MaaS service.

Any technology that cannot be used from the user's perspective can be destined to fail. Although current targeted customers of MaaS are mainly the young generations who wish to adopt a sustainable lifestyle, MaaS may have great benefits to elderly travelers and play an important role to meet travel demand of the aging society in rural areas, Li & Voege (2017). Therefore a customization will be one of the most important MaaS parameters.

Considering experience of RUMOBIL project we suppose that an application of MaaS in rural areas of Central Europe is still limited especially by the following factors:

- Willingness of responsible transport authorities to implement new technologies;
- Legal barriers for service implementation (e.g. UBER);
- The complexity of transportation operational systems and diversity of technology platforms;
- Responsibilities of different actors to operate joint MaaS services;
- Implementation costs.

However we suppose that MaaS, when coupled with conventional public transport services, could give rural areas the catalyst they need to deliver substantive reform, providing a cost-effective and reliable alternative to car use. As one of the fundamental recommendations for Maas implementation is to start to build organizational structure of MaaS based on entities who hold control over the integration of mobility services and to set up legislative framework and rules.

FUTURE RESEARCH DIRECTIONS

Current research conducted in the area of concept of smart city and its mobility, or mobility in megacities has discovered the basic condition of managing huge passenger flow in various forms. Solving the problem has been discussed in various studies, e.g. Canitez, F. (2019), or Ghosh, Schot, (2019). In particular the focus on mobility opens a variety of approaches as described in the study by Ismagilova et al. (2019). This study provides a valuable synthesis of the relevant literature by analyzing and discussing the key findings from existing research on issues related to smart cities from an Information Systems perspective. The research focuses on many aspects of smart cities including smart mobility highlighting the limitations of current developments and potential future directions.

Based on these results, the authors agree that there is still a lack of research dealing with the implementation and examination of how new innovative technologies can enhance mobility in rural areas. Also, the perspective of demographical and economic changes in rural regions has been described in various studies, however, there is still a problem of how to assess their consequences. According to material "*Demographic trends in EU regions*" provided by European Parliamentary Research Service (2019), the EU response to the demographic challenge is currently limited and poorly developed and indicates that cohesion policy should play a more critical role in this respect. It suggests that the EU should seek to mainstream demographic considerations across all policy areas and include budget headings to enable further development of these policies. The opinion calls for measures to fight transport isolation and the digital backwardness of rural, peripheral and remote areas.

Economists often use the argument that the operation of public transport should be done efficiently. But cost constraints are not as important as the other factors or added values which are generated by providing a transport service in rural areas. Due to its capacity and ecological acceptability, public transportation still serves as a framework for mobility in rural regions. An interesting topic for further research is an examination of different types of rural areas varying from country to country as written in the study of Li, Westlund, Liu, (2019). Further research work in the field of mobility in rural areas should be examined using adequate data such as data on population (e.g. demography, age-groups, etc.), travel behavior, and demographic changes. It should be easier to gain such data in the future thanks to new and innovative technologies not only dedicated to Smart Cities, but also to Smart Rural Areas.

Some former RUMOBIL partners are continuing their work in a new project. The YOUMOBIL project launched in April 2019 aims to enhance the passenger transport system for young people living in rural areas and their access to the European and national transport networks. Poor mobility options other than the personal car is among the most frequent reasons why young people choose to leave their native rural area and to migrate to larger cities or even beyond, intensifying the demographic change troubling most rural areas in Central Europe. Some tools including shared mobility and demand-oriented flexible collective transport services have emerged across Europe but only a few have been introduced in the rural areas of Central Europe. This has led to a lack of knowledge and data among the relevant transport stakeholders and researchers as well. Moreover, public transport and related infrastructure suffer from a rather poor image among Central European youth. Therefore, YOUMOBIL partners together with rural areas' local youth test the potential of new demand-driven service features through interfaces for mobile devices. Moreover they explore how youth initiatives can revitalize disused rail infrastructure to enhance the attractiveness and image of public transport. Five technology-driven pilots will demonstrate the use of modern ICT (Information communication technologies) solutions for smart youth-oriented transport in rural areas while one investment and five feasibility studies will deal with upcycling and adaptive reuse of rail infrastructure.

Pilots will test solutions based on the MaaS concept, combining different services and providing a platform, where multimodal journey planning and booking are integrated. The RUMOBIL pilot test realized in Italy has confirmed the feasibility of such a service. A website (www.prontobus-rumobil.eu) and a mobile app (Prontobus-RUMOBIL) have been developed with the purpose of giving users real-time information about the DRT service in Castelfranco Emilia. The results were very satisfying especially regarding the increase in travelers that occurred, which showed the importance of infomobility. The state of Saxony-Anhalt supported the two Citizens' Bus initiatives in the municipalities of Möser and Osterburg by procuring the vehicles and relevant infrastructure, and in planning and implementing the schedule. Both citizens' buses were provided with tablets, ticket printers and distribution software enabling the bus drivers to sell tickets directly on-board. Furthermore technical equipment is being used to track the vehicle and inform potential passengers in real-time through INSA, the public transport information system of Saxony-Anhalt. The success of the creation of a small multimodal rural hub in the Žilina region will be accomplished by the new app integrating ticketing of main public transport services (rail, bus) and other services focusing on individual mobility (bicycle parking, bicycle sharing, etc.).

New MaaS solutions planned in the YOUMOBIL project will provide new options or mobility packages. Those will be predefined with the help of a survey among young people and feedback from them, taking into account several rural specific parameters (e.g. sociodemographic, economic, or environmental). The focus of the YOUMOBIL tool-box will be on innovative approaches and involving local youth (gamification apps, on-demand collective mobility for rural areas). The concrete results of these on-field experiences allow the enrichment of the tool-box with practical guidelines for its transfer and implementation. During the test phase, all available data on usage of the pilot service is aggregated (without violation of data protection legislation) for careful analysis. The quantitative data demonstrates the added value of the tested services to decision-makers in public transport and allows detailed estimation of the financial costs to continue the service beyond YOUMOBIL's pilot phase. This represents a further value-added that will strengthen the reliability of the partner areas' transport strategies. The data will be also used for further research activities and transport demand prognoses.

ACKNOWLEDGMENT

This chapter is published based on the results of the project RUMOBIL (CE55), co-funded by the European Regional Development Fund under Interreg CENTRAL EUROPE.

REFERENCES

Barreto, L., Amaral, A., & Baltazar, S. (2018). Mobility as a Service (MaaS) in rural regions: An overview. *Proceedings of International Conference on Intelligent Systems.* Doi:10.1109/IS.2018.8710455

Canitez, F. (2019). *Pathways to sustainable urban mobility in developing megacities: A socio-technical transition perspective.* Technological Forecasting and Social Change. Doi:10.1016/j.techfore.2019.01.008.2019

Eckhardt, J., Nykänen, L., Aapaoja, A., & Niemi, P. (2018). MaaS in rural areas - case Finland. *Research in Transportation Business & Management, 27*(June), 75–83. doi:10.1016/j.rtbm.2018.09.005

European Parliamentary Research Service. (2019). *Demographic trends in EU regions.* Available at: https://ec.europa.eu/futurium/en/system/files/ged/eprs-briefing-633160-demographic-trends-eu-regions-final.pdf

Geurs, K. T., Gkiotsalitis, K., Fioreze, T., Visser, G., & Veenstra, M. (2018). The potential of a Mobility-as-a-Service platform in a depopulating area in The Netherlands: An exploration of small and big data. *Advances in Transport Policy and Planning, 2*, 57-79.

Ghosh, B., & Schot, J. (2019). Towards a novel regime change framework: Studying mobility transitions in public transport regimes in an Indian megacity. *Energy Research & Social Science, 51*, 82–95.

Hennig, B. D. (2018). The growth and decline of urban agglomerations in Germany. *Environment, and Planning A: Economy and Space.* Doi:10.1177/2F0308518X18798835

Ho, Ch. Q., Hensher, D. A., Mulley, C., & Wong, Y. Z. (2018). Potential uptake and willingness-to-pay for Mobility as a Service (MaaS): A stated choice study. *Transportation Research Part A, Policy and Practice, 117*, 302–318.

Ismagilova, E., Hughes, L., Dwivedi, Y. K., & Raman, K. R. (2019). Smart cities: Advances in research—An information systems perspective. *International Journal of Information Management*, *47*, 88–100. doi:10.1016/j.ijinfomgt.2019.01.004

Leibert, T., & Golinski, S. (2016). Peripheralisation: The Missing Link in Dealing with Demographic Change. *Comparative Population Studies*, *41*(3-4). doi:10.12765/CPoS-2017-02en

Li, Y., & Voege, T. (2017). Mobility as a Service (MaaS): Challenges of Implementation and Policy Required, January 2017. *Journal of Transportation Technologies*, *07*(02), 95–106. doi:10.4236/jtts.2017.72007

Li, Y., Westlund, H., & Liu, Y. (2019). Why some rural areas decline while some others not: An overview of rural evolution in the world. *Journal of Rural Studies, 68*, 135-143. doi:10.1016/j.jrurstud.2019.03.003

Lyons, G., Hammond, P., & Mackay, K. (2019). The importance of user perspective in the evolution of MaaS. *Transportation Research Part A, Policy and Practice*, *121*(March), 22–36. doi:10.1016/j.tra.2018.12.010

McDonagh, J. (2006, September). Transport policy instruments and transport-related social exclusion in the rural Republic of Ireland. *Journal of Transport Geography*, *14*(5), 355–366. doi:10.1016/j.jtrangeo.2005.06.005

Milbourne, P., & Kitchen, L. (2014). Rural mobilities: Connecting movement and fixity in rural places. *Journal of Rural Studies*, *34*, 326–336. doi:10.1016/j.jrurstud.2014.01.004

O'Donnell, P., O'Donovan, D., & Elmusharaf, K. (2018). Measuring social exclusion in healthcare settings: A scoping review. *International Journal for Equity in Health*, *17*(1), 15. doi:10.118612939-018-0732-1 PMID:29391016

Ranković Plazinić, B., & Jović, J. (2013) *Transportation demand management in rural areas*, 4th International Conference "Towards a Humane City", Novi Sad, Serbia.

Rogers, D., Dufty-Jones, R., & Steele, W. (2015). Housing and Socio-Spatial Inclusion. *Social Inclusion*, *3*(2), 1–51. doi:10.17645i.v3i2.264

RUMOBIL. (2018). *D.T1.3.5. Adopted RUMOBIL strategy*. Available at: https://www.interreg-central.eu/Content.Node/RUMOBIL-WPT1-T1.3.5-Adopted-RUMOBIL-Strategy-v06-2018-03-12.pdf

RUMOBIL. (2019a). *D.T2.6.2. Quantitative evaluation*. Author.

RUMOBIL. (2019b). *DELIVERABLE D.T1.6.1 – Work paper summarizing the learning from RUMOBIL pilots*. Author.

Smith, N., Hirsch, D., & Davis, A. (2012). Accessibility and capability: The minimum transport needs and costs of rural households. *Journal of Transport Geography, 21*, 93–101. doi:10.1016/j.jtrangeo.2012.01.004

Stastna, M., Vaishar, A., & Stonawská, K. (2015). Integrated Transport System of the South-Moravian Region and its impact on rural development. *Transportation Research Part D, Transport and Environment, 36*(May), 53–64. doi:10.1016/j.trd.2015.02.012

UITP (Union Internationale des Transports Publics). (2016). *Statistics Brief, 'Local public transport in the European Union'*. Retrieved from https://www.uitp.org/sites/default/files/cck-focus-papers-files/UITP_Statistics_PT_in_EU_DEF_0.pdf

Wang, Ch., Quddus, M., Enoch, M., Ryley, T., & Davison, L. (2015). Exploring the propensity to travel by demand responsive transport in the rural area of Lincolnshire in England. *Case Studies on Transport Policy, 3*(2), 129-136.

Wollner, M. (2015). *Today's concepts for a sustainable and mobile tomorrow in rural areas*. Available at: http://www.casi2020.eu/blog/posts/today-s-concepts-for-a-sustainable-and-mobile-tomorrow-in-rural-areas/

Wolny, A. (2019). Are suburban commuters confined to private transport? A case study of a medium-sized functional urban area (FUA) in Poland. *Cities (London, England), 92*(September), 82–96. doi:10.1016/j.cities.2019.03.013

ENDNOTES

1. https://ec.europa.eu/eurostat/web/products-eurostat-news/-/EDN-20180917-1
2. https://www.slov-lex.sk/pravne-predpisy/SK/ZZ/2012/56/ Act also determines the basic conditions and rules for public bus transport organization in Slovakia.
3. https://mlv.sachsen-anhalt.de/themen/nahverkehr/plan-fuer-den-oeffentlichen-personennahverkehr-des-landes-sachsen-anhalt-2020-2030/
4. https://www.delijn.be/de/belbus/
5. https://www.landbusunterland.at/anrufbus/
6. https://www.nidirect.gov.uk/articles/dial-a-lift

Section 3
Emerging Trends and Innovations in Mobility Systems

The final section explores future trends and innovations exploring the most likely paths where MaaS can take modern societies, providing new perspectives and presenting the latest mobility improvements regarding this global challenge based on the adoption of competitive solutions for transport efficiency and sustainability.

Chapter 8
Sustainable Mobility in Smart Cities:
The Key Role of Gamified Motivational Systems for Citizens' Engagement and Behavior Change

Annapaola Marconi
https://orcid.org/0000-0001-8699-7777
Fondazione Bruno Kessler, Italy

Enrica Loria
Fondazione Bruno Kessler, Italy

ABSTRACT

MaaS, with its user-centric vision aiming at more flexible, personalized, and on-demand mobility services, has a perspective to address the key sustainability challenges of the mobility of the future: transport solutions that are integrated, inclusive, and time-efficient, while containing the economic, social, and environmental costs. Civic engagement is a key success factor towards this objective: to fully embrace MaaS transformation, citizens need to feel part of the process. Gamification proved to be effective to raise citizens' awareness, encourage their participation, and promote a gradual but profound behavior change. These characteristics can be exploited by MaaS solutions to attract users and encourage the acceptance of tailored mobility plans. The chapter presents two successful cases of gamified systems aiming at promoting a more sustainable mobility, Play&Go and Kids Go Green; investigates the potential of gamified systems, in combination with MaaS solutions, in terms of citizens' engagement and behavior change; and discusses current limitations and future challenges.

DOI: 10.4018/978-1-7998-1614-0.ch008

INTRODUCTION

Within a Smart City, mobility plays a fundamental role: the way in which citizens experience the city, access its core services and participate in the city life strongly depends on its mobility organization and efficiency (Vesco, 2015). In this context, the challenge that cities are facing is very ambitious: on the one hand, administrators must guarantee to their citizens the right to mobility and to easily access local services, on the other hand they need to minimize the economic, social and environmental cost of the mobility system.

Dealing with this challenge requires a holistic approach that allows to efficiently harness existing mobility resources while integrating and promoting emerging mobility services to enable an integrated, efficient and sustainable mobility ecosystem. To this end, cities are planning and implementing interventions at the level of infrastructures, services and mobility policies.

The inadequacy of traditional transportation models is proven by the proliferation of alternative, social and grassroots initiatives aiming at a more flexible, customized and collective way of organizing transport (e.g., car pooling, ride and park sharing services, flexi-buses). Some of these attempts have become popular (e.g., Uber, BlaBlaCar, FlixBus), even though, in most cases, these are isolated solutions targeting specific mobility target groups and are not part of the city mobility eco-system, which is still mainly based on traditional public and private transport facilities.

An attempt of re-thinking the way mobility is managed and offered is represented by the Mobility as a Service (MaaS) paradigm. MaaS solutions (e.g., MaaS Global: http://maas.global) aim at arranging the most suitable transport solution for their customers thanks to cost effective integrated offer of different multi-modal means of transportation. MaaS also foresees radical changes in the business landscape, with a new generation of mobility operators emerging as key actors to manage the increased flexibility and dynamism offered by this new concept of mobility. The MaaS paradigm aims at changing mobility habits by empowering the users in finding customized answers to their own specific mobility needs (Jittrapirom et al., 2017). Today, the user has to plan the route from source to destination (i.e. use Google maps), select the mobility modes (e.g. bus, train, car etc.) for the various trip legs and acquire the resources needed (reservations/tickets etc.). With MaaS, mobility will become a service: the user will get from a single mobility operator the best "all-inclusive" solution according to his preferences and to the current availability and traffic/environmental conditions; the user will buy a single ticket for the whole trip, with the mobility operator taking care of re-routing the user in real time to different modes in case of delays, changed conditions or new better traveling options.

MaaS essential characteristic is its user-centric vision, with mobility services that are flexible, personalized and on-demand. By enabling seamless multi-modal travels with accessible and affordable costs, MaaS can play a key role in shifting the interest of end-users from private car to alternative transport modes (Chowdhury and Ceder, 2016; CIVITAS, 2016; König et al., 2016). MaaS has thus the potential to contribute towards the strategic goals of Smart Cities: a more efficient, sustainable and accessible mobility (Jittrapirom et al., 2017).

Availability of real-time mobility information and interoperable inter-modal solutions (e.g., for travel planning, booking and ticketing) are certainly key ingredients offered by the MaaS paradigm; but another very important aspect of MaaS, as a socio-technical phenomenon, is users' acceptance and adoption (Giesecke et al., 2016; König et al., 2016). Innovative infrastructures, applications and services are liable to fail if they are not combined with actions aimed at making citizens aware and involved in this process and to influence their mobility habits in a gradual but profound way (Vesco, 2015).

In most cases, citizens' daily mobility choices are driven by habits and are based on wrong or outdated beliefs (Hamari et al., 2014). Citizens need to be aware of the mobility services offered by their city and of their actual value (in terms of time, cost, and environmental impact). They need to be conscious of the impact of their individual daily choices (in terms of traffic, greenhouse gas emissions and social cost). Most importantly, they need to feel part of a community that, through daily individual choices, can play a key role towards the fulfillment of city-level mobility strategic objectives. In other words, individuals and communities have to learn to take responsible actions and this can only be achieved through the development of a new culture for urban and rural mobility.

In recent years, a significant effort has been undertaken to understand how interactive technologies can be leveraged to raise citizens' awareness, encourage their active participation, break bad habits and promote behavior change towards a more sustainable lifestyle. To this extent, gamification is emerging as a persuasive technology with significant potential (Koivisto et al., 2019; Di Dio et al., 2018) and opportunities for application in the mobility domain (Hamari et al., 2014; Gabrielli et al., 2013; Holleis et al., 2012; Kazhamiakin et al., 2015; Marconi et al., 2018; Briones et al., 2018), as well as in several other domains in the environmental sustainability realm (Cowley et al., 2011; Orland et al., 2014; Shiraishi et al., 2009; Lee et al., 2013). The key idea is to leverage on the motivational and persuasive power of games by designing motivational systems that properly exploit and embed game concepts and elements (Weiser et al., 2015; Nakamura et al., 2014) to pursue a change in the player's behavior. Gamified systems motivate people to make certain decisions or carry out certain tasks that are important in order to reach valuable objectives, even when they are somehow unattractive, by turning them into fun and rewarding experiences.

The aim of this chapter is to investigate in detail the application of gamified systems in the mobility domain and to explore the potential impact of their integration within MaaS based solutions, in terms of end-user's engagement, motivation and retention, towards the goal of a more efficient and sustainable mobility of the future.

The chapter starts with an overview of existing solutions, discussing their application context and impact. Then, in the following Sections, two particularly significant experiences that we have developed and experimented on the field for several years, are analyzed in detail: Play&Go and Kids Go Green. The former is a gamified system targeting citizens and promoting sustainable home-work mobility, the latter is a playful education system that promotes active and sustainable mobility in primary schools. In both cases, these initiatives have proven to be effective in terms of retaining participants' engagement on a long term (several months), raising their awareness, and inducing a change their mobility habits.

Play&Go is a long-running gamified urban mobility campaign promoting voluntary travel behavior change, which has been active in Trentino (Italy) since 2015 (through four yearly editions). In Play&Go players exploit a mobile app to track their sustainable journeys, which are subjected to an automatic validation procedure and are rewarded with virtual points and allows the player to proceed in the competition. Play&Go combines standard game elements (e.g., points, badges, leaderboards, real prizes) with personalized game content (i.e., weekly challenges) that is tailored to the player's profile and is specifically focused on encouraging and rewarding a positive change in the player behavior. Field trials have achieved significant results. In 2019 edition, in six months 590 players actively participated to the game and tracked 64.251 trips, of which 45.238 with zero impact (by foot or by bike), and more than 250,000 kilometers were traveled sustainably. Furthermore, 66% of players declared to have adopted more sustainable mobility habits thanks to the game and 45% experienced, thanks to the game, a new sustainable mode of transport. These experiments have shown both the ability of the approach to support citizen participation in long-term games, and the ability to break habits and change the behavior of players towards more sustainable mobility habits.

Kids Go Green involves the entire school community (children, teachers, and families) in a gameful educational journey to promote sustainable and active home-school mobility for primary and middle school age children. In Kids Go Green, the sustainable kilometers made by each child on their trips from home to school (e.g., on foot, by walking bus, by bike, with the school bus) contribute to the progress of the whole group of children on a virtual educational journey to places in the real world that is personalized according to the group interests and capabilities. Kids Go Green has been adopted for three consecutive school years in Trento and Ferrara

(Italy), involving a total of 16 schools, 1844 children, and more than 100 teachers. After three years of research and development, Kids Go Green has demonstrated strong impact in terms of sustainable mobility, with increased awareness and understanding of the topic by the children and an impressive impact on the mobility behavior lasting beyond the conclusion of the trials. This solution also proved to be effective in terms of educational value, offering innovative, interdisciplinary learning opportunities integrated into the existing academic program and the creation of a sense of community as the students are motivated by a common goal and move together virtually through the space of the interactive, educational game.

Despite the success of these experiences, there are still some key aspects that need to be investigated to fully exploit the potential of gamified motivational systems: reach also those citizens that are not particularly sensitive to sustainability issues; reward the actual behavior improvement, rather than the player's mobility performances; provide an accurate measure of the impact of the mobility campaigns. The chapter ends with a discussion on lessons learned, best practices and future challenges, that might guide researchers and practitioners in this field.

BACKGROUND

The idea of using game design elements in non-game contexts (Deterding et al., 2011) to motivate and increase user activity and retention (Ryan et al. 2006; Johnson et al., 2016) has rapidly gained traction in interaction design and digital marketing. This approach, also known as gamification or gameful design, fosters motivation by employing elements and characteristic proper of entertainment games in other systems (Ryan et al. 2006; Johnson et al., 2016). The individual's personal inner motivation (Glans, 2010; Michie et al., 2010) and intention (Fogg, 1998) guide the process of behavior change, leading to a positive influence (Hamari et al., 2014; Johnson et al., 2016). One of the key application domains are technologies that aim at persuading to assume a positive behavioral change (PBC). Designers can exploit the symbolic nature of games (Salen et al., 2004; De la Hera Conde-Pumpido, 2013) to guide the players during their experience so as to foster a behavioral change (Sicart, 2010). Moreover, the global and massive usage of smartphones, can be exploited as a tool to give immediate feedback on users behaviors by sensing and predicting how the way they act impacts the surrounding environment. Fogg (1998) defined the concept of persuasion as "an attempt to shape, reinforce or change behaviors, feelings or thoughts about an issue, object or an action". Hence, making users understand and internalize complex world events, even though being a challenging task, can be eased through simulation in games. In fact, according to the semiotic principle, meaning that users perception of signs in the system is influenced by the context in which they encounter them (Salen et al., 2004). Proceduralists share this line of thought as they assess that, due to their procedural nature, games can address serious topics (Sicart, 2010). The designers can embed meaning in the games' elements to guide players during their experiences and, more importantly, to persuade them to assume a certain behavior by making them actually experience the intended message (Sicart, 2011). In other words, games can be used as a Persuasive Technology (PT) since they can function as cognitive frames to guide the interpretation of the signs embedded in them (De la Hera Conde-Pumpido, 2013). However, despite Wright and Bogost's (2007) belief that forcing constraints contribute to the creative process, Sicart (2011) argues that proceduralism hinder players from the ability of expressing themselves. This limitation is dangerous, since the behavior change is guided by the individual's personal inner motivation (Glanz et al., 2010; Michie et al., 2011) and intention (Fogg, 1998). Another great supporter in building intrinsic motivation in performing the task in question (e.g. eating healthily or adopt sustainable mobility habits) is relatedness (Zhang, 2008). In fact, one of people's basic needs is the sense of belonging and of being in connection with other human beings (Deci et al., 2000). Gameful systems, by building collective intentions (we-intentions), make true cooperation happen (Searle, 1990; Tuomela, 2011), which leads to a higher commitment in the activity that elicits this feeling of collectivity (Morschheuser et al., 2017).

As a consequence, using game-like systems to assist the development of a Smart City, with the common benefits that derive from it, appears to be a natural evolution. Among the many aspects that can be treated, sustainable mobility takes a big slice of the interest. In fact, there is a plethora of applications which, with a diverse extent of complexity, employ game elements. In London, the gameful system of Chromaroma (The Guardian, 2010) incentive users to visit new areas to earn virtual points and win challenges. In the city of Bogota, Colombia, gamification has been used to change people's behavior (Castellanos, 2016), by pushing them to use ecological transportation means, like walking and biking. To this purpose, both monetary and non-monetary incentives have been delivered. Game-like mechanics have also been used to discourage bad driving behaviors, for instance by Volkswagen (Volkswagen, 2011), which awarded prizes to conscientious drivers by rewarding them with money collected from hazardous drivers. Gamification can also be employed to promote a particular transportation service and hence to have an impact in terms of mobility, such as bike sharing systems - which are usually a cheap and ecological choice. An example of an attempt to foster the usage of bike sharing, despite the cold season approaching, was in Washington DC, USA, where the city started a contest with both real and virtual prizes to avoid people abandoning the service. The campaign was so successful that the rides increased by 67% in two months (Capital.com, 2011).

However, setting up an incentive policy to push certain behaviors is challenging. This is particularly true when behavior-based rewards are introduced since they may function only in the short-term. All the systems above-mentioned rely on the willingness to move ecologically. If the users are reluctant to abandon the car some other strategies should be employed. An alternative is eco-driving, which is a way of driving with the purpose of reducing CO2 emissions, which could be extremely beneficial for the environment (Barkenbus, 2010); in the literature there are already some examples of the usage of game elements in this context (Froehlich, 2015). Metropia (Metropia, 2017) is a mobile platform that assists users in finding alternative to their travels, to give an optimized solution. This is achieved by predicting the traffic exploiting real-time data. Drivers are rewarded if they shift their journey in less busy timeframes, and the reward strategy is informed by behavioral economics strategies and gamification elements. The positive aspects also involve environmental benefits in terms of CO2 emissions. On the side of academic experiments, there is "Spitsmijden" or "traffic avoidance" (Knockaert et al., 2012), which took place in The Netherlands in 2006. The aim was to encourage users to avoid traveling in the peak hours through virtual points, which could be used to obtain a smartphone. Despite having impressive initial results, the main drawback was that once the incentives were revoked, the mobility habits returned as those pre-intervention. This is mostly due to the nature of the gameful strategy employed, heavily relying on physical prizes. However, this does not stimulate players to choose other means of transport instead of personal transportation. Carpooling is a system through which users with similar routes can use one car. Its main goal is to match people who commute to work. Olszewski et al. (2018) propose a variation of urban game in the form of a serious game to help to significantly reduce the number of cars in Warsaw. Another example of gamification implemented by the transport administration is described by Shreenath et al. (2015). They conducted a feasibility study for the implementation of gamification within Trafikverket (the Swedish transport administration), by augmenting with game elements the existing processes. They came up with guidelines on how and when gamification should be applied. UbiFit (Consolvo, 2007) is a mobile app that gives users feedback on the level of "greenness" of their transportation behaviors by sensing their actions through both the smartphones and wearable devices, and self-declarations (Choudhury et al., 2008). The feedback is in the form of an abstract visualization. This visualization exploited the psychological theory that employing animals is helpful in fostering a connection between humans and nature (Myers et al., 2002). Moreover, adopting a green behavior resulted in positive rewards - earning points. UbiFit's focus was on making users aware of their habits and of the impact that their actions have on the environment. Another example of a mobile system employing game-like mechanics to promote an ecologic mobility behavior is Bella Mossa, a game active in the city of Bologna (Italy). The game awards physical and virtual prizes to citizens that track their journeys through the app, by giving a higher importance to the more sustainable means. However, “we are at the Atari 2600 stage of persuasive technology and gamification” (Froehlich, 2015) and still plenty of studies have to be held so as to understand how to exploit its true potential.

PLAY&GO: AN URBAN MOBILITY GAME

Play&Go (Figure 1a) is a mobile app fostering a positive behavioral change, built upon the framework of Kazhamiakin et al. (2016). Players are involved in a gameful environment in which they can track their daily journeys (Figure 1b) and get points for undertaking a green behavior. This initiative was born in 2016 from a partnership of Fondazione Bruno Kessler and the Municipality of Trento (Italy), whose purpose was and is to move towards a green and smart city. Every year, the Municipality set a 6-month game, structured in weeks, in which also physical prizes are put out for grabs. Examples of prizes are public transit passes, museum or event tickets, or fitness classes. Since the first release of the game, which is currently in its 4th edition, the dialog among the three main stakeholders - designers, municipality and citizens - has contributed to significantly improve the system.

The core mechanic of the game is that players can track their movement through the app by specifying whether they are walking, riding a bike or using public transportation (bus or train). At the end of the travel, they receive an amount of Green Leaves points proportional to the number of kilometers travelled and the level of sustainability of the means employed (e.g., walking confers more points than using the bus). However, before having awarded any point, the trip is subjected to a semi-automatic validation process, which assess the veracity of the declaration. For example, when a player registers a bus trip, the system looks for matches between the bus timetable, and the time and geo-localization of the tracked trip. Players can also review the actions they performed in their *Diary* (Figure 1c).

The gameplay has a weekly structure so as to have weekly leaderboards, which can be used to assign the weekly prizes offered by local sponsors. Moreover, weekly leaderboards (Figure 1d), that consider the performance achieved in a small window of time, allow newcomers to have a fair opportunity to compete with veterans in the game. The seniority of players is represented by a level associated to their profile. The level (e.g., Green Soldier, Green Warrior, Green Guru) is calculated in function of the total amount of points gathered and is a measure of the players’ experience.

In the game particular attention is paid to keep users in a state of flow. To this extent, players are motivated to improve (or maintain) their performance through weekly challenges, with care to not frustrate them with unreachable targets. Challenges

Figure 1a. Play&Go screenshots: Homepage

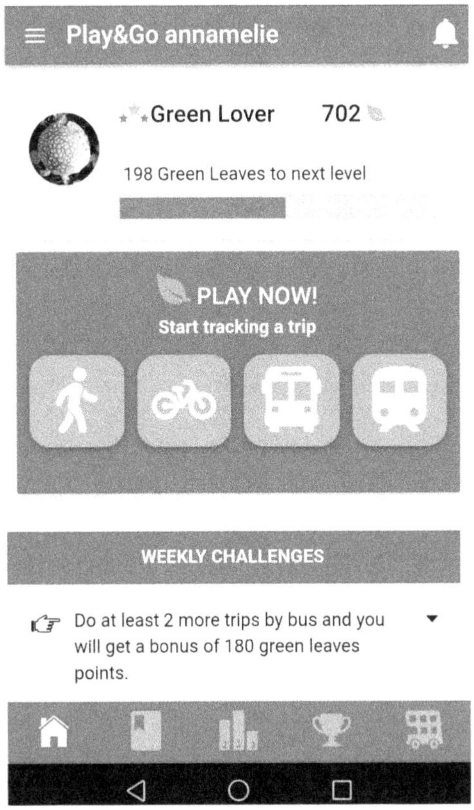

are generated by a Recommender System (RS) so to be tailored to each user, considering player history, preferences and skills. The rewards for such challenges are computed in function of the estimated difficulty for the challenge considered the player in terms of the effort she is forecasted to make to win it. Example of an individual challenge is the following: "Do at least 3 kilometers by bike to win 120 Green Leaves points".

In Play&Go multiplayer challenges are also available (Figure 2), so as to allow players to build a sense of relatedness and community by engaging in a sustainable behavior. The multiplayer challenges always involve two players and can be one of the following kinds: cooperative, time-based competitive and performance-based competitive. Both the target and the reward are automatically calculated by the system, according to the history and preferences of the two players involved in the challenge. The target is always the same for the two players, while the reward may differ due to the estimated difficulty for each player. In the cooperative mode, the target is cumulative, and players have to work together to reach it, before the

Figure 1b. Play&Go screenshots: Tracking a trip

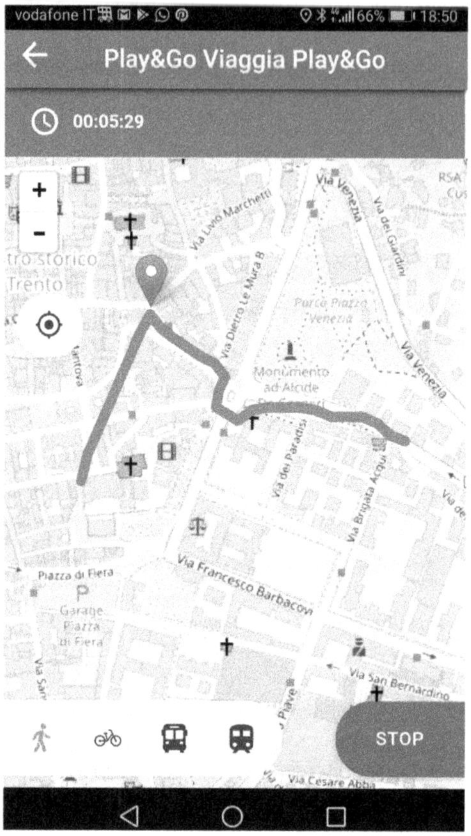

challenge expires. The individual contribution is not important and, in case of win, they both receive the same prize. An example of cooperative challenge is: "Join forces: bike at least 10 km between you and Player 2. If you win, you'll both get 120 green leaves". The other two multiplayer challenge modes are both competitive, but the mechanic is slightly different. In the time-based competitive mode, there is a fixed target and the winner is the player that is fastest in reaching it. The target is calculated to be achievable for both players, without being too easy to reach. It could happen that players have different experience level, and hence for one of them the effort to employ is higher. Therefore, the reward for the two players may be different, since it is proportional to the level of difficulty calculated for each party: - i.e., the strongest players would get a lower reward than the weakest. An example of such a challenge is: "Bike 4 km before your opponent Player 2. If you win, you'll get 100 green leaves. If Player 2 wins, she-he will get 150 green leaves". The last competitive mode is the performance-based competitive challenges. This time there is no target,

Figure 1c. Play&Go screenshots: Diary of actions

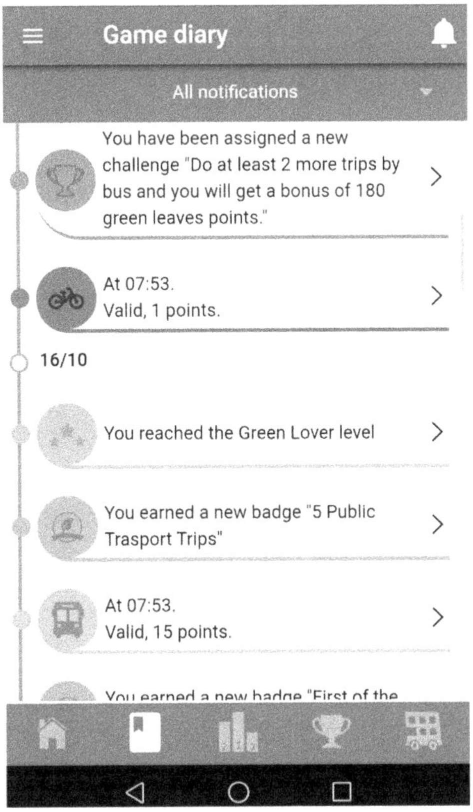

but the challenge will be active for the whole week. Upon expiration, the player that has the highest performance wins. To avoid players setting for a low performance, the reward is a booster calculated on the weekly points obtained - e.g., "Walk more km than your opponent Player 2. The winner will get a 40% bonus of all the points got during the week".

Since the game not only implements the "blue-print triad" – e.g., points, badges and leaderboards – but also more articulated game elements, some of the features are made available to players later on in the gameplay. In other words, unlockable content is associated to the reaching of every level.

Figure 1d. Play&Go screenshots: Leaderboard

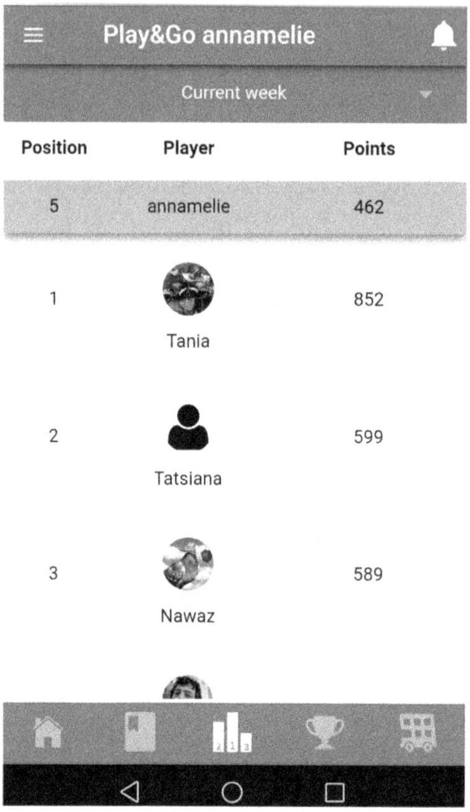

Level 1: Players receive their first customized individual challenge.
Level 2: Players can choose one among two customized individual challenges.
Level 3: Players can choose one among three customized individual challenges.
Level 4: Players can unlock one among the three modes of multiplayer challenges available (Figure 2a) - cooperative, time-based competitive and performance-based competitive challenges. From this point onward, they can invite other players to a multiplayer challenge (Figure 2b).
Level 5 and 6: Players can unlock another multiplayer challenge mode.

In conclusion, the more players advance in the game, the vaster are their possible choices. It must be noticed that each week, players can only have one active challenge, no matter whether it is single player or multiplayer.

At the beginning of every week, players have a small window of time (a couple of days) when they can program the challenge for the following week. Programming a challenge means selecting whether they want to engage in an individual challenge

Figure 2d. Play&Go - multiplayer game elements: Accept an invite

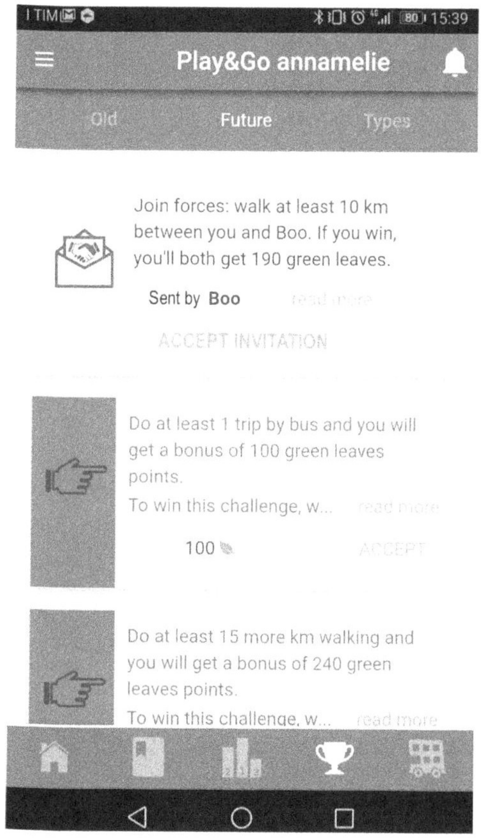

240k kilometers to 250k kilometers (Figure 3). Of those trips, the vast majority were tracked by foot, followed by train, bus and bike. Therefore, the transportation means we call "zero impact" (i.e., by foot or by bike) were the favored ones. This result should be interpreted considering also the background of players, in terms of preferred transportation means used. The majority of our users (35%) declared to mostly use public transportation means, while the remaining selected as primary means car, walk and bike (25%, 21% and 18% respectively). Therefore, the 75% of our users already moved in a green way, with 39% of them assuming a "zero impact" behavior. Despite this, the achievements are still impressive, considering that before the game 48% of our players declared to use the car about 3 times per day. Moreover, 66% of our users declared to have changed their mobility behavior, thanks to the app, to a medium-to-high extent.

Figure 2a. Play&Go - multiplayer game elements: Unlockable content

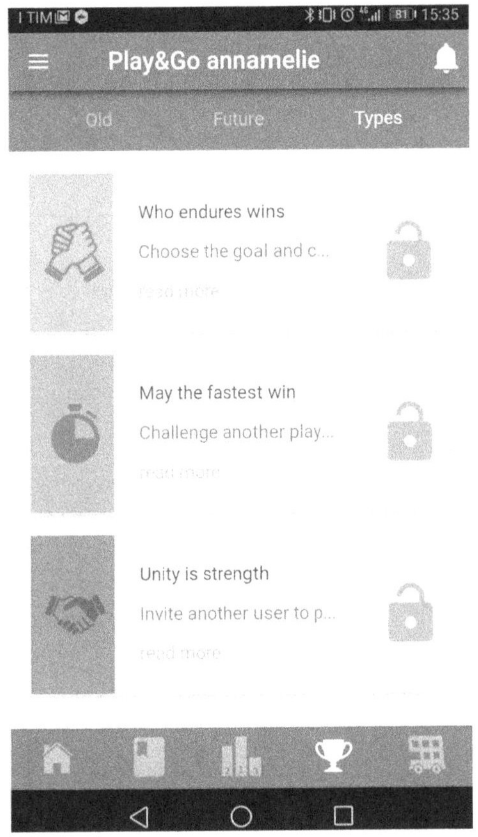

(and choose one, if they are at least at level 2) or they want to engage in a multiplayer challenge. To engage in a multiplayer challenge, players have to reach at least level 4. At that point, they can unlock one multiplayer mode. To initiate a multiplayer challenge, players can select one of the modes that they have unlocked and, then, invite another user. Prior to the invitation, the initiator of the challenge must specify the type of challenge they want to activate (Figure 2c), and hence choose among bike, walk or a general challenge on the amount of Green Leaves points obtained. The RS will calculate the best target depending on the two players involved and the mode selected and, then, will generate the appropriate reward(s). During this step the more competitive players could actuate a strategy, meaning that they could play with the invitation parameters – i.e., challenge mode and opponent – so as to have the best chances to win or the highest reward. Players can make only one request at a time. Once an invite has been sent, the player who sent the invite can still cancel the request to make another one, while the invited player can either accept it or reject

Figure 2b. Play&Go - multiplayer game elements: Multiplayer Challenges

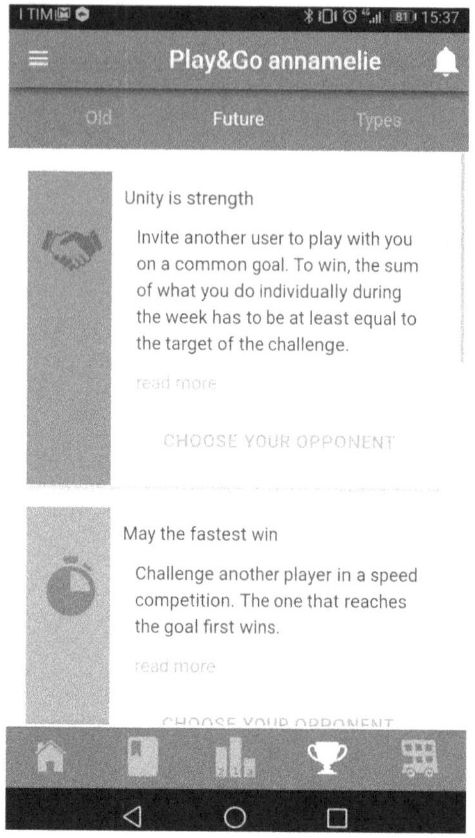

it (Figure 2d). When the invite has been accepted, the players involved cannot make other invitations and/or accept other challenges until the following week.

Every gamification campaign has been followed by the analysis of the data retrieved throughout the six-month period. We exploited two kinds of data sources which helped in conducting both quantitative and qualitative studies. The first and most prolific data source is gameplay logs, in which we stored every in-game action of the players, examples being the trips tracked, information on the challenges received and the points gained. The second kind of data source are surveys. We submitted players an initial survey, to collect information on their preference and personality; a survey inspecting players response to the multiplayer challenges; and a final survey inquiring players about their game experience. The main difference between the two data sources lies in the amount of data obtained, in that, on the one hand, we have logs for all players, while, on the other hand, not every user answered to every (or any) questionnaire. In particular, we had 480 respondents for the initial

Figure 2c. Play&Go - multiplayer game elements: Send an invite

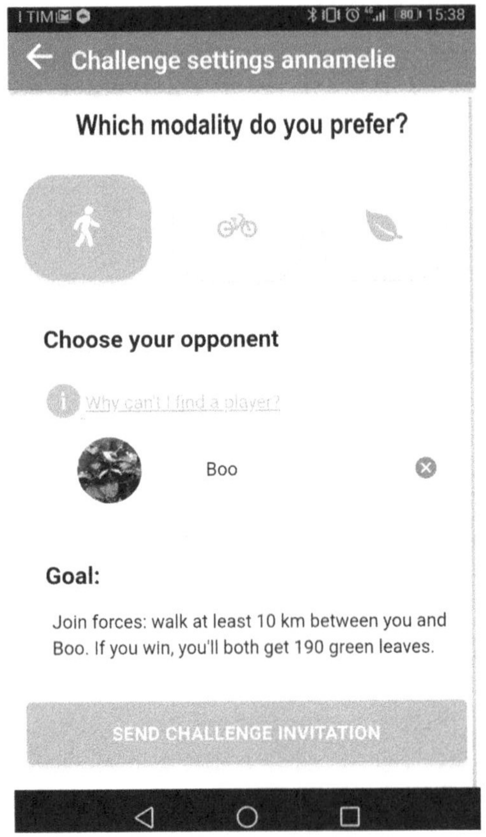

survey, 81 respondents for the questionnaire about the multiplayer cl
87 respondents for the final survey. The insight gained during such
and will lead the design of further versions of the game.

In our most recent edition of the game, which run from the end o
to April 2019, we observed a more committed audience of players.
to the previous edition, where the amount of time spent in the ga
modest despite having a higher rate of registered players, we found
users were fewer but more constant in the usage. In fact, while on 20
we counted 1079 registered players (of which 747 were active), t
834 registered players (590 active). By active players we mean pla
at least one trip in the game. It follows that the rate of active pl
registered ones increased. Moreover, even though the active play
less in this year edition we found an increase in the total amo
(and trips) tracked – i.e., from about 53k trips to about 64k tri

Figure 3. Amount of trips and kilometers travelled, divided by transportation means, in Play&Go, October 2018 - April 2019

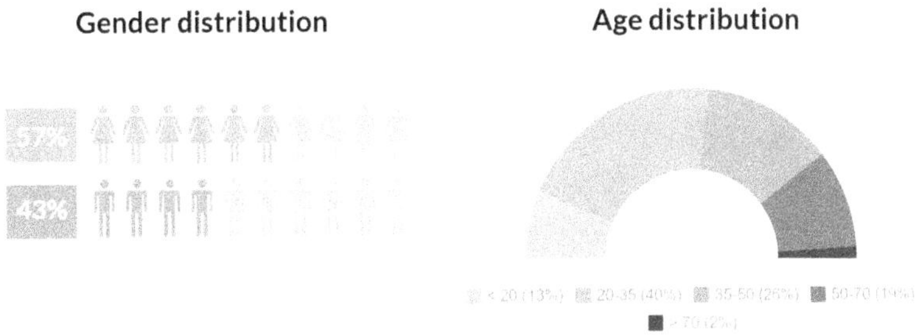

Figure 4. Distribution of the number of tracked trips over time in Play&Go, October 2018 - April 2019

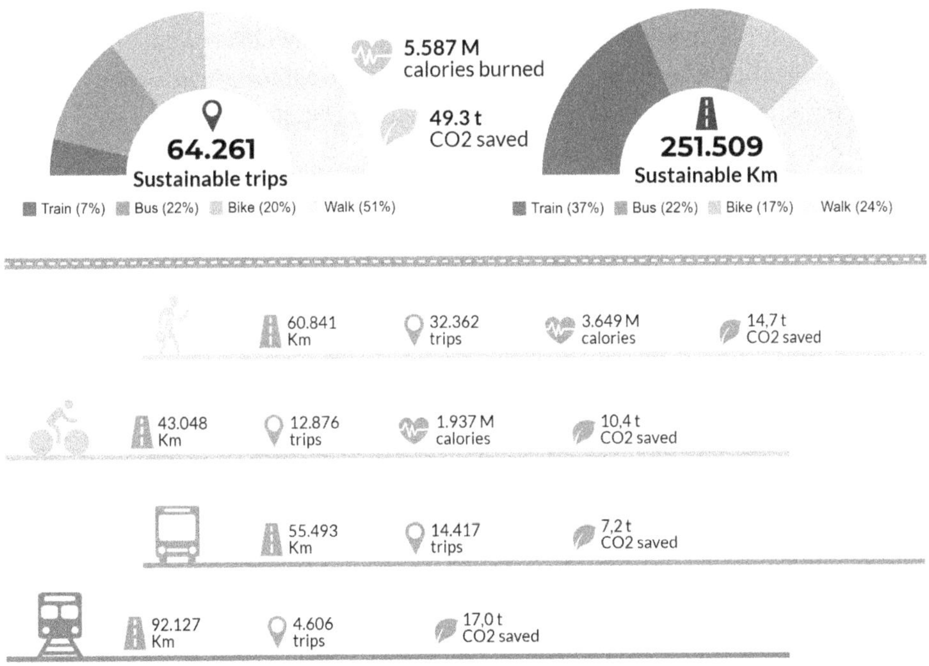

Players participation and effort can also be measured through the outcomes of the challenges. Among the 5.9k challenges assigned, players won half of them. Players were also well disposed to contribute to our understanding of their experience by answering to in-game surveys: 483 of them answered to the initial survey and 86 to the final survey. On the overall, player experience was rated positively: 85% of users

Figure 5. Gender and age distribution in Play&Go, October 2018 - April 2019

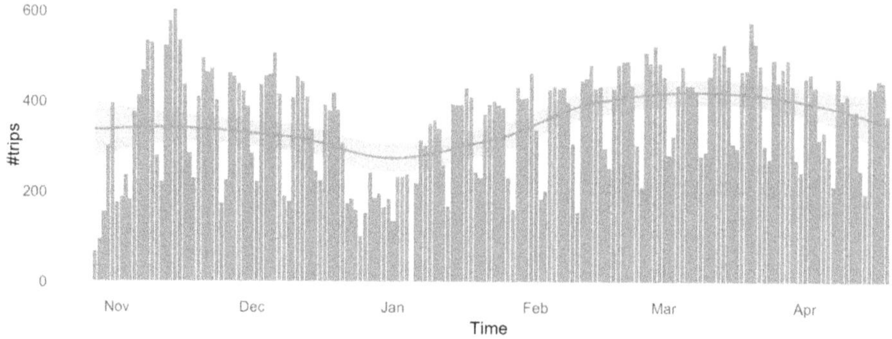

gave a rate higher than 4, with the scale going from 1 to 6 and 6 representing the best experience. The game also had an impact in terms of mobility, since a fair number of users (40%) tried a new means of transportation, which was public transportation for half of them. It is also interesting to notice that we did not observe drastic low peak in the participation rate, in spite of holidays and bad weather (Figure 4), apart from a lower participation that has been registered during Christmas holidays (December 22, 2018 to January 6, 2019).

In terms of demographic, we observed that our players are quite heterogeneous (Figure 5) in terms of gender (57% female), which is probably due to the variety of game elements and mechanics implemented. Moreover, the distribution of the age range follows the shape of a gaussian curve slightly skewed towards the lowest extreme: 13% of our users are younger than 20 years old, 40% are between 20 and 35 years old, 25% are between 35 and 50 years old, 19% are between 50 and 70 years old and 2% are older than 70. This age range distribution is in-line with similar initiatives: all the age range are well represented, exception made for the younger users (less than 20 years old, 13%) and the elderly users (more than 70 years old, 2%). The participation of younger users is limited by the age restrictions imposed by the national regulation for prize-based competitions (players younger than 16 cannot register). The low rate of players over 70 years old can be related to elderlies' general resistance to smartphone applications: even when owning a smartphone, elderlies mainly exploit its basic functionalities (e.g., calls and messages).

Figure 2a. Play&Go - multiplayer game elements: Unlockable content

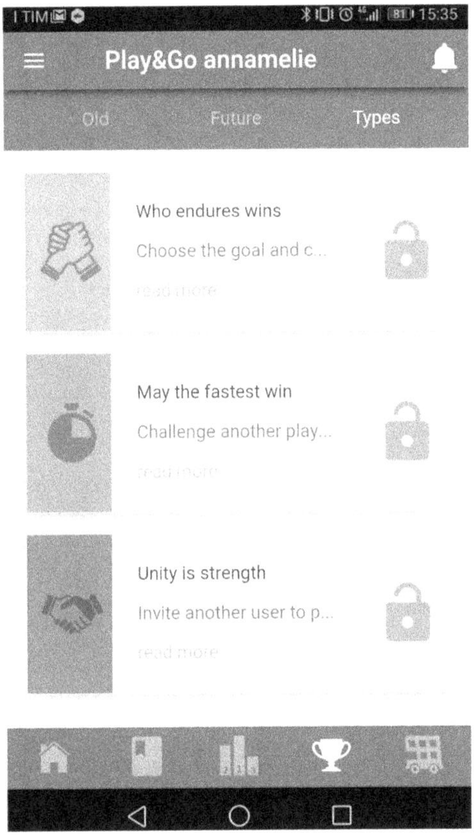

(and choose one, if they are at least at level 2) or they want to engage in a multiplayer challenge. To engage in a multiplayer challenge, players have to reach at least level 4. At that point, they can unlock one multiplayer mode. To initiate a multiplayer challenge, players can select one of the modes that they have unlocked and, then, invite another user. Prior to the invitation, the initiator of the challenge must specify the type of challenge they want to activate (Figure 2c), and hence choose among bike, walk or a general challenge on the amount of Green Leaves points obtained. The RS will calculate the best target depending on the two players involved and the mode selected and, then, will generate the appropriate reward(s). During this step the more competitive players could actuate a strategy, meaning that they could play with the invitation parameters – i.e., challenge mode and opponent – so as to have the best chances to win or the highest reward. Players can make only one request at a time. Once an invite has been sent, the player who sent the invite can still cancel the request to make another one, while the invited player can either accept it or reject

Figure 2b. Play&Go - multiplayer game elements: Multiplayer Challenges

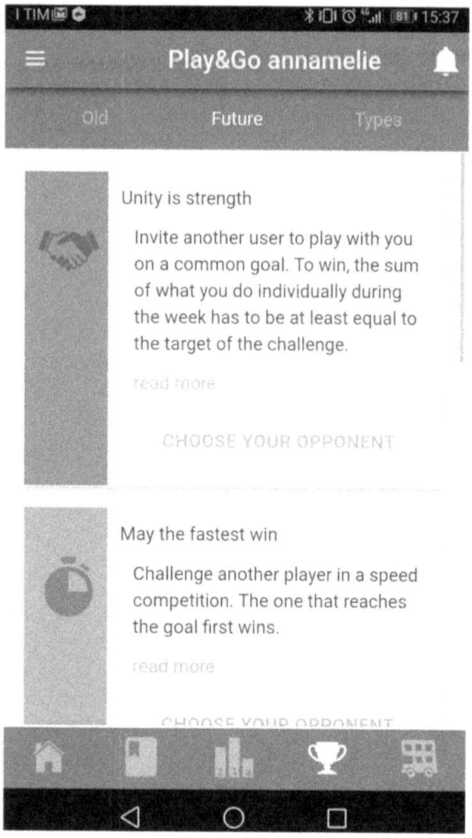

it (Figure 2d). When the invite has been accepted, the players involved cannot make other invitations and/or accept other challenges until the following week.

Every gamification campaign has been followed by the analysis of the data retrieved throughout the six-month period. We exploited two kinds of data sources which helped in conducting both quantitative and qualitative studies. The first and most prolific data source is gameplay logs, in which we stored every in-game action of the players, examples being the trips tracked, information on the challenges received and the points gained. The second kind of data source are surveys. We submitted players an initial survey, to collect information on their preference and personality; a survey inspecting players response to the multiplayer challenges; and a final survey inquiring players about their game experience. The main difference between the two data sources lies in the amount of data obtained, in that, on the one hand, we have logs for all players, while, on the other hand, not every user answered to every (or any) questionnaire. In particular, we had 480 respondents for the initial

Figure 2c. Play&Go - multiplayer game elements: Send an invite

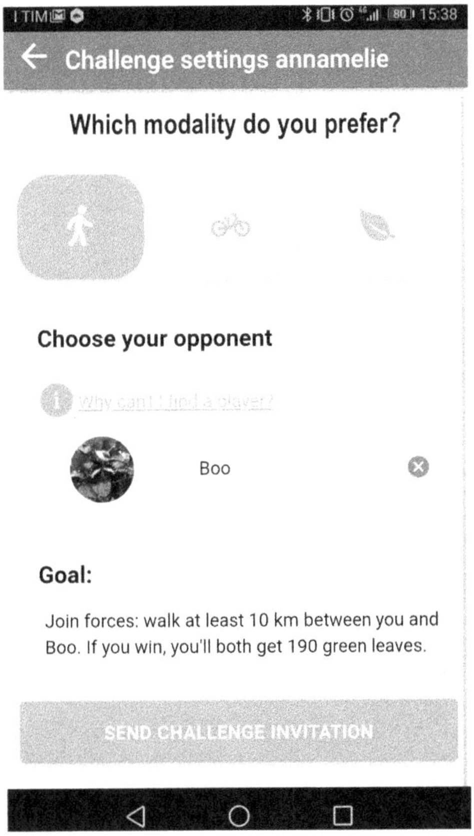

survey, 81 respondents for the questionnaire about the multiplayer challenges, and 87 respondents for the final survey. The insight gained during such analysis have and will lead the design of further versions of the game.

In our most recent edition of the game, which run from the end of October 2018 to April 2019, we observed a more committed audience of players. In comparison to the previous edition, where the amount of time spent in the game were more modest despite having a higher rate of registered players, we found that the active users were fewer but more constant in the usage. In fact, while on 2017-2018 edition we counted 1079 registered players (of which 747 were active), this year we had 834 registered players (590 active). By active players we mean players that tracked at least one trip in the game. It follows that the rate of active players among the registered ones increased. Moreover, even though the active players were slightly less in this year edition we found an increase in the total amount of kilometers (and trips) tracked – i.e., from about 53k trips to about 64k trip and from about

Figure 2d. Play&Go - multiplayer game elements: Accept an invite

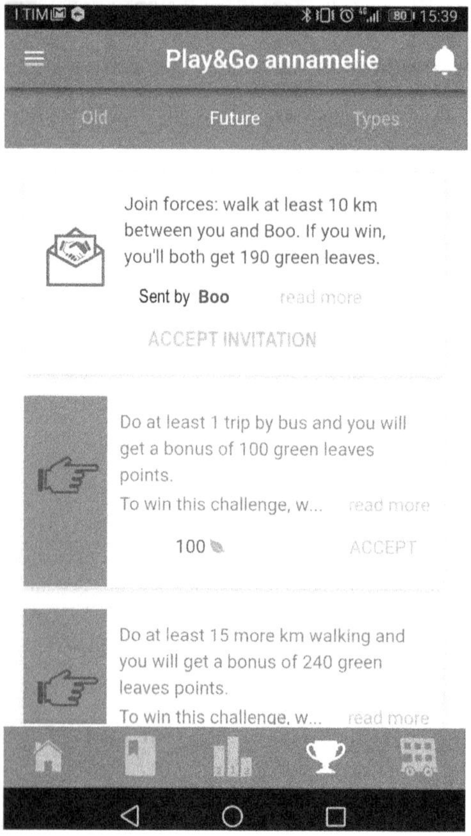

240k kilometers to 250k kilometers (Figure 3). Of those trips, the vast majority were tracked by foot, followed by train, bus and bike. Therefore, the transportation means we call "zero impact" (i.e., by foot or by bike) were the favored ones. This result should be interpreted considering also the background of players, in terms of preferred transportation means used. The majority of our users (35%) declared to mostly use public transportation means, while the remaining selected as primary means car, walk and bike (25%, 21% and 18% respectively). Therefore, the 75% of our users already moved in a green way, with 39% of them assuming a "zero impact" behavior. Despite this, the achievements are still impressive, considering that before the game 48% of our players declared to use the car about 3 times per day. Moreover, 66% of our users declared to have changed their mobility behavior, thanks to the app, to a medium-to-high extent.

Figure 3. Amount of trips and kilometers travelled, divided by transportation means, in Play&Go, October 2018 - April 2019

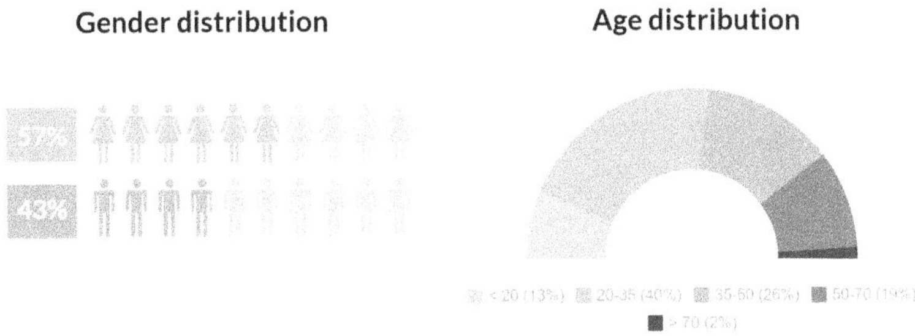

Figure 4. Distribution of the number of tracked trips over time in Play&Go, October 2018 - April 2019

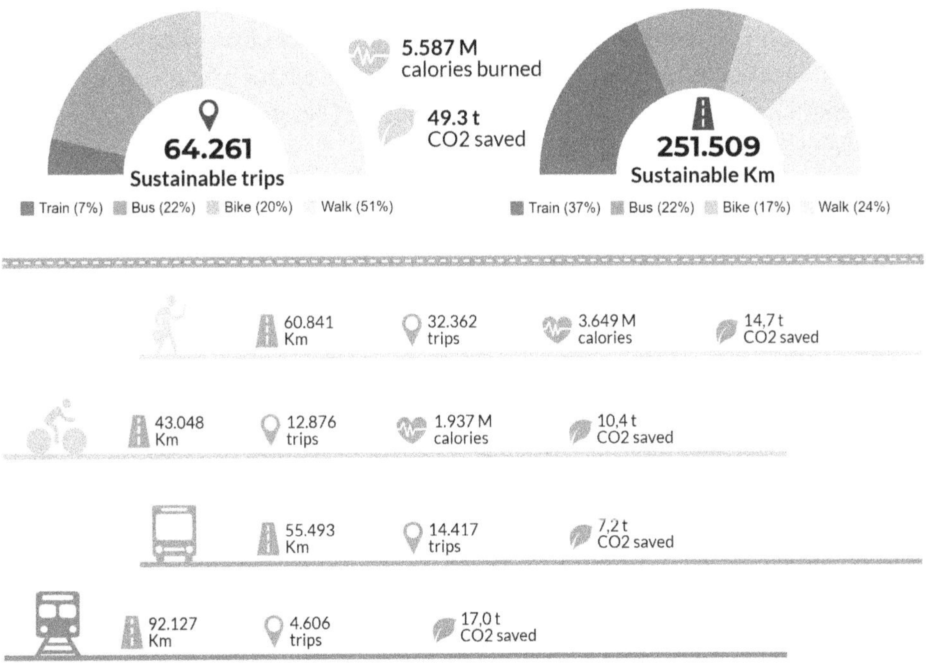

Players participation and effort can also be measured through the outcomes of the challenges. Among the 5.9k challenges assigned, players won half of them. Players were also well disposed to contribute to our understanding of their experience by answering to in-game surveys: 483 of them answered to the initial survey and 86 to the final survey. On the overall, player experience was rated positively: 85% of users

Figure 5. Gender and age distribution in Play&Go, October 2018 - April 2019

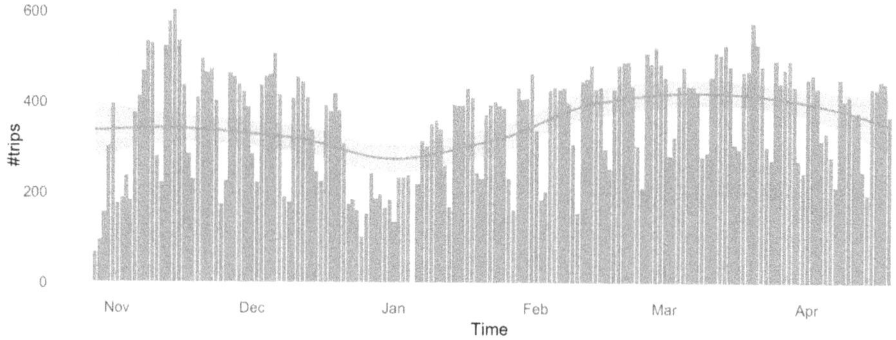

gave a rate higher than 4, with the scale going from 1 to 6 and 6 representing the best experience. The game also had an impact in terms of mobility, since a fair number of users (40%) tried a new means of transportation, which was public transportation for half of them. It is also interesting to notice that we did not observe drastic low peak in the participation rate, in spite of holidays and bad weather (Figure 4), apart from a lower participation that has been registered during Christmas holidays (December 22, 2018 to January 6, 2019).

In terms of demographic, we observed that our players are quite heterogeneous (Figure 5) in terms of gender (57% female), which is probably due to the variety of game elements and mechanics implemented. Moreover, the distribution of the age range follows the shape of a gaussian curve slightly skewed towards the lowest extreme: 13% of our users are younger than 20 years old, 40% are between 20 and 35 years old, 25% are between 35 and 50 years old, 19% are between 50 and 70 years old and 2% are older than 70. This age range distribution is in-line with similar initiatives: all the age range are well represented, exception made for the younger users (less than 20 years old, 13%) and the elderly users (more than 70 years old, 2%). The participation of younger users is limited by the age restrictions imposed by the national regulation for prize-based competitions (players younger than 16 cannot register). The low rate of players over 70 years old can be related to elderlies' general resistance to smartphone applications: even when owning a smartphone, elderlies mainly exploit its basic functionalities (e.g., calls and messages).

KIDS GO GREEN: GAMEFUL EDUCATION FOR SUSTAINABLE MOBILITY

Being an independent and active road user is fundamental for the physical, social, cognitive and emotional development of children and adolescents (Beunderman, 2010; Brown et al., 2008; Kolotkin et al., 2001). Nevertheless, in recent decades there has been an increase in the number of parents placing restrictions on - and thereby limiting - the active mobility of their children (Björklid, 2002; O'Brien et al., 2000; Hillman et al., 1990; Hillman, 1993; Rissotto and Tonucci, 2002) and the expression "back-seat generation" is sadly becoming well known in international debates (Child Friendly Cities Initiative, 1996; Child in the City Foundation, 2002). The reasons why parents choose to drive their children to school vary. For some cases it is due to lack of traffic safety measures for children as pedestrians or cyclists. For others it is related to practical considerations (Fotel and Thomsen, 2004; SKL, 2012), convenience and time saving (McDonald and Aalborg, 2009), since driving the children to school makes it easier to manage the daily family juggle.

Considering that children's mobility accounts for approximately 20% of the total daily travelling population in the EU (European Commission, 2002) and that chauffeuring children accounts for a large proportion of cars on the road, the impact of this ever growing phenomena is huge in terms of traffic, safety and pollution in our cities.

Plenty of grassroots and participatory initiatives are being launched at a local level providing bottom-up solutions supporting children's sustainable and active mobility (e.g., walking buses, children bike-trains, volunteers at crosswalks). However, most of these initiatives struggle to scale up and often do not survive long after their bootstrapping. This is mainly due to their fragmentation and isolation, lack of reliability, convenience and flexibility of the offered mobility services, especially if compared to the use of private cars, but also due to the lack of awareness and difficulty in breaking away from bad habits.

Education is our best hope and most effective means to cope with environmental sustainability challenges, since it can empower a new generation of citizens with values, knowledge and skills to critically reflect, make informed decisions and act in a sustainable way. Childhood is a privileged time for environmental education: it creates a new attitude of care, respect and sustainable life habits and establishes a reinforcement loop spreading the educational impact to their families and the communities they live in. Children are the citizens of the future and will bring their habits into adulthood, thus shaping our future cities.

Kids Go Green (Marconi et al, 2018; Kids Go Green Website, 2016) aims at changing the attitude of children, their families and their community towards environmental sustainability challenges, by making the adoption of eco-friendly mobility habits a fun and social experience. Kids Go Green aims at reducing GHG emissions by inducing a collective modal shift from cars to clean sustainable mobility (walking, biking, public transport) in home-school trips for primary and secondary school pupils.

To this extent, Kids Go Green involves the entire school community (children, teachers, and families) in a gameful educational activity where the sustainable kilometers made by each child on their daily trips from home to school contribute to the progress of the whole group of children on a virtual educational journey to places in the real world. Stops along the way allow the discovery of cities, countries, and places, conveying the value of the collective sustainable kilometers travelled to date. The route, the final destination and the stops of the virtual journey are customized according to the interests and the educational needs of the teachers and can cover interdisciplinary subjects (e.g., a trip around Europe discovering culture, natural landscapes, history and architecture) or target a specific topic of interest (e.g., Italian National Parks, the journey of Ulysses, a journey discovering children's countries of origin). Teachers associate to each stop multimedia educational material such as videos pictures, documents or web-links. Stops are unlocked when a sufficient number of sustainable kilometers is reached by the entire class, and thus can be exploited for in-class learning and for motivating children towards sustainable mobility transportation. The game is supported by a platform that captures the sustainable kilometers of each child, managing progress in the game and, through a Web App, displaying the achievements on an interactive map.

Before launching the game in the school, a set of preparatory activities is performed in order to design and calibrate the virtual journey to the needs and objectives of the participants. Teachers, involving the students and their families, collect information about current home-school mobility habits (i.e., how many children walk and cycle to school, how many are taking the schoolbus, how many are chauffeured by car) and define behavior change objectives (i.e., number of children that might adopt for more sustainable habits and modal shift). Starting from predefined ranges, a constant value is associated to each transport modality (e.g., 400 meters for walk trips, 1.000 meters for school bus trips). From these information (transport habits, behavior change targets, transport distances) the daily distance travelled by the group is calculated and the design and calibration of the journey can start. This is a co-creation activity that mediates didactical interests and needs (in terms of educational content, intermediate stops and final destination to be reached), distances to be covered by the group and the duration of the game (typically from 2 months to the whole school year). Then, the teacher associates multimedia educational material (e.g. pictures, video, web-links, documents) to each stop. Teachers exploit the Kids Go Green Web-based management application that allows them to upload their own educational content, search material from predefined online educational sources, or

Figure 6a. Kids Go Green Web App: Daily Mobility Journal

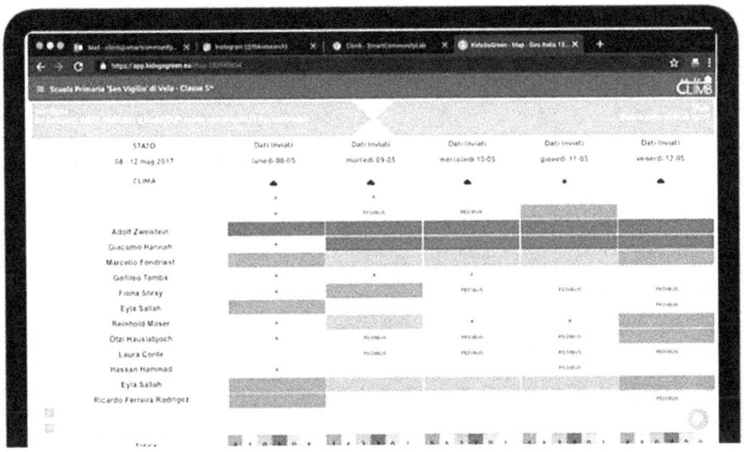

Figure 6b. Kids Go Green Web App: Daily Mobility Journal

reuse geo-referenced educational content previously uploaded by other members of the Kids Go Green community.

During game execution, every day each class exploits the Kids Go Green Web application and fill out the Mobility Journal (see Figure 6). Through a color code, each child indicates the way she reached the school on that day. Supported transport mode are walking, cycling, by school bus or public transport, by foot from a parking lot in proximity of the school (identified by teachers and families during the game

Figure 7. Kids Go Green web app: Interactive digital map

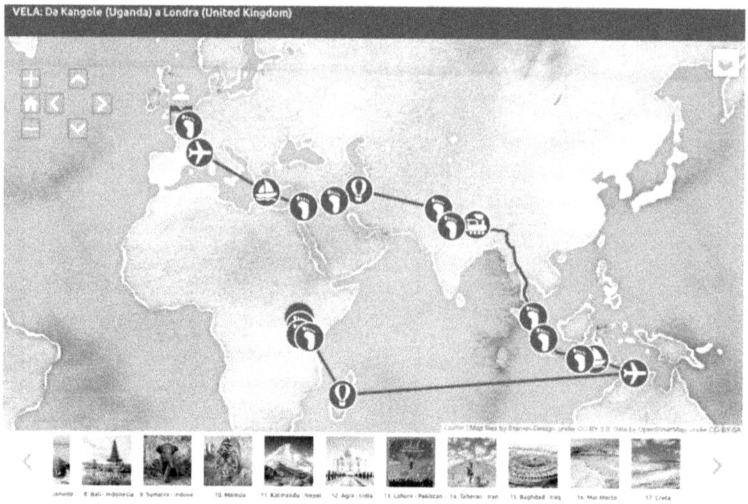

Figure 8. Kids Go Green web app: Multimedia educational material

set-up), or by car. The number of kilometers traveled in a sustainable way by the group (calculated on previously defined constant values in meters) in their daily trips to school contributes to the progress on the virtual journey. Classes also indicate the daily weather and, in case of rain or snow, trips by foot and by bike are incentivized through bonus meters rewards. Classes can also enter their walking field trips done as part of the school activities.

Figure 9. Kids Go Green web app: Mobility statistics

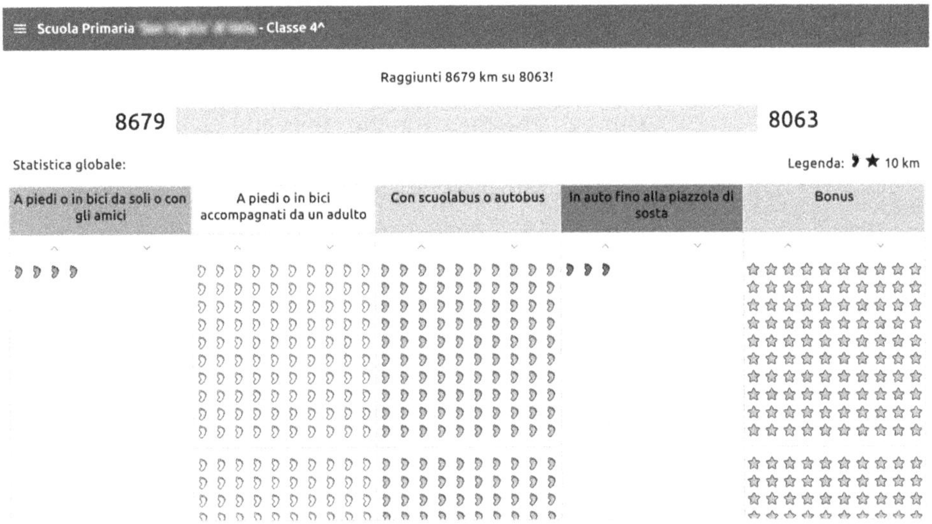

Kids Go Green also supports for various class-level and team-level challenges: the group, upon completion of a specific achievement (e.g., zero impact day, no-car week), benefit from virtual prizes that can be exploited to cover long distances (e.g., a cruise ticket along a river, a train ticket, a hot-air balloon ticket to reach an oversea stop) or to speed up games (and promote a change in mobility habits) where the group is proceeding slower than planned.

Children can see their journey progress through an Interactive Digital Map (see Figure 7), usually displayed in class on an interactive whiteboard or large display. Participants can inspect their current position and the reached stops. Upon arrival to a stop, all the associated educational material is unlocked and can be exploited for in-class lessons (see Figure 8). The Interactive Digital Map is publicly accessible: in this way, children can share the progress with their families and can enjoy and benefit from the educational material also from home.

The Mobility Statistics page allows to view the amount of sustainable distances covered by the group with different transport means (Figure 9). Given the age range of target users, distances are shown as ideograms (each “foot” symbol corresponds to 10 kilometers).

Kids Go Green has been adopted for three consecutive school years in various Italian cities, involving a total of 16 schools, 72 classes, 1844 children, and more than 100 teachers. In the 2016-2017 school year, the S. Vigilio school in Trento (Italy), experimented with the first version of the system, developing a virtual path from Trento (Italy) to Kangole (Uganda). In just four months 87 children traveled more than 8800 sustainable kilometers, crossing all of Italy and passing through Greece, Egypt and other countries of the African continent.

The data collected for the evaluation included: the Kids Go Green game logs, depth-interviews with teachers (n=3), as well as two questionnaires for the families (n=87 for both surveys). During these three months, the children took 4400 home-school trips and only 31 were by car. The impact was impressive, teachers confirmed that cars disappeared from the school surroundings and a post experiment evaluation, carried out six months after the game end through a survey subministered to families (n=85), revealed that 86% of the children maintained the adopted sustainable habits.

The success of this first experimentation and the enthusiasm of children, parents and teachers enabled 10 Kids Go Green games to be activated in six schools in Trento during the 2017-2018 school year, for a total of over 600 children and 100 teachers involved. During school year 2018-2019, 25 games have been activated in 18 schools in various Italian cities, for a total of 1189 children.

These experiments have shown an impact on home-school mobility that has exceeded all expectations: during the experiments run in 2017-2018 children have traveled more than 20,000 sustainable kms; and of the 26,000 home-school trips, only 5.9% were done by car. Applied solutions proved to be effective not only in ideal settings (e.g., in schools with small communities in safe suburbs), where percentages of car-trips was almost null (0.3 – 0.6% of total trips), but also in schools located in the city center, where the maximum percentage of car trips was 15.6% (from an initial baseline of 40-50% over the total number of home-school trips). Similarly to the previous school year, the evaluation was based on the Kids Go Green game logs, depth-interviews with teachers (n=13), as well as pre and post surveys for the families (n=436).

Opinions collected from questionnaires with parents show a positive environmental effect in terms of children's and families' behavioral change as well as the educational value of Kids Go Green. For 42% of the participants, Kids Go Green had a positive impact on the way children reached school and for 26% the game changed their mobility habits also during free time and family trips, highlight that sustainable behavior was also generalized to other contexts.

Although not being a primary objective of the proposed solution, Kids Go Green resulted to have a significant value also in terms of didactics. Teachers report that Kids Go Green is an innovative educational tool that, through the principles of gameful education, promotes the definition of interdisciplinary, inclusive and participatory educational paths. Interdisciplinary, because it enables the co-creation of personalized thematic paths that integrate topics from different disciplines into a choral narration. Inclusive, because thanks to the multimedia material associated with the stops, it allows to exploit various communication channels (images, drawings, videos, websites and material produced by students and families) that offer children a varied learning experience. Participatory, since it provides for an active collaboration of teachers, students, families, educators and researchers, both in the co-creation of educational paths, and during the execution of the game which, through cooperative mechanisms, promotes a collective commitment and a concrete alliance.

Kids Go Green supports and sustains over time, through technology, the sustainable and active mobility of children, together with their peers, making them fun, participatory and safe. The game is well integrated into the educational path, nourishing and supporting the project over time, avoiding the loss of motivation. Thanks to its simplicity and immediacy, the Kids Go Green game has demonstrated a high impact in terms of didactic and socio-educational terms, significantly affecting the participation and the involvement of pupils and families in solving the problems related to traffic, pollution and safety in proximity of schools.

BEST PRACTICES, LESSONS LEARNED AND FUTURE CHALLENGES

As it clearly emerges from the presented case studies, game-based motivational systems can play a key role for promoting the engagement of citizens' and transform them into active players for addressing mobility-related sustainability challenges. Gamified systems can be exploited to engage end-users and retain them on the long-term. Gamified systems are also an effective means to support reflection on the impact of citizens' current mobility choices, in that they can be exploited to encourage a break of bad habits and to discover and appreciate more sustainable, efficient, or newly introduced mobility services. If carefully designed, they can induce a gradual, but profound and long-lasting change in citizens' mobility habits.

These characteristics of gamified systems can become key ingredients of MaaS solutions. Within the MaaS paradigm, gamification and game-based motivational systems could be seen as a new layer specifically targeting end-user's engagement, motivation and retention.

Gamified sustainable mobility campaigns could be exploited to launch MaaS-based applications, attracting new users and creating an active and committed community of customers. Game-based elements could be easily integrated in MaaS solutions to promote, among all available integrated and tailored mobility options, those that are more efficient and sustainable, or to advertise newly integrated mobility services and to increase their chance of being tried and appreciated by customers. Through a combination of virtual and real incentives, game-based systems could also be exploited to retain MaaS application customers, especially for what concerns booking and payment services.

In the following, we discuss the lessons learned, current limitations and recommendations for the design and operation of gamified motivational systems in the mobility domain, as emerged from the presented case studies.

Unfortunately, it is difficult to design a fun and entertaining environment, especially when it is tied up to a "serious" application domain, such as positive behavioral change. Retaining players on the long-term (several months) is particularly important within games with a purpose, since sustainable behaviors induced by the game need to be reinforced long enough to become durable mobility habits. To this extent, following the design principles of persuasive technologies (Fogg, 2003), personalization and dynamicity are key features of an effective gamified system. The dynamicity of game mechanics (such as personalized challenges and missions, or level-based unlock of game features and virtual resources) avoids player's boredom and supports retention. Personalization is a key feature that allows to tune game objectives and goals to the player's profile and performances, promoting and rewarding behaviors that are in-line with the game ulterior motive and that are challenging but feasible for the player (thus avoiding player's frustration).

In both our case studies, personalized, dynamic content has been the most appreciated motivational affordance by end-users. In Play&Go, weekly challenges have been the most enjoyed feature throughout the four yearly editions. In Kids Go Green, the personalization of the virtual journey, that can be designed according to the interests and didactical objectives of the group of pupils, has been the distinguishing factor for the retainment of children and teachers. Moreover, requiring a gradual feasible behavior change to players has proven to be successful both in Play&Go and in Kids Go Green initiatives. This has been achieved in Play&Go through weekly personalized challenges that required players to have a small, but incremental and continuous improvement. In Kids Go Green, one of the winning strategies has been the identification of a parking place in proximity of the school (distance from 300 to 500 meters) that parents could reach by car and from which children could walk to school (either accompanied by adults or in small groups, dependently on the child age and autonomy).

The capability to identify and block cheaters (or killers - disruptors in taxonomies of player types) is another fundamental factor to tackle in gamification campaigns. This allows to establish a sense of trust among all players, who are strongly demotivated when witnessing unfair rewards. The accuracy of the automatic trip validation algorithms in Play&Go has been a key success factor of this initiative. In Kids Go Green, this problem is widely limited by the application context, where informal social control prevents children from cheating (i.e., teachers and children, as a group, know how each kid reached the school).

Another success factor of any sustainable mobility campaign is community building. Making citizens an integral part of the innovation process and building a sense of relatedness with the rest of the community is what truly fosters motivation to use the system, and thus, to pursue a positive behavior change. This is particularly relevant for those citizens that are sensitive to sustainability issues and whose motivation can be leveraged to spread the campaign's impact and reach thousands of users (e.g., in Play&Go, 44% of participants have come to know the initiative through their friends). In the presented case studies this has been achieved through a continuous dialogue with end-users and a participatory co-design approach. End-users (citizens, teachers, children and their parents) through the years, thanks to their feedback and suggestions, had a prominent role in the enhancement of Play&Go and Kids Go Green solutions. This is a win-win situation, since organizers can gain valuable insights on the current strengths and limitations of their solution, while end-users strongly appreciate that their suggestions are appreciated and implemented.

Despite the proven benefits, there are still some limitations that emerged from the presented case studies, which constitute future challenges for practitioners and researchers in the field.

The first challenge concerns the capability of gamified motivational systems of rewarding the actual behavior improvement, rather than the participant performances. In Play&Go, this has been partially achieved through weekly challenges, which provide a reward to those players achieving the challenge goal. Nevertheless, bonuses gained when a challenge is won contribute only to a limited part of the virtual points (Green Leaves) that define leaderboards, and thus, to the assignment of real prizes. A considerable amount of virtual points is gained by players proportionally to the kilometers traveled during each game week and to the sustainability of the used transport means. As a consequence, despite in-game daily and weekly limitations to the number of trips and kilometers that can be tracked, top players in the game were mostly citizens traveling long distances by foot or by bike, independently from the induced changed by the game in their mobility habits. This aspect, related to a well-known limitation of competitive games exploiting leaderboards as a motivational affordance (Hamari et al., 2014), can result in a demotivation of all those players that are committed to the game objectives but cannot sustain top players' performances.

Another limitation concerns the capability of accurately measuring the impact, in terms of behavior change, of gamification campaigns. The main challenge in this case is related to the possibility of obtaining a baseline. In Play&Go, as in most similar initiatives, the pre-game behavior is obtained through players' self-assessment (i.e., through in-game surveys), where players declare the number of kilometers traveled per day and the main transport mode. A more reliable strategy for obtaining a precise baseline, might be to track all participants trips within a limited timeframe at the beginning of the game (i.e. one week, exploiting automatic functionalities embedded in the mobile App). However, this solution presents several drawbacks in terms of mobile phone battery consumption and end-user privacy protection and might thus result in a low acceptance from players, especially within the initial phase of the game, that is already the most critical phase in terms of churn rate. Kids Go Green does not suffer from this issue: children report all their daily trips, independently from the transport mode, from the very beginning of the game. The collected data, together with the information about mobility habits and game objectives defined by teachers and families, allow to have an accurate measure of the change induced by the game.

Last but not least, a well-known and ambitious challenge of sustainable mobility campaigns (and motivational - behavior change systems in general) is to reach those end-users that are not sensitive to the targeted issue or are particularly resistant to changes (i.e., inveterate drivers, in the mobility domain). This is a current limitation of Play&Go, where, based on the registration survey, only 25% of registered players declared to use the car as main transport mode. Kids Go Green, however, proved to be a successful case, also with respect to this challenging and widely unresolved issue. Kids Go Green is an activity that is integrated within the school programme and thus all families are involved in the game, independently from their sensitivity to environmental sustainability issues and from their practical (and in most cases understandable) resistance against adopting more sustainable mobility habits. Being a cooperative game, where the group wins only if there is a collective change in terms of mobility behavior, peer pressure is particularly effective in Kids Go Green, where children play a prominent role in convincing their schoolmates in breaking bad habits. Moreover, parents are proven to be particularly sensitive to their children's request for a change (Damerell et al., 2013). And thus, within this setting, behavior change naturally spreads through children's peers and up to their families. These characteristics of Kids Go Green, that proved to be able to tackle a major challenge of gamified systems, can inspire the design of future campaigns that are pursuing similar objectives.

CONCLUSION

Game-based motivational affordances can be successfully exploited for civic engagement and to develop a new culture for urban mobility. Through a combination of game elements, virtual and real incentives, gamified systems have proven to be effective in promoting an active participation, breaking bad habits and inducing long-lasting changes in citizens' mobility habits. These characteristics of gamified systems can be exploited by MaaS solutions to attract new users, retain them on a long-term, and encourage the acceptance of tailored mobility plans suggested by

MaaS-based applications. MaaS systems, offering innovative, tailored and convenient mobile solutions to citizens, combined with the motivational and persuasive power of game-based systems, have a perspective to address the key sustainability challenges of the mobility of the future and to achieve the goal of replacing private vehicles with alternative transport models.

Two application cases, targeting home-work and home-school mobility, applied for several consecutive years and involving thousands of citizens, are presented in detail. The key success factors of these case studies are analyzed, and best practices are derived to guide the design and development of similar initiatives.

Despite the proven benefits, in terms of engagement and behavior change, the investigated case studies still suffer from some limitations: being able of rewarding the actual behavior improvement, rather than personal mobility performances; providing an accurate measure of the impact of the sustainable mobility campaigns; engaging citizens' that are not sensitive to the theme, such as inveterate drivers. These constitute important challenges to be addressed in the future by practitioners and researchers in the field.

ACKNOWLEDGMENT

This research was supported by EIT Climate KIC [grant numbers TC_2.2.1_190152_ P420-1B, TC2018A_3.1.9-SMASHb_P420-1B].

REFERENCES

Barkenbus, J. N. (2010). Eco-driving: An overlooked climate change initiative. *Energy Policy*, *38*(2), 762–769. doi:10.1016/j.enpol.2009.10.021

Beunderman, J. (2010). People make play: the impact of staffed play provision on children, families and communities. A research report by Demos for Play England National Children's Bureau, London, UK.

Björklid, P. (2002). Parental restrictions and children's independent mobility. In *Proceedings of IAPS Conference*. The Stockholm Institute of Education, Department of Educational Research.

Bologna Municipality. (2017). *Bella Mossa*. Available online: https://www.bellamossa.it/

Briones, A. G., Chamoso, P., Rivas, A., Rodríguez, S., De La Prieta, F., Prieto, J., & Corchado, J. M. (2018). Use of gamification techniques to encourage garbage recycling. A smart city approach. In *International Conference on Knowledge Management in Organizations* (pp. 674-685). Springer. 10.1007/978-3-319-95204-8_56

Brown, T. & Summerbell C. (2008). *Systematic review of school-based interventions that focus on changing dietary intake and physical activity levels to prevent childhood obesity: an update to the obesity guidance*. National Institute for Health and Clinical Excellence.

Capitalbikeshare. (2011). *Capital bikeshare boasts near record ridership in February.* Available online: https://www.motivateco.com/capital-bikeshare-boasts-near-record-ridership-in-february/

Castellanos, S. (2016). Delivering modal-shift incentives by using gamification and smartphones: A field study example in Bogota, Colombia. *Case Studies on Transport Policy*, *4*(4), 269-278.

Child Friendly Cities Initiative. (1996). Retrieved from https://childfriendlycities.org

Child in the City Foundation. (2002). Retrieved from https://www.childinthecity.org

Choudhury, T., Borriello, G., Consolvo, S., Haehnel, D., Harrison, B., Hemingway, B., & Lester, J. (2008). The mobile sensing platform: An embedded system for capturing and recognizing human activities. *IEEE Pervasive Computing*, *7*(2), 32–41. doi:10.1109/MPRV.2008.39

Chowdhury, S., & Ceder, A. (2016). Users' willingness to ride an integrated public-transport service: A literature review. *Transport Policy*, *48*, 183–195. doi:10.1016/j.tranpol.2016.03.007

CIVITAS. (2016). *Mobility-as-a-Service: A new transport model.* Retrieved from http://civitas.eu/content/civitas-insight-18-mobility-service-new-transport-model

Consolvo, S. (2007). *UbiFit*. Available online: https://www.consolvo.org/ubifit

Cowley, B., Moutinho, J. L., Bateman, C., & Oliveira, A. (2011). Learning Principles and Interaction Design for 'Green My Place': A Massively Multiplayer Serious Game. *Entertainment Computing*, *2*(2), 103–113. doi:10.1016/j.entcom.2011.01.001

Damerell, P., Howe, C., & Milner-Gulland, E. J. (2013). Child-orientated environmental education influences adult knowledge and household behaviour. *Environmental Research Letters*, *8*(1), 15-16. doi:10.1088/1748-9326/8/1/015016

De la Hera Conde-Pumpido, T. (2013). A conceptual model for the study of persuasive games. *DiGRA Conference 2013.*

Deci, E. L., & Ryan, R. M. (2000). The" what" and" why" of goal pursuits: Human needs and the self-determination of behavior. *Psychological Inquiry, 11*(4), 227–268. doi:10.1207/S15327965PLI1104_01

Deterding, C. S., Dixon, D., Khaled, R., & Nacke, L. (2011). From Game Design Elements to Gamefulness: Defining Gamification. *Proceedings of the 15th International Academic MindTrek Conference*, 9–15. 10.1145/2181037.2181040

Di Dio, S., Lissandrello, E., Schillaci, D., Caroleo, B., Vesco, A., & D'Hespeel, I. (2018). MUV: A Game to Encourage Sustainable Mobility Habits. In *International Conference on Games and Learning Alliance* (pp. 60-70). Springer.

European Commission. (2002). *Kids on the Move. Luxembourg: Environment Directorate-General*. Office for Official Publications of the European Communities.

Fogg, B. J. (1998). Persuasive computers: perspectives and research directions. In *Proceedings of the SIGCHI conference on Human factors in computing systems* (pp. 225-232). ACM Press/Addison-Wesley Publishing Co. 10.1145/274644.274677

Fogg, B. J. (2003). *Persuasive Technology: Using Computers to Change What We Think and Do*. Morgan Kaufmann. doi:10.1016/B978-155860643-2/50011-1

Fotel, T., & Thomsen, T. U. (2004). The Surveillance of Children's Mobility. *Surveillance & Society, 1*(4), 535–554.

Froehlich, J. (2015). Gamifying Green: Gamification and Environmental Sustainability. In *The Gameful World*. MIT Press.

Gabrielli, S., Maimone, R., Forbes, P., Masthoff, J., Wells, S., Primerano, L., ... Pompa, M. (2013). Designing Motivational Features for Sustainable Urban Mobility. In *CHI '13 Extended Abstracts on Human Factors in Computing Systems (CHI EA '13)* (pp. 1461–1466). New York: ACM. doi:10.1145/2468356.2468617

Gibson, D. V., Kozmetsky, G., & Smilor, R. W. (Eds.). (1992). *The technopolis phenomenon: Smart cities, fast systems, global networks*. Rowman & Littlefield.

Giesecke, R., Surakka, T., & Hakonen, M. (2016). Conceptualising Mobility as a Service. A user centric view on key issues of mobility services. *Eleventh International Conference on Ecological Vehicles and Renewable Energies*.

Glanz, K., & Bishop, D. B. (2010). The role of behavioral science theory in development and implementation of public health interventions. *Annual Review of Public Health, 31*(1), 399–418. doi:10.1146/annurev.publhealth.012809.103604 PMID:20070207

Hamari, J., Koivisto, J., & Sarsa, H. (2014). Does Gamification Work? A Literature Review of Empirical Studies on Gamification. HICSS, 14, 3025-3034.

Hillman, M. (1993). *Children, Transport and the Quality of Life*. London: Policy Studies Institute.

Hillman, M., Adams, J., & Whitelegg, J. (1990). *One False Move... A Study of Children's Independent Mobility*. London: Policy Studies Institute.

Holleis, P., Luther, M., Broll, G., Cao, H., Koolwaaij, J., Peddemors, A., ... Raaphorst, S. (2012). TRIPZOOM: A System to Motivate Sustainable Urban Mobility. *1st Int. Conf. on Smart Systems, Devices and Technologies*.

Jahangiri, A., & Rakha, H. A. (2015). Applying machine learning techniques to transportation mode recognition using mobile phone sensor data. *IEEE Transactions on Intelligent Transportation Systems, 16*(5), 2406–2417. doi:10.1109/TITS.2015.2405759

Jittrapirom, P., Caiati, V., Feneri, A.-M., Ebrahimigharehbaghi, S., Alonso González, M. J., & Narayan, J. (2017). Mobility as a Service: A Critical Review of Definitions, Assessments of Schemes, and Key Challenges. *Urban Planning, 2*(2), 13–25. doi:10.17645/up.v2i2.931

Johnson, D., Deterding, S., Kuhn, K. A., Staneva, A., Stoyanov, S., & Hides, L. (2016). Gamification for health and wellbeing: A systematic review of the literature. *Internet Interventions, 6*, 89-106.

Kazhamiakin, R., Marconi, A., Martinelli, A., Pistore, M., & Valetto, G. (2016). A gamification framework for the long-term engagement of smart citizens. In *2016 IEEE International Smart Cities Conference (ISC2)* (pp. 1-7). IEEE.

Kazhamiakin, R., Marconi, A., Perillo, M., Pistore, M., Valetto, G., Piras, L., . . . Perri, N. (2015). Using gamification to incentivize sustainable urban mobility. *Proceedings of IEEE International Smart Cities Conference*.

Kids Go Green Website. (2016). Retrieved from http://www.kidsgogreen.eu/

Knockaert, J., Tseng, Y. Y., Verhoef, E. T., & Rouwendal, J. (2012). The Spitsmijden experiment: A reward to battle congestion. *Transport Policy, 24*, 260–272. doi:10.1016/j.tranpol.2012.07.007

Koivisto, J., & Hamari, J. (2019). The rise of motivational information systems: A review of gamification research. *International Journal of Information Management*, *45*, 191–210. doi:10.1016/j.ijinfomgt.2018.10.013

Kolotkin, R. L., Meter, K., & Williams, G. R. (2001). Quality of life and obesity. *Obesity Reviews*, *2*(4), 219–229. doi:10.1046/j.1467-789X.2001.00040.x PMID:12119993

König, D., Eckhardt, J., Aapaoja, A., Sochor, J., & Karlsson, M. (2016). Business and operator models for Mobility as a Service (MaaS). Deliverable 3 to the MAASiFiE project, Brussels, Belgium.

Lee, J. J., Matamoros, E., Kern, R., Marks, J., de Luna, C., & Jordan-Cooley, W. (2013). Greenify: Fostering Sustainable Communities via Gamification. In *CHI '13 Extended Abstracts on Human Factors in Computing Systems (CHI EA '13)* (pp. 1497–1502). New York: ACM. doi:10.1145/2468356.2468623

Marconi, A., Schiavo, G., Zancanaro, M., Valetto, G., & Pistore, M. (2018). Exploring the world through small green steps: improving sustainable school transportation with a game-based learning interface. *Proceedings of ACM International Conference on Advanced Visual Interfaces*. 10.1145/3206505.3206521

McDonald, N. C., & Aalborg, A. (2009). Why Parents Drive Children to School: Implications for Safe Routes to School Programs. *Journal of the American Planning Association*, *75*(3), 331–342. doi:10.1080/01944360902988794

Metropia. (2017). *Metropia App*. Available online: http://www.metropia.com/

Michie, S., Van Stralen, M. M., & West, R. (2011). The behaviour change wheel: A new method for characterising and designing behaviour change interventions. *Implementation Science; IS*, *6*(1), 42. doi:10.1186/1748-5908-6-42 PMID:21513547

Morschheuser, B., Riar, M., Hamari, J., & Maedche, A. (2017). How games induce cooperation? A study on the relationship between game features and we-intentions in an augmented reality game. *Computers in Human Behavior*, *77*, 169–183. doi:10.1016/j.chb.2017.08.026

Myers Jr, O. E. & Saunders, C. D. (2002). Animals as links toward developing caring relationships with the natural world. *Children and nature: Psychological, sociocultural, and evolutionary investigations*, 153-178.

O'Brien, M., Jones, D., Sloan, D., & Ristin, M. (2000). Children's independent spatial mobility in the urban public realm. *Childhood*, *7*(3), 257–277. doi:10.1177/0907568200007003002

Olszewski, R., Pałka, P., & Turek, A. (2018). Solving "Smart City" Transport Problems by Designing Carpooling Gamification Schemes with Multi-Agent Systems: The Case of the So-Called "Mordor of Warsaw". *Sensors (Basel), 18*(1), 141. doi:10.339018010141 PMID:29316643

Orland, B., Ram, N., Lang, D., Houser, K., Kling, N., & Coccia, M. (2014). Saving Energy in an Office Environment: A Serious Game Intervention. *Energy and Buildings, 74*(C), 43–52. doi:10.1016/j.enbuild.2014.01.036

Patsakis, C., Venanzio, R., Bellavista, P., Solanas, A., & Bouroche, M. (2014). Personalized medical services using smart cities' infrastructures. In *2014 IEEE International Symposium on Medical Measurements and Applications (MeMeA)* (pp. 1-5). IEEE.

Rissotto, A., & Tonucci, F. (2002). Freedom of Movement and environmental knowledge in elementary school children. *Journal of Environmental Psychology, 22*(1-2), 65–77. doi:10.1006/jevp.2002.0243

Ryan, R. M., Rigby, C. S., & Przybylski, A. (2006). The motivational pull of video games: A self-determination theory approach. *Motivation and Emotion, 30*(4), 344–360. doi:10.100711031-006-9051-8

Salen, K., Tekinbaş, K. S., & Zimmerman, E. (2004). *Rules of play: Game design fundamentals.* MIT Press.

Sassi, A., & Zambonelli, F. (2014). Coordination infrastructures for future smart social mobility services. *IEEE Intelligent Systems, 29*(5), 78–82. doi:10.1109/MIS.2014.81

Searle, J. R. (1990). Collective intentions and actions. *Intentions in Communication, 401.*

Shiraishi, M., Washio, Y., Takayama, C., Lehdonvirta, V., Kimura, H., & Nakajima, T. (2009). Using Individual, Social and Economic Persuasion Techniques to Reduce CO2 Emissions in a Family Setting. In *Proceedings of the 4th International Conference on Persuasive Technology (Persuasive '09).* ACM.

Shreenath, V. M., Kornevs, M., Raghothama, J., & Meijer, S. (2015). A Feasibility Study for Gamification in Transport Maintenance: Requirements to implement gamification in heterogeneous organizations. In *2015 7th International Conference on Games and Virtual Worlds for Serious Applications (VS-Games)* (pp. 1-7). IEEE.

Sicart, M. (2011). Against procedurality. *Game Studies, 11*(3), 209.

SKL (2012). *Why do parents drive their children to school.* SKL rapport.

The Guardian. (2010). *Chromaroma Adds Some Color to London Commutes.* Retrieved from https://www.theguardian.com/media/pda/2010/nov/30/chromaroma-oyster-transport-gaming

Tuomela, R. (2011). Cooperation as joint action. *Analyse & Kritik, 33*(1), 65–86. doi:10.1515/auk-2011-0106

Vesco, A. (2015). *Handbook of research on social, economic, and environmental sustainability in the development of smart cities.* IGI Global. doi:10.4018/978-1-4666-8282-5

Volkswagen. (2011). *The Fun Theory.* Available online: https://goodvertising.site/the-fun-theory/

Wright, W., & Bogost, I. (2007). *Persuasive games: The expressive power of videogames.* MIT Press.

Zanella, A., Bui, N., Castellani, A., Vangelista, L., & Zorzi, M. (2014). Internet of things for smart cities. *IEEE Internet of Things Journal, 1*(1), 22-32.

Zhang, P. (2008). Motivational affordances: Reasons for ICT design and use. *Communications of the ACM, 51*(11), 145–147. doi:10.1145/1400214.1400244

KEY TERMS AND DEFINITIONS

Game Elements: These are the atoms of games and build the foundations of the game itself. Examples of game elements are points, badges, challenges, and leaderboards.

Game Mechanics: This term refers to the ensemble of rules that regulates players activities, and that define the allowed actions. Game mechanics are built over game elements.

Gamification: The employment of game-like elements in contexts other than games, generally, to pursue a certain goal (e.g., a positive behavioral change).

Participation: This is a measure to evaluate the response of users (or players) to the system in question. By analyzing the intensity and the commitment of participants, a hint on the goodness of the system's design can be grasped.

Persuasive Technology: The technology designed with the underlying motive of modifying a certain attitude of behavior, exploiting psychological and sociological theories, such as persuasion and social influence.

Smart City: An urban environment that employs technology with the main motive of enhancing citizens' quality of life. It widely employs sensors to gather and analyze data, to handle the city's resources efficiently.

Sustainable Mobility: Moving in a sustainable way refers to the choice of preferring ecological transportation means (e.g., walking or biking) and public transportation to reduce CO_2 emissions.

Chapter 9
The Future of Mobility as a Service (MaaS): Driving Through the Internet of Mobility (IoM)

Sara Baltazar
Escola Superior de Ciências Empresariais (ESCE), Instituto Politécnico Viana do Castelo (IPVC), Portugal & Research Center for Systems and Technologies (SYSTEC), University of Porto, Portugal & Institute of Systems and Robotics (ISR), University of Coimbra, Portugal

António Amaral
https://orcid.org/0000-0002-7910-2418
Escola Superior de Tecnologia e Gestão, Polytechnic Institute of Porto, Portugal

Luís Barreto
https://orcid.org/0000-0001-6173-433X
Applied Research Center for Digital Transformation (ARC4DigiT), Escola Superior de Ciências Empresariais, Instituto Politécnico Viana do Castelo, Portugal & Instituto de Telecomunicações (IT), Universidade de Aveiro, Portugal

João Pedro Silva
https://orcid.org/0000-0003-4206-243X
Escola Superior de Tecnologia e Gestão, Politécnico de Leiria, Portugal

Luísa Gonçalves
https://orcid.org/0000-0002-6265-8903
Instituto Politécnico de Leiria, Portugal

ABSTRACT

The environmental concerns together with social inclusion issues and the need to promote economic equity in the society have profound implications regarding the sustainable mobility concept. This allied to a technological (r)evolution leads to the path of the internet of mobility (IoM). On the other hand, we are witnessing the prosperity of mobility associated with services, mobility as a service (MaaS), which also aims at the integration of different transport modes. Linking together IoM and MaaS, the internet of mobility as a service (IoMaaS) concept is introduced, which can learn from the end user experiences and behaviors, enabling the reduction of ease of use and sustainable mobility, while supporting a much-needed cultural shift regarding mobility habits.

DOI: 10.4018/978-1-7998-1614-0.ch009

INTRODUCTION

The beginning of the XXI century brought us an overwhelming variety of challenges, most of which are related to the dynamic context of markets and of their social and economic consequences. Others are related with the global climate movement, the focus on green alternatives and the search for sustainability practices. More recently, new challenges have arisen related with the digitization of society and its broad implications within our way of living. These issues have had an evident preponderance in the collective effort to develop new solutions and approaches which might properly address the future development.

Since the dawn of globalization, the mobility of people and goods has played a major role in the economic dynamics of countries and their overall functioning. Hence, there is a need to shift and redefine mobility, the concept and expenses of owning a car, and the scarce use of public and shared transport modes, along with a global strategy to integrate all the different means of transportation among countries. Thus, it is inevitable to drive cars with cleaner energy consumption (Travelspirit Foundation, 2019a) as well as to prepare cities/regions for the vital transformations and societal changes to ensure the cultural changes necessary for the adoption of new mobility practices (Travelspirit Foundation, 2019b). In recent years, the technological boom has been considered to be one of the main triggers of society's digitization. It has produced unprecedented options and features for citizens' mobility, with an impact on the wellbeing and the quality of life of the population. This means that Information and Communications Technologies (ICT), which continue to be improved at every minute and hour of every day, are becoming fundamental to ensure that this new paradigm of mobility is achieved and that new mobility systems are developed. Thus, emerging technologies – automated vehicles, peer-to-peer sharing applications and the Internet of Things (IoT) – are transforming and revolutionizing how individuals and society in general face the mobility evolvement (Docherty, Marsden & Anable, 2018). Hence, in this era of digitalization, mobility issues need to be properly addressed ensuring a social levelling mechanism – in order to guarantee a universal access to all citizens, especially those who suffer from any type of impairment, as well as children and elderly people.

Urban mobility is rapidly evolving, due to an increasing number of new entrants demanding high standard experiences and rising consumer expectations (Travelspirit Foundation, 2019a). However, still many smaller cities have not yet been reached by the 'wave' of transportation technology revolution, due to a lack of investment and absence of the proper infrastructures (Hallock & Inglis, 2015). Notwithstanding, the trend of urbanization is noticeable, and it is intrinsically related to younger people moving to growing, developing cities, as opposed to elderly people who are remaining in rural areas (Diewald, Möller, Roalter & Kranz, 2012). Thus, in order to respond to the variety of mobility features and conditions, we are observing an emergency of providing new mobility models in cities, all over the world (Travelspirit Foundation, 2019a), which require decentralized, multimodal, automated approaches (Cohen, 2018a), shared and on demand mobility (Shaheen, Cohen, Yelchuru, & Sarkhili, 2017). In the urban areas, it is imperative to incorporate new mobility trends in the, foreseen and current, sustainable urban mobility plan (SUMP) and consider their impacts, options and level of integration. Having various options improves mobility in general, allowing transportation systems to become far more flexible and resilient (Hallock & Inglis, 2015). The integration of diverse transport modes and connecting them to other services is a current challenge. Nevertheless, these challenges along with the appearance of new laws and regulations, will have to operationally tackle the regional differences, the market size asymmetries, the current market shares, the stakeholders and key actors in the market, which might constrain the full potential of developing the Mobility as a Service (MaaS) business models (Eckhardt, Aapaoja, Nykänen & Sochor, 2017). Despite the increasing adoption level of technology in mobility, it needs to ensure the required data anonymization in order to guarantee the freedom of movement, as is required by the General Data Protection Regulation (Cohen, 2018a; GDPR, 2016). Beyond the creation of a technological infrastructure with its functional requirements, too much of the city's development and planning is based on common habits and behaviors, which oftentimes neither address the challenges, nor promote happiness among the citizens (Travelspirit Foundation, 2019a). As pointed out by Baltazar, Amaral & Barreto (2018) happiness and the prosperity indexes are strongly related, i.e., the greater prosperity levels a country has, the happier the people of that country tend to be.

The way we travel from point A to point B is changing dramatically. This transformation is creating a new ecosystem focused on the personal mobility, with strong implications that could affect several stakeholders. The way the future of mobility practice could embrace the entire mobility ecosystem or actively shape its emergence (Corwin & Pankratz, 2017) will promote the proliferation of the Internet of Mobility (IoM) – that would maximize the full potential of MaaS. This shall be accomplished in a decentralized manner and democratizing the access to mobility users for benefits gained through shared mobility startups and allowing IoM to, somehow, enhance the MaaS customization through individuals' own travel patterns (Cohen, 2018c). Notwithstanding, the volatile, uncertain, complex and ambiguous (VUCA) context of the world we are living is continuously challenging the future's definition, creating serious doubts about how things could be organized, substantially increasing the difficulty of the mid-term prospects of how things could be shaped or even the different types of options that we might have under assessment. Therefore, this chapter intended to present and discuss the main mobility challenges and features in future implications regarding MaaS adoption in a broader context, and also to present some discussion of new opportunities that can be defined and explored. Nevertheless, the authors are conscious of the difficulties of establishing future milestones due to the dynamic nature of our context, as well as the rhythm of technological change that might increase the subjectivity of our view.

This proposition chapter is organized as follows: the research methodology followed in the development of this work; the mobility challenges are described and linked to its future implications in terms of MaaS adoption; a discussion of the main determinants of mobility in the future and new opportunities is presented; and finally some closing remarks and future research trends in a near future are indicated.

RESEARCH METHODOLOGY

According to Saunders, Lewis and Thornhill (2016), who proposed the definition of a multilayer research methodology, illustrated by an onion's layers. These layers corresponded to the philosophy of research, the approach, the strategy, the choice of method, the definition of the time horizon and the selected techniques and procedures.

Regarding the philosophy selection, there are four types of common philosophies: positivism, realism, interpretivism, and pragmatism (Saunders Lewis & Thornhill, 2016). In this research work, we have followed an interpretivist philosophy, where it argues that we are in a continuous process of interpretation of the social world around us, which leads us to the adjustment and definition of meanings and actions.

Regarding the approach, it can be either deductive or inductive in nature. This research work followed a deductive approach, which happens when a theory could be derived from the literature review and in which the developed hypothesis might be tested or proved in the future. Regarding the research strategy, the authors present several options whose selection depends, namely, on the research objectives, the existing knowledge and the amount of time and resources available. The most common strategies are: experimentation, surveys, action research, ethnography, documentary research, and case studies.

This research work will follow a case study approach, since it presupposes the development of detailed knowledge about a particular case that occurs in a certain context. Commonly used techniques include interviews, observation and questionnaires. In our particular case we've used observation in order to be able to envision some future MaaS scenarios. These techniques used can be useful for exploring or questioning an existing theory and provide new hypotheses. Regarding the choice of method, this can be unique when opting for a quantitative or qualitative method; or multimethod, if several exclusively quantitative or exclusively qualitative methods are selected; and finally, mixed, when combining quantitative and qualitative methods. This research will follow a qualitative approach.

The time horizon of the investigation may be transversal or longitudinal. It is characterized as transversal when the phenomenon is studied at a given time, and longitudinal when the same phenomenon is studied over time on different occasions. This research will be transversal. As for techniques and procedures, they can be categorized as qualitative and quantitative. Only qualitative techniques will be applied in this research.

MOBILITY: A FRONTVIEW MIRROR OF ACTUAL CHALLENGES AND MAIN FEATURES

This section presents some actual challenges that mobility and MaaS are facing in this (r)evolutionary digitalization era.

The purpose of interconnecting societal needs and ambitions, regarding mobility, with the environment is a current challenge. It should be guaranteed that the development of new mobility models is proprietary, non-monopolistic, and closed to outsiders, in the public or private sector (Travelspirit Foundation, 2019a). It urges the appearance of mobility providers to move towards innovative, interconnected, and seamless service offerings (Deon Digital, 2019). Thus, a challenge to achieve a unified mobility involves providing an equilibrium between demand responsive transport (DRT), autonomous vehicles, sharing systems, public transport (PT) and active modes, enhanced and supported by technological tools and trends. The optimal mobility business models will have to blend public and private approaches towards achieving sustainability outcomes (Cohen, 2018a). The modern mobility should be offering higher levels of flexibility and appropriateness (Li & Voege, 2017). This new mobility concept that is being shaped will in all likelihood affect the automotive, insurance, health care, energy and media industries which should reconsider how they could create value in this emerging environment (Corwin & Pankratz, 2017). Currently, younger generations are not so interested in ownership and are more focused on enhancing the number and quality of experiences (Transport Systems Catapult, 2016). Therefore, all decision-makers should benefit from the potential inclusion of new tools to reduce the burdens of vehicle ownership (Hallock & Inglis, 2015). Today's traditional players may benefit, but some of them may be weakened if they do not proactively take part in these new shaping mobility ecosystems (Van Audenhove *et al.*, 2018). Travelers and other transport users increasingly view transport as a service which is vital to revitalize (Transport Systems Catapult, 2016) the mobility marketplace (Cohen, 2019). Thus, the MaaS paradigm arises in order to unify all the stakeholders and users in a general concept of mobility, which offers them shapeable market conditions, where the challenge of achieving the balance between the development of policies, the impacts of which are still unknown, that catalyzes the MaaS marketplace and the planning for any possible negative impacts of MaaS (Transport Systems Catapult, 2016). Since 2016, MaaS has emerged offering travelers a new alternative to car ownership and use in the form of seamless multimodal transport. It provides a range of mobility offers across modes and locations, all linked through an integrated user interface (Travelspirit Foundation, 2019a) – Integrated Mobility Platforms (IMPs). These IMPs are being launched, tested and used all over the world, with different kinds of integration levels and value propositions (i.e., Citymapper, GoEuro, Google Maps, Moovel, Moovit, Qixxit, Rome2Rio). Nevertheless, none of these players has yet emerged as a prevailing provider with a superior offering – allowing to recognize key success factors for shaping the future of integrated mobility (Baron, Zinte, Schemken & Uferer, 2018) to be recognized.

Public agencies have been slow to integrate the new mobility systems into their planning and policies toolbox (Hallock & Inglis, 2015). Another crucial point evolves from the paradigm of integrated mobility – the relationship between transit and other shared mobility modes of travel, and how technology can enhance this relationship (Kimley-Horn & IBI Group, 2019). The inclusion of adequate technology is adding substantial value to travelers in terms of the appearance of new applications. As a result of the IoT, transport data will potentially become more widespread and ubiquitous allowing the combination of autonomous transport systems with the Artificial Intelligence (AI) applied to big data, increasing opportunities to innovate and to reinforce the MaaS paradigm (Transport Systems Catapult, 2016). Thus, representing a real revolution for public transport system (PTS) and making it urgent to address technical and social implications of IoT in order to enhance all the potential benefits that can be effectively gained (Melis, Prandini, Sartori & Callegati, 2016). Currently, the infrastructure and data for travel booking, payment and information is locked away in silos, which means a slight interoperability between transport modes, across borders or between competing transport companies (Travelspirit Foundation, 2019a).

Since the MaaS concept is a holistic one and is still under development, it can be defined and approached from many different points of view (Eckhardt, Aapaoja, Nykänen & Sochor, 2017). Thus, global trends towards developing the integrated mobility are stimulating cities/regions and governments with challenges related to public safety and security, the supply and consumption of energy, traffic management and social integration. Table 1 presents a brief sum of mobility trends, with the concept associated and the current and coming challenges.

Despite the development of MaaS as an innovative approach towards enhancing global mobility, there are also several challenges that need to be properly handled so as to ensure that the MaaS ecosystem is able to reach its full potential. Firstly, it is critical to identify and characterize all stakeholders in the system, trying to detect

Table 1. Mobility trends, associated concepts / players, and actual and future challenges

Mobility Trends	Concepts / Players	Challenges Features
Digitalization Data Science Interoperability Integration MaaS IoM Apps	Consumers Providers Public partners Multimodality Campaigns Technological Literacy	Provide real-time information Social inclusion Consumers needs and frustrations Economical viable transport system to the end user Integrated Mobility Platforms (IMPs) Integrate the new mobility systems into the planning and policy toolbox Ensure the data anonymization, and the systems integrity without disclosing critical data

their particularities and vicissitudes. This task is critical to reduce the asymmetries as well as the misassumptions that could impact the adoption and the full integration of all the stakeholders involved. Therefore, some of the barriers that could be detected by each type of stakeholder are the following (Transport Systems Catapult, 2016):

- Lack of options for the consumer. This means the consumer needs to feel that he/she is free to make his or her own choices and the absence of them would limit the MaaS value, and therefore, the societal benefits derived from it;
- The MaaS marketplace should be controlled by local authorities in order to ensure ensuring that the population´s rights are being respected and that the benefits retained from the MaaS ecosystem are not concentrated;
- The development of systems' interoperability is critical but, for that, it is necessary to ensure a secure linkage between data providers, transport operations and other players in any MaaS ecosystem;
- The transport operators need to explore new business models to be fully engaged and aligned with the MaaS ecosystem.

As presented by Sochor, Arby, Karlsson, and Sarasini (2018), despite the MaaS general definition, the value preposition of the MaaS common layer must be based on offering a service aligned with customer/user/traveler/consumer transport needs; involving multimodal options – mobility rather than transport; and integration of transport services, without disregarding information, payment and ticketing availability. Together with future mobility, the tendency is to involve the adoption of active modes and new mobility models, significantly reducing the actual mobility model based on car ownership (Transport Systems Catapult, 2016). The economic strength and efficiency should be combined with the different modes to ensure the best, the fastest and the cheapest option, bringing their strengths altogether (Baldwin, 2018). A future integration of all mobility system modes will be characterized by a continuous and widespread collaboration, where resources value and data can be seamlessly exchanged, with the proper technological support, ensuring the privacy of all members (Cohen, 2018a). There are multiple services which individually help travelers. Notwithstanding the fact that they are properly combined, they will become "more than the sum of their parts" (Hallock & Inglis, 2015). In this sense, the MaaS paradigm could evolve to include services (i.e. museums, gym, concerts, cinema or restaurant bookings) and, in parallel, integrating these with the transport service needed to easily access them (Pangbourne, Stead, Mladenovic & Milakis, 2018). Thus, an integrated transport master plan to be adopted and implemented in a city/region or even at a national level will be a key imperative towards improving the overall mobility performance and its attractiveness (Van Audenhove *et al.*, 2018), involving traditional transport companies to be open to and prepared for competition and accepting the fact that they do nott own the complete customer journey (Travelspirit Foundation, 2019b), and challenging individual modes of transportation as being credible alternatives, especially in terms of short-distance trips or even in areas not well covered by public transport (Van Audenhove *et al.*, 2018). The IMPs can benefit from a wide range of stakeholders (Corwin, Dinamani & Pankratz 2019) and could be a key solution to urban traffic management and to the integration of different modes of transport, since it simplifies the route planning and guarantees a more efficient journal, while being able to provide highly customer-tailored solutions (Baron, Zinte, Schemken & Uferer, 2018). The integration of traffic and network management is increasing significantly, in order to address people and goods mobility requirements, due to the ability to create interface systems across different modes and, therefore increase the connectivity between people, vehicles and goods – which will be responsible for boosting the overall performance of the system. According to Lyons, Hammond, and Mackay (2019), Durand, Harms, Hoogendoorn-Lanser, and Zijlstra, (2018), and Sochor, Arby, Karlsson, and Sarasini (2018), the operational integration has to do with low interchange penalties and a "seamless" door-to-door journey experience. The level of information integration is concerned with producing an interface for the different modes where the journey planning and execution information is available; and the transactional integration can be defined as a common interface for ensuring booking, ticketing and payments. The currently accepted 5 levels of MaaS's integration are defined according to Durand, Harms, Hoogendoorn-Lanser & Zijlstra (2018), namely: (Level 0) no integration; (Level 1) basic integration (information integration), (Level 2) Limited Integration (i.e. ticketing and payment integration); (Level 3) Partial Integration; (Level 4) Full integration under certain conditions; and (Level 5) Full integration in any situation. As pointed in the following Figure 1, the levels of integration present the roadmap of

Figure 1. MaaS integration levels and cognitive user effort
Source: Adapted (Durand, Harms, Hoogendoorn-Lanser & Zijlstra, 2018, Lyons, Hammond & Mackay, 2019 and Sochor, Arby, Karlsson & Sarasini, 2018).

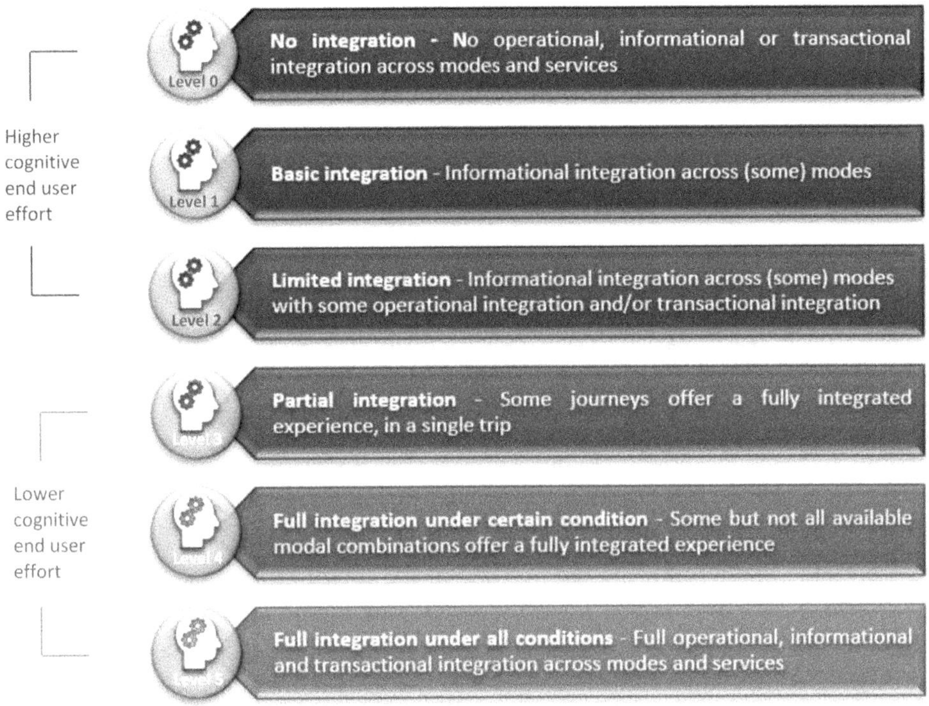

the development of the MaaS paradigm as a whole and include the necessary steps and alignments towards the achievement of the maximum value and maturity of all mobility systems. However, in the future, it could be interesting to present other frameworks detailing the integration between the different stakeholders as well as the inclusion of services and the appearance of new business models.

The customer experience and his personal expectations can shape and contribute to the redefinition of the IMPs features and of the overall mobility system progress. Thereby, the data sharing/access process allows end users to adopt a best fit approach to their particular service and therefore contribute to the system as a whole, to the overall increasement of its efficiency and satisfaction (Cohen, 2018a). The user should be at the center of the mobility system, providing and ensuring the unification of all of its components to design the future mobility system. Simultaneously, the IMPs should guarantee that the cognitive effort applied by the end user effort is minimum, to ensure the user-friendliness of the whole technological ecosystem for mobility as well as working as a counter-measure to combat the technological literacy of some potential users. In this specific manner it is important to point out several possibilities that could be adopted in the future regarding the reduction of the impacts of this issue. Namely, the human-machine interface (HMI) systems could be developed in order to enhance the quality of interaction as well as to maximize the value of the user experience. Other high-tech approaches could be incorporated based on augmented reality or mixed reality, which could be created to support a personal assistant for mobility or even as a hologram/avatar to ensure the easiness and maximum enjoyability of the experience.

A cultural shift is also required to increase the awareness and level of understanding of personal mobility choices and their impacts on and implications for the citizens' quality of life, environment, costs, health & happiness among others. The MaaS concept/paradigm is leading the establishment of mobility milestones, suitably adapted to the requirements of smart cities' mobility patterns and the sustainability of all systems, as well as the full integration and commitment of all the stakeholders involved (Barreto, Amaral & Baltazar, 2019b). It must be noticed, that in several regions, all over the world, including Europe, a low level of technological literacy per inhabitant is denoted– technological illiteracy, thus affects the use and implementation of the smart transportation and MaaS concepts (Baltazar, Barreto and Amaral, 2018) as a present paradigm for the redefinition of the actual mobility systems. In areas where the age of the users is high, the population has low income levels and the number of users is low and is expected to have a low return on investment. Thus, the transport companies must have incentives to operate in those areas. Otherwise, these areas are continuously being ignored and the populations are limited as regards their mobility options (Barreto, Amaral, & Baltazar, 2019a). This cannot happen, since for the MaaS paradigm to attain its maximum potentiality, it presupposes the mobility users' universal access to the different transport options as well as to its associated services.

Another important dimension is associated with the cultural vision that needs to be included, especially in some regions where the tourism activity represents a huge amount of income. Thus, regarding the sustainable development of tourism, MaaS should be able to contribute to a vibrant and creative local community that continuously nurtures, enhances and celebrates its diversity as well as its own cultural heritage and modernity, which strengthens community identity, and which is a growing attraction for visitors (Richins, 2009).

As depicted in Figure 1, the cognitive effort levels evolution regarding the MaaS integration level, where according to Lyons, Hammond & Mackay (2019) the cognitive user effort is the effort involved in relying on the mobility system beyond the private car to fulfil mobility goals. Consumer markets, especially those of the millennials, have moved away from a desire to own cars to being in control of their digitalized lives: leading to a global demand for MaaS – an expected market valued of £1 trillion by 2030 (Travelspirit Foundation, 2019a). The aim is that the end users can use a single technological interface to plan and pay for their trips across all available modes of transport (Corwin, Dinamani & Pankratz 2019) – anticipating travel choices and providing suitable payment mechanisms–, together with a dynamic journey management service – providing real-time information for the end user – and a personalized service – building a relationship between the end user and the MaaS provider (Transport Systems Catapult, 2016). Simultaneously, it is important to start working with young people for the adoption and dissemination of this new mobility reality and of its main implications for the territory sustainability. This awareness-raising work is essential to produce an effective cultural change that stimulates the community to adopt these sustainable mobility practices, as opposed to the current mobility paradigm.

Digitalization will be one of the main drivers for the mobility system progress, in order to achieve a complete and connected mobility system, enabling the creation of new business models and allowing new players to be part of the mobility system as service providers (Van Audenhove *et al.*, 2018). Therefore, public and private partnerships will favor the development of common approaches to standardize, share and protect data for its potential trigger in increasing mobility's impact on the transportation network and encouraging the appearance of innovative solutions (Kimley-Horn & IBI Group, 2019). This broad collaboration – data sharing, whereby the MaaS provider shares data on the mobility needs of end users – will lead to a comprehensive "MaaS Offer", where a diverse mix of operators including automotive firms, bus companies, taxis, bike share and trains will work closely towards helping transport operators to improve their service (Transport Systems Catapult, 2016). Thus, the future MaaS market can be foreseen as fragmented and struggling to reconcile the private automotive world with the world of public transit (Travelspirit Foundation, 2019a). For that reason, without the creation of a shared infrastructure and the adoption of a collaborative approach mindset, the siloed operating systems and data will definitely condition the transformational impacts of multimodal mobility, and therefore the cities will remain tied to mobility models which do not guarantee their full development and sustainability (Cohen, 2018a). Another important issue that needs to be properly tackled is the overall security, namely of the personal data. The amount of data available, the increasing processing power and the almost ubiquitous use of mobile phones have enabled travelers and their needs to be better understood (Transport Systems Catapult, 2016). Hence, with the proliferation and volume of data sources increasing, questions arise as to who will control the end-to-end flow of data and who has the right to use it and for what types of purpose. The proliferation of data also generates questions regarding regulatory concerns, and the required necessity to include smart governance authorities to act according to the best interests. The data security will be of paramount concern throughout the ecosystem and across the entire technology stack, so who will be accountable for any leaks or misuses? Furthermore, how would the involvement of players attempting to resolve these issues on a city/region, country, regional, or global level be managed and centralized? (Heineke, Ménard, Södergren & Wrulich, 2019)

The future mobility will continue to gain in importance in the way urban metabolism is designed and planned. The way cities will be influenced by the concepts of smart, sustainable and resilient will completely alter the living standards in urban and rural places. MaaS is encouraging transportation to universally affordable and inclusive, environmentally friendly, personalized and flexible to deal with the complexities and dynamics inherent in the mobility systems. AI technology will form an intrinsic part of any type of mobility solution. It will be designed to provide a personalized travel experience, easing the interaction between users and APP/technological support systems in an integrated manner. The usage of natural language voice recognition will therefore, help with the decision process of planning the user´s own travel in real-time with all the peculiarities and vicissitudes any daily journey may have, making sure that the users' preferred settings are properly considered in order to select the best mobility options at any time, without conditioning the personal privacy or other important data that might cause any embrace or inappropriate/non-authorized usage. In the cities of the future, the amount of information available will increase immensely. Especially by using technologies like sensor-ingested maps, sending real-time incident data or other road attributes to the vehicles themselves, moving towards breakthroughs in autonomy and connectivity, as well as deploying key information about the cost, time, transport types alternatives, amount of CO_2 discharged, among others. Therefore, it will be extremely important to select the mobility pattern according to the user's personal desire, physical conditions, environmental friendliness, health and age restrictions. In terms of the conditions of the cities of tomorrow, it will be also of considerable importance to work the interconnectivity between cities and rural places as well as between countries. An upper level of MaaS might need to be considered to enhance this necessity and urgency to be able to easily connect foreign countries with different mobility ecosystems in an integrated way (Barreto, Amaral & Baltazar, 2019b).

FROM MaaS REARVIEW MIRROR TOWARDS FUTURE MOBILITY

This section presents future mobility overview, based on the MaaS paradigm, its features and IoM.

The future of urban and regional mobility should be planned and developed as an entire system, where mobility features could be viewed as positive points to link the different contexts. MaaS is not only about providing access to multimodal transportation, customized and digitized mobility, it is also about enabling open collaboration and competition amongst all legal and shared private mobility service providers (Cohen, 2018a), as well as including door-to-door travel with one ticket, last mile integration, failure management, and seamless ancillary services integration (Deon Digital, 2019). MaaS needs to be socially inclusive and permeable to national borders, transport modes, governance types, and other types of boundaries (cognition, cultures, languages, and currencies) (Giesecke, Surakka & Hakonen, 2016), which constitutes an outstanding challenge of cooperation and standardization so as to be able to fulfil that goal. Nevertheless, the time required to attain that purpose and the policy convergence between different countries raises several obstacles and resistance. Therefore, that could be easier to accomplish in a common economic space.

To explore the determinants of MaaS growth, it is necessary to understand the relationships between all stakeholders, technologies and capabilities that are involved in delivering the broad view of MaaS (Transport Systems Catapult, 2016), as well as to enhance a fast adaptation to the different environments, characteristics and features. There is very limited insight about the behavioral impact of MaaS and of its future potential (Lyons, Hammond & Mackay, 2019). Nevertheless, the MaaS system is an example of a systemic innovation, where mobility services are incorporated taking into account customers' transportation needs with real-time traveler information, ticketing, and payment services, among other possibilities (Surakka *et al.*, 2018). Uniformization of MaaS concepts adoption is still an open challenge, as referred to by Sochor, Arby, Karlsson, and Sarasini (2018). Together with the creation of the proper technological infrastructure and cultural shift it ensures the suitable conditions to implement a global MaaS. This approach needs all the requirements and functionalities so as to be established as the future mobility paradigm and to make a decisive contribution to the attainment of the sustainable development. MaaS is still a very new mobility concept, which has the potential to contribute to a more sustainable transport system. Based on MaaS central elements, a future should be defined, regarding the evolution of the society and its desires and characteristics, acquiring a deeper understanding of the key actors' motives, expectations, perceptions and concerns (Polydoropoulou, Pagoni & Tsirimpa, 2018).

Figure 2. MaaS evolution
Source: Adapted from (Cohen, 2018a)

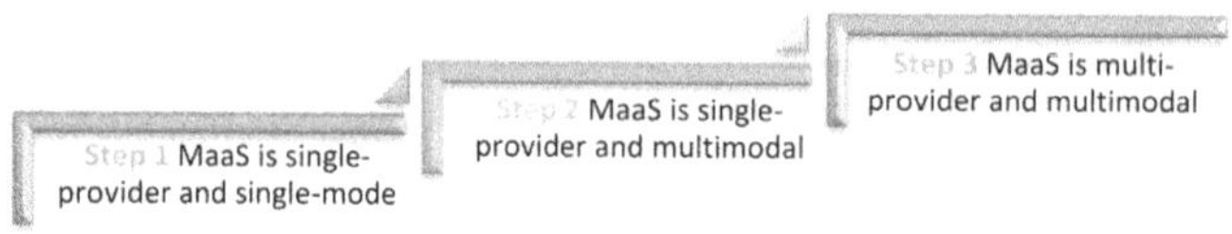

Therefore, there is a temporal dimension that needs to be properly considered in order to be able to achieve the full efficiency of the MaaS paradigm.

According to Cohen (2018a), MaaS requires a developmental roadmap to achieve its full potential, as can be seen in Figure 2. Therefore, throughout the MaaS paradigm development and maturation, a continuous level of integration between different transport modes, a cultural shift among users and the population in general, an increased stakeholder engagement level through the introduction of new services and possibilities as well as the development of a common technological infrastructure which guarantees a full integration among all the mobility ecosystem will be require.

This MaaS paradigm offers a transformative approach towards dealing with the fragmentation problem still plaguing most of the current urban mobility ecosystems (Cohen, 2018a). Nevertheless, there are undeniable trends that support the future MaaS adoption and growth. Consumers are increasingly expecting their mobility experiences to be delivered as a 'service', and to get the highest value as a result (Transport Systems Catapult, 2016). Despite the improvements in urban mobility ecosystems that MaaS has embodied, it still presents significant shortcomings that need to be overcome in order to provide an advancement in MaaS and thus ensure a seamless and interconnected way, that of the IoM truly serving people (Cohen, 2018a). The role of ICT is clearly of an utmost importance for ensuring the necessary integration of all information and its convergence between users, transport providers, and services, in a dynamic and interactive process. The IoT, in the future, could ensure the collection of the necessary information, as well as it processing and continuous availability (Florence School of Regulation, 2015 and Nemtanu *et al.*, 2016).

The rapid technological advances achieved in the last five years have enabled the creation of new transportation tools (Hallock & Inglis, 2015), which have been spreading rapidly across cities/regions. The increased access to the Internet, the significant growth of the smartphone's proliferation, the usage of social media and the shared economy are revolutionizing the world of transportation by creating new possibilities for getting around that are accessible, convenient and affordable (Hallock & Inglis, 2015). Thus, the IoM concept was born as an open and secure network, exploring new revenue streams through seamless collaborations (Deon Digital, 2019) and by transforming future mobility to contribute to a superior end

Figure 3. Open architecture built by IoMob
Source: (Cohen, 2018b).

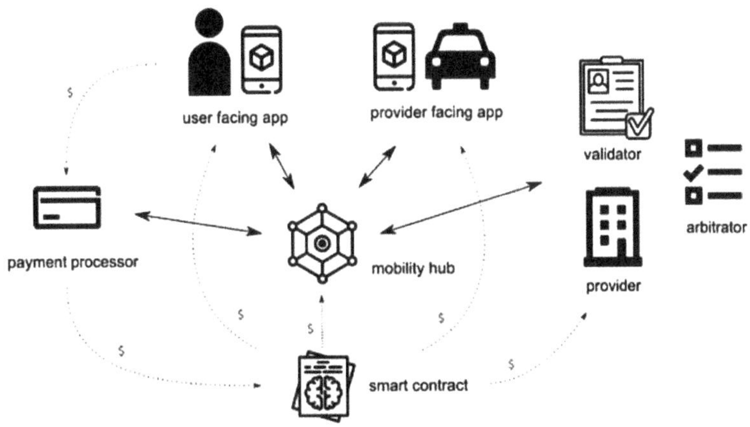

user experience – meeting their needs in a wider array of mobility services (Cohen, 2018b), beyond the optimization of public transit, together with a more inclusive access to the mobility system for other stakeholders like startups, small operators and even independent ones (Cohen, 2019).

In the sense of being able to provide individualized customer-centric services universally available to everyone (Cohen, 2018b), the IoM can be seen as a smart connection between MaaS and Connected & Autonomous Vehicles (CAV) – new transport modes, software and information streams – enabling the benefits of both, according to an open source architecture and blockchain protocols (Travelspirit Foundation, 2019a). According to Figure 3, an important point addressed in IoM is ensuring a valid mobility service, allowing any user the ability to use it at his or her personal convenience, contributing to the overall resilience of the mobility ecosystem (Cohen, 2018b).

At a higher level of the MaaS maturity ladder, according to Figure 4, the IoM incorporation into MaaS is disclosed. Therefore, in times to come, any city/region will need to be connected to a broader mobility operating system – an integrated global platform that brings together physical infrastructure (roads, railways, and so on), modes of transportation (cars, public transit, ridesharing, bikesharing, and so on), and service providers (aggregators, PTS) which could create greater throughput, as well as being able to optimize the system through market clearing mechanisms (Corwin, Dinamani & Pankratz 2019). The evolution of MaaS through the adoption of IoM could speed up the level of integration of the multi-service providers and of the multimodal transport's possibilities, as well as its geographical coverage so that it is not only concentrated in big cities or even in developed countries.

Figure 4. From MaaS to IoM evolution
Source: Adapted from (Cohen, 2018a).

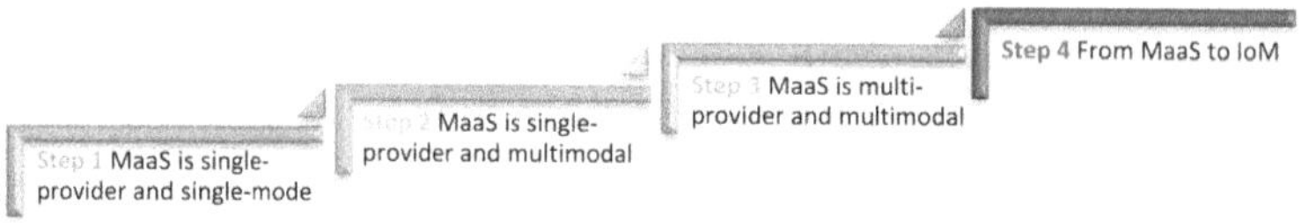

A broad change needs to be performed in mobility in general to be fulfilled towards rapidly respond to the societal, environmental, economic and technological challenges predicted in the sustainable development agenda of 2030 (UN, 2016). Thus, the integration of all the aforementioned key mobility components is needed and, in some cases, is already in progress. The challenge lies in developing adequate strategies and aligning them to the different contexts and realities in a dynamic way.

In order to properly respond to the future mobility challenges, cities' local government bodies and mobility solutions providers must first adopt a thorough and transparent analysis of the real dimension of the mobility challenge in the context in which they are inserted, so as to be fully aware and to make a concerted effort to develop management strategies and proper mechanisms to attain the desired goals and develop proactive approaches to mobility demand management in order to better influence behaviors in space and time. It is almost a common view that the mobility systems of tomorrow will be intermodal, personalized, convenient, and connected, and encourage the usage of more sustainable modes of transport (PT, cycling, walking) while integrating new mobility solutions and services (Van Audenhove *et al.*, 2018). Thus, the involvement from operators is critical to ensure the best technical solutions, resulting from new mobility ecosystems that could offer a faster, cheaper, cleaner, safer, more efficient, and more customized travel or service (Corwin & Pankratz, 2017). Notwithstanding, there are mature cities that do not yet have clear visions of what their mobility systems should look like in the future, and do not have coherent strategies for attaining it (Van Audenhove *et al.*, 2018). A set of strategies must be developed to answer key features and attributes of future mobility based on MaaS and IoM adoption, as disclosed in Figure 5.

The promotion of intermodality will require the expansion and availability of shared-use transportation modes and other technology-enabled tools (Hallock & Inglis, 2015) which will enhance the role of mobility coordination. Global coordination is a principal attribute, especially with the increasing integration of stakeholders in the mobility ecosystem over time. Therefore, it should be properly handled and managed otherwise the lack of coordination could make the problem worse instead of helping (Corwin, Dinamani & Pankratz 2019).

Figure 5. Future mobility key features

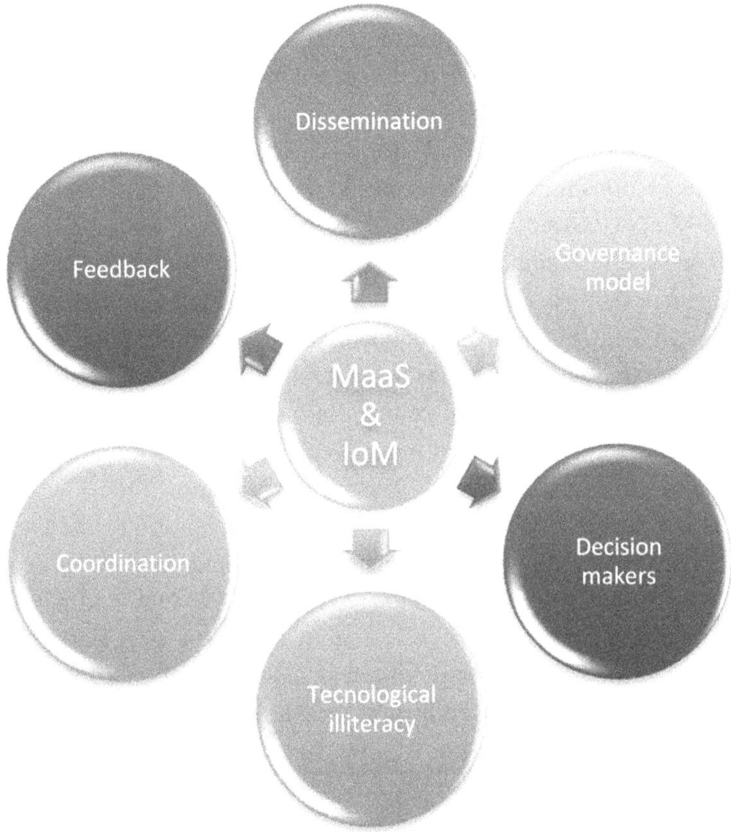

A sustainable mobility index should be created and monitored, based on sustainability indicators, promoting a unified global monitoring and creating clear benchmarking roadmaps to achieve the establishment of sustainable mobility policies (Baltazar, Amaral & Barreto, 2018). Together with a balanced and inclusive governance model which proclaims the proper decision-making influence of each stakeholder ensuring the right equilibrium of the mobility ecosystem (Travelspirit Foundation, 2019a).

The inclusion of new connected services into the mobility context can promote cooperative work among all players in order to perform a clear definition of the benefits expected for users, third-party providers, and the city governance (Corwin, Dinamani & Pankratz 2019). Such a series of recommendations and an implementation plan to manage stakeholders will allow them to discuss the goals and objectives as well as to voice specific concerns and research interests (Kimley-Horn & IBI Group, 2019) which can also contribute to densify the dissemination process.

The technological dimension of the binomial MaaS/IoM, as it was previously mentioned, is critical to obtain the overall benefits of the mobility ecosystem. Nevertheless, an important dimension must be considered which is the technological illiteracy felt by the elderly and the disabled people which, perhaps, could be diminished by introducing personal assistance avatars in a "augmented reality" approach to speed up the universal usability of the system, as well as providing important suggestions for other services and realities that could work jointly in the adoption of MaaS (Barreto, Amaral, & Baltazar, 2019a).

Nevertheless, the technological breakthroughs like the advent of autonomous self-driving vehicles will, without any doubt, favour the appearance of new business models through the development of innovative services in the mobility ecosystem. Therefore, some trade-offs will have to be determined based on the appearance of this new form of mobility alternative that claims lower costs per kilometer, lower accident rates, less congestion and more urban space, which can be gained from other active transport modes (Webb, Wilson, & Kularatne, 2018).

The development of new technologies for supporting mobility, the establishment of commercial partnerships and the creation of trusted data systems is vital for realizing both the pro-social and economic potential of MaaS and the adoption of the CAV, creating a more resilient, sustainable, and connected world (Travelspirit Foundation, 2019a). The experts point out that MaaS will gain global scalability with the emergence of CAV, which the forecasts project at being around 2030's (Travelspirit Foundation, 2019a).

DISCUSSION

In this section, a discussion of ideas and possible future scenarios is presented, leading us to possible routes and allowing us to embrace the different ways of constructing a future mobility and opening the door to new options. Some barriers are also presented in this discussion.

The mobility evolution and its future trends will have major impacts on any city/region's organization and planning and also on the social inclusion and the population's well-being. The impact of all the above-mentioned initiatives should be national – urban, suburban and rural – and, moreover, international so that its impact could be well known. To achieve an integrated mobility system, the support of all the stakeholders, business, the public sector and individuals as well as the major companies and the researchers, is fundamental (Barreto, Amaral, & Baltazar, 2019a).

The city/region or countries have been experiencing different mobility demands and it is imperative to learn from other countries the good practices they have implemented and how their plans resulted in success. The result of this endeavor could be an increased level of awareness of the strengths and opportunities that should be kept in sight to ensure the overall operationality of the mobility system and its security level so as to minimize the systems' vulnerabilities as well as their vulnerability to cyberattacks, reducing the overall effects and impacts (Pangbourne, Stead, Mladenovic & Milakis, 2018).

An important output and functionality of MaaS' IMP will be to guarantee the easiness of booking, the ticketing system, travel selection and payment methods within a single customer order record and process, holding all data elements obtained throughout a specific time frame and detecting the best billing conditions based on the history of transport usage (Travelspirit Foundation, 2019b). An integrated perspective together with the knowledge of diverse consulting/interviews and studies done by the few leaders in the MaaS field, which provide knowledge and diverse points of view, with the aim of a global sustainable and inclusive mobility in all its domains, e.g. MaaS4EU (MaaS4EU, 2019).

Without a doubt, the IMP will play a vital role into the future mobility design, and it will enhance several potentialities with the inclusion of autonomous vehicles which will be an undeniable reality in a non-distant future.

As the mobility system matures, cities and the private sector could face a broad array of consequences and implications. Cities themselves could become mobility platforms upon which entrepreneurs and innovators can build new services and applications, ensuring a good possibility of retaining their talented citizens through the creation of this new reality (Corwin, Dinamani & Pankratz 2019). Simultaneously, new opportunities will appear along with the full development of the Maas/IoM paradigm integration that could be reinvented throughout the development of new policies and services. This will assure a continuous redefinition of the mobility ecosystem in order to guarantee its monitoring in a dynamic way.

Based on the full integration of MaaS and IoM concepts, it is possible to predict the (r)evolutionary appearance of the Internet of Mobility as a Service – IoMaaS. This may be based on actions such as those explored by Smith, Sochor and Karlsson (2019), where an attempt to establish a national Intermediary MaaS Integrator, in Sweden, is fully evaluated. The Intermediary MaaS Integrators can be defined as the intermediate actors that collect the financial gains from Transport Service Providers and distribute them amongst MaaS Operators in order to facilitate their development, operation and management (Smith, Sochor & Karlsson, 2019).

Other social demographic variables like the different needs of gender or age perspectives could be taken into consideration when designing MaaS in terms of its easiness and safeness in public spaces (Melis, Prandini, Sartori & Callegati, 2016), and reshaping its redesigning in a concrete and smart manner.

A broad plan needs to be fully developed in order to meet the need to integrate all populations into the digitalization era, including those who have less technological/informatic knowledge, in other words technological illiteracy. The virtual mobility assistant (VMA) could have access to users' contacts, agenda, and location, as well as information about weather, traffic, and mass transit, adjusting when to alert user's upcoming appointment (Bodart & Condon, 2007). The VMA could also provide real-time instructions to handicapped people to follow the trip status. It could also suggest options to complete the trip with complementary services.

There are several limitations that could create barriers and difficulties in terms of the MaaS paradigm adoption and in the attainment of its full potential, namely: the entrance of new mobility actors with little know-how about mobility in general, as well as an unclear definition of their roles. Therefore, the importance of establishing a clear set of overarching goals and a commitment to act in ways which will enable innovation to flourish but which simultaneously ensures the enhancement of public value is essential (Docherty, Marsden & Anable, 2018). Moreover, there is a substantial level of uncertainty related with the MaaS adoption and with its capacity to endure that, could somehow restrict or delay its implementation on a large scale, such as the required collaboration among actors (due to uncertainties about the right business models) and to customers' acceptance of MaaS (Jittrapirom, Marchau, van der Heijden & Meurs, 2018). In the European Union, there is still a low level of integration among multimodal transport systems, as well as the inexistence of integrated policies, regarding complementary training within the scope of a technological literacy context. This situation is specially felt in regions where soft modes are less used which could somehow be connected with cultural factors. The rhythm of technological sophistication and of customers' demand for viable solutions will only be possible with the proper balance between stimuli and penalties to reduce the current externalities of transport in general (Barreto, Amaral, & Baltazar, 2019a). Communication between the public sector and consumer can be a weakness of MaaS technology (Eckhardt, Nykänen, Aapaoja & Niemi, 2018). Nevertheless, the barriers to an open MaaS system are not limited to technology but also lie in the creation of standards, regulations, business models and in the political environment in order to be able to positively regulate and integrate new forms of transport, which could ensure the interoperability between modes, and providers to enable people to make their journeys and swap easily between different modes (Travelspirit Foundation, 2019b).

CONCLUSION AND FUTURE WORK

In this digitalization era, MaaS is a complex and multidimensional recent concept. Cities/regions and public authorities, private innovators, transport operators, investigators, users, in other words, all the actors, need to know and understand the different perspectives and collaborate full.

Another important point regarding the mobility system improvement is the need to share experiences from diverse geographical contexts allowing a collaborative approach and classifying the mobility system evolution levels as well as setting the standard by example. Therefore, taking advantage of the current MaaS development stage and integrating the potentialities of IoM an IoMaaS or a similar future mobility solution can be achieved that guarantees the best quality mobility service, which can be accessed and used by every citizen in all regions.

The future is always uncertain and brings, to a certain extent, new possibilities that are difficult to detect and whose impacts are difficult to estimate in the present. Nevertheless, it is clear that some future developments should be pursued to guarantee the adoption of MaaS supported by the IoM. There are some main ideas to follow up in the near future to increase the robustness of the MaaS paradigm, especially through:

- The full integration among all regional and urban mobility systems. This aspect will be of utmost importance and will limit the adoption of some of the transport possibilities;
- The determination of services' monetization, ensuring that the benefits are equally distributed among all the MaaS IoM' members and, of course, among all the different stakeholders, as well as the trade-offs being clear among all the members;
- Developing the best approaches to combat the technological illiteracy some mobility system members could have through the inclusion of virtual/augmented reality personal assistants, to guarantee the universal usability of the IoM system;
- Reassess the main implications, barriers and good practices achieved during the steps to achieve the MaaS maturity.
- Defining a MaaS roadmap and a work methodology, including focus groups, workshops, interviews, surveys and meeting with transport authorities and investigators to validate and define future strategies.

ACKNOWLEDGMENT

This research was supported by FEDER (Fundo Europeu de Desenvolvimento Regional) through Programa Operacional Competitividade e Internacionalização, in the scope of the Project "ALTO MINHO. SMOB - Mobilidade Sustentável para o Alto Minho", Ref. POCI-01-0145-FEDER-024043.

The authors are very grateful to the reviewers for their insightful comments that greatly contributed to improve this work.

REFERENCES

Baldwin, M. (2018). Keynote speech on Year of Multimodality. Brussels: Academic Press.

Baltazar, S., Amaral, A., & Barreto, L. (2018). Happiness, Prosperity and Quality of Life: A Relation Towards Future Mobility. In *4th International Conference on Production Economics and Project Evaluation (ICOPEV 2018)*, (pp. 93-98). Academic Press.

Baltazar, S., Barreto, L., & Amaral, A. (2018). Future Mobility in Alto Minho Region - Towards the path of sustainability. In *2018 IEEE International Conference on on Engineering, Technology and Innovation (ICE/ITMC)*, (pp. 361-368). IEEE.

Baron, R., Zinte, M., Schemken, N., & Uferer, C. (2018). *Integrated Mobility Platforms: How to win the customer in tomorrow's world of smart mobility*. Retrieved from https://www.adlittle.com/sites/default/files/viewpoints/adl_integrated_mobility_platforms.compressed.pdf

Barreto, L., Amaral, A., & Baltazar, S. (2019a). Flexible on Demand Transport Services (FDTS): The future of mobility systems. In *25th ICE/IEEE International Technology Management Conference (ITMC)*. Nice, France: IEEE. 10.1109/ICE.2019.8792649

Barreto, L., Amaral, A., & Baltazar, S. (2019b). Urban Mobility Digitalization: Towards Mobility as a Service (MaaS). *9th International Conference on Intelligent Systems 2018: Theory, Research and Innovation in Applications, IS 2018 - Proceedings*, 850–855.

Bodart, A. J., & Condon, C. E. (2007). 11/206,829. Academic Press.

Cohen, B. (2018a). *Future transport: From MaaS to the Internet of Mobility.* Retrieved July 24, 2019, from https://www.smartcitiesworld.net/opinions/opinions/future-transport-from-maas-to-the-internet-of-mobility?source=post_page

Cohen, B. (2018b). *How the Internet of Mobility can make Urban Mobility More Resilient.* Retrieved July 24, 2019, from https://medium.com/iomob/how-the-internet-of-mobility-can-make-urban-mobility-more-resilient-5697194ff7bc

Cohen, B. (2019). *How the Internet of Mobility will Transform Mobility Ecosystems.* Retrieved July 24, 2019, from https://medium.com/iomob/how-the-internet-of-mobility-will-transform-mobility-ecosystems-6c83c15e1553

Corwin, S., Dinamani, A., & Pankratz, D. (2019). *Toward a mobility operating system: Establishing a lingua franca for urban transportation.* Deloitte Center for Integrated Research.

Corwin, S., & Pankratz, D. M. (2017). *Forces of change: The future of mobility.* Deloitte Insights. Retrieved August 01, 2019, from https://www2.deloitte.com/content/dam/insights/us/articles/4328_Forces-of-change_FoM/DI_Forces-of-change_FoM.pdf

Deon Digital. (2019). *Internet of Mobility.* Retrieved July 24, 2019, from https://www.deondigital.com/mobility/

Diewald, S., Möller, A., Roalter, L., & Kranz, M. (2012). MobiliNet: A Social Network for Optimized Mobility. In *4th International Conference on Automotive User Interfaces and Interactive Vehicular Applications (AutomotiveUI 2012)* (pp. 145–150). Academic Press.

Docherty, I., Marsden, G., & Anable, J. (2018). The governance of smart mobility. *Transportation Research Part A, Policy and Practice, 115*, 114–125. doi:10.1016/j.tra.2017.09.012

Durand, A., Harms, L., Hoogendoorn-Lanser, S., & Zijlstra, T. (2018). *Mobility-as-a-Service and changes in travel preferences and travel behaviour: a literature review.* Ministry of Infrastructure and Water Management. KiM Netherlands Institute for Transport Policy Analysis.

Eckhardt, J., Aapaoja, A., Nykänen, L., & Sochor, J. (2017). *Mobility as a Service business and operator models. In 12th ITS European Congress on iItelligent Transportation Systems* (pp. 19–22). Retrieved from https://www.researchgate.net/publication/316243907%0AMobility

Eckhardt, J., Nykänen, L., Aapaoja, A., & Niemi, P. (2018, Aug.). MaaS in rural areas - case Finland. *Research in Transportation Business & Management*, 1–9. doi:10.1016/j.rtbm.2018.09.005

Giesecke, R., Surakka, T., & Hakonen, M. (2016). Conceptualising Mobility as a Service - A User Centric View on Key Issues of Mobility Services. *11th International Conference on Ecological Vehicles and Renewable Energies, EVER 2016*. 10.1109/EVER.2016.7476443

Hallock, L., & Inglis, J. (2015). *The Innovative Transportation Index: The Cities Where New Technologies and Tools Can Reduce Your Need to Own a Car*. Academic Press.

Heineke, K., Ménard, A., Södergren, F., & Wrulich, M. (2019). *Development in the mobility technology ecosystem — how can 5G help?* McKinsey & Company.

Jittrapirom, P., Marchau, V., van der Heijden, R., & Meurs, H. (2018). *Future implementation of mobility as a service (MaaS): Results of an international Delphi study*. Travel Behaviour and Society.

Kimley-Horn & IBI Group. (2019). *DRPT Statewide Integrated Mobility Initiative*. Retrieved from http://www.censusindia.gov.in/2011census/PCA/PCA_Highlights/pca_highlights_file/Tamil_Nadu/6.Figures_Glance_Tamil Nadu.pdf

Li, Y., & Voege, T. (2017). Mobility as a Service (MaaS): Challenges of Implementation and Policy Required. *Journal of Transportation Technologies*, *7*(02), 95–106. doi:10.4236/jtts.2017.72007

Litman, T. (2017). *Public Transportation's Impact on Rural and Small Towns: A Vital Mobility Link*. Community Transportation Association of America.

Lyons, G., Hammond, P., & Mackay, K. (2019). The importance of user perspective in the evolution of MaaS. *Transportation Research Part A, Policy and Practice*, *121*, 22–36. doi:10.1016/j.tra.2018.12.010

MaaS4EU. (2019) *End-to-End Approach for Mobility-as-a-Service tools, business models, enabling framework and evidence for European seamless mobility*. Retrieved from http://www.maas4eu.eu/

Melis, A., Prandini, M., Sartori, L., & Callegati, F. (2016). Public transportation, IoT, Trust and Urban habits. In *International Conference on Internet Science* (pp. 318–325). Springer. 10.1007/978-3-319-45982-0_27

Pangbourne, K., Stead, D., Mladenovic, M., & Milakis, D. (2018). The case of Mobility as a Service: a critical reflection on challenges for urban transport and mobility governance. In *Governance of the smart mobility transition* (pp. 33–48). Emerald Publishing Limited. doi:10.1108/978-1-78754-317-120181003

Polydoropoulou, A., Pagoni, I., & Tsirimpa, A. (2018). *Ready for Mobility as a Service? Insights from stakeholders and end-users*. Travel Behaviour and Society; doi:10.1016/j.tbs.2018.11.003

Richins, H. (2009). Environmental, cultural, economic and socio-community sustainability: A framework for sustainable tourism in resort destinations. *Environment, Development and Sustainability*, *11*(4), 785–800. doi:10.100710668-008-9143-6

Saunders, M., Lewis, P., & Thornhill, A. (2016). *Research methods for business students, (no. Book, Whole)*. Harlow, UK: Pearson Education Limited.

Shaheen, S., Cohen, A., Yelchuru, B., & Sarkhili, S. (2017). *Mobility on Demand*. Report No. FHWA-JPO-18-611, U.S. Department of Transportation.

Smith, G., Sochor, J., & Karlsson, I. C. M. (2019). Intermediary MaaS Integrators : A case study on hopes and fears. *Transportation Research Part A, Policy and Practice*. doi:10.1016/j.tra.2019.09.024

Sochor, J., Arby, H., Karlsson, I. C. M. A., & Sarasini, S. (2018). A topological approach to Mobility as a Service: A proposed tool for understanding requirements and effects, and for aiding the integration of societal goals. *Research in Transportation Business & Management*, *27*(March), 3–14. doi:10.1016/j.rtbm.2018.12.003

Surakka, T., Härri, F., Haahtela, T., Horila, A., & Michl, T. (2018). Regulation and governance supporting systemic MaaS innovations. *Research in Transportation Business & Management*, *27*, 56–66. doi:10.1016/j.rtbm.2018.12.001

Transport Systems Catapult. (2016). *Mobility As a Service: Exploring the Opportunity for Mobility as a Service in the UK*. Retrieved from https://ts.catapult.org.uk/wp-content/uploads/2016/07/Mobility-as-a-Service_Exploring-the-Opportunity-for-MaaS-in-the-UK-Web.pdf?fbclid=IwAR2lTnX2KEKwp5Fd0xjcsMpr7rlDD-oQtHCcVXwxhc2cBQigL8_qoQ1e5t4

Travelspirit Foundation. (2019a). *Internet of Mobility*. Retrieved from doi:10.1177/002205740105302515

Travelspirit Foundation. (2019b). *An Open Future for Cities*. Author.

Van Audenhove, F.-J. A., Bettati, A., Smith, A., Rominger, G., Steylemans, N., Haon, S., . . . Zintel, M. (2018). The Future of Mobility 3.0: Reinventing mobility in the era of disruption and creativity. *Arthur D. Little*. Retrieved from http://www.adlittle.com/futuremobilitylab/assets/file/180330_Arthur_D.Little_&_UITP_Future_of_Mobility_3_study.pdf

Webb, J., Wilson, C., & Kularatne, T. (2018). Will people accept shared autonomous electric vehicles? A survey before and after receipt of the costs and benefits. *Economic Analysis and Policy*.

KEY TERMS AND DEFINITIONS

Future Mobility: An integrated and multimodal mobility, using technology and data to learn about and suggest mobility solutions/services in a friendly and universal environment, thereby contributing to a superior user experience and also to a sustainable and inclusive mobility.

Integrated Mobility: A connection between all transport modes, allowing the user to shift between modes and create new routes according to his/her habits and preferences.

Internet of Mobility (IoM): An open and secure network connected to promote mobility, meeting users' needs and optimizing public transit, together with a more inclusive access to the mobility system.

Internet of Mobility as a Service (IoMaaS): The combination of IoM and MaaS, bringing together the benefits of both – creating a mobility service using internet connections (ubiquitous) and providing also other additional and complementary services.

Internet of Things (IoT): A system connecting any device to the Internet, with the seamless ability to transfer data over a network without requiring human-to-human or human-to-computer interaction.

Mobility as a Service (MaaS): A mobility management and distribution system which aims to integrate all transport modes (public and private) in a single platform, allowing users to seamlessly plan and pay for their mobility.

Smart Mobility: A mobility system/services supported by the use of technology, smartphones and data, creating connections between people, places and goods with all transport modes and learning about the users' habits.

Related Readings

To continue IGI Global's long-standing tradition of advancing innovation through emerging research, please find below a compiled list of recommended IGI Global book chapters and journal articles in the areas of green cities, environmental management, and sustainable urban development. These related readings will provide additional information and guidance to further enrich your knowledge and assist you with your own research.

Ahuja, K., & Khosla, A. (2019). A Framework to Develop a Zero-Carbon Emission Sustainable Cognitive City. In K. Ahuja & A. Khosla (Eds.), *Driving the Development, Management, and Sustainability of Cognitive Cities* (pp. 1–26). Hershey, PA: IGI Global. doi:10.4018/978-1-5225-8085-0.ch001

Akkucuk, U. (2016). SCOR Model and the Green Supply Chain. In U. Akkucuk (Ed.), *Handbook of Research on Waste Management Techniques for Sustainability* (pp. 108–124). Hershey, PA: IGI Global. doi:10.4018/978-1-4666-9723-2.ch006

Albuquerque, C. M. (2017). Cities Really Smart and Inclusive: Possibilities and Limits for Social Inclusion and Participation. In L. Carvalho (Ed.), *Handbook of Research on Entrepreneurial Development and Innovation Within Smart Cities* (pp. 229–247). Hershey, PA: IGI Global. doi:10.4018/978-1-5225-1978-2.ch011

Angelucci, F., Di Girolamo, C., & Zazzero, E. (2018). New Designing Codes for Urban Infrastructures: A Hypothesis of a Transdisciplinary Approach. In G. Carlone, N. Martinelli, & F. Rotondo (Eds.), *Designing Grid Cities for Optimized Urban Development and Planning* (pp. 209–237). Hershey, PA: IGI Global. doi:10.4018/978-1-5225-3613-0.ch012

Avdimiotis, S., & Tilikidou, I. (2017). Smart Tourism Development: The Case of Halkidiki. In L. Carvalho (Ed.), *Handbook of Research on Entrepreneurial Development and Innovation Within Smart Cities* (pp. 491–513). Hershey, PA: IGI Global. doi:10.4018/978-1-5225-1978-2.ch021

Ayvaz, B., & Görener, A. (2016). Reverse Logistics in the Electronics Waste Industry. In U. Akkucuk (Ed.), *Handbook of Research on Waste Management Techniques for Sustainability* (pp. 155–171). Hershey, PA: IGI Global. doi:10.4018/978-1-4666-9723-2.ch008

Bagnato, V. P. (2018). Technology and Urban Structure: The Grid City Between Technological Innovation and New Public Space System. In G. Carlone, N. Martinelli, & F. Rotondo (Eds.), *Designing Grid Cities for Optimized Urban Development and Planning* (pp. 238–253). Hershey, PA: IGI Global. doi:10.4018/978-1-5225-3613-0.ch013

Bala, I., & Singh, G. (2019). Green Communication for Cognitive Cities. In K. Ahuja & A. Khosla (Eds.), *Driving the Development, Management, and Sustainability of Cognitive Cities* (pp. 87–110). Hershey, PA: IGI Global. doi:10.4018/978-1-5225-8085-0.ch004

Bernardino, S., & Santos, J. F. (2017). Building Smarter Cities through Social Entrepreneurship. In L. Carvalho (Ed.), *Handbook of Research on Entrepreneurial Development and Innovation Within Smart Cities* (pp. 327–362). Hershey, PA: IGI Global. doi:10.4018/978-1-5225-1978-2.ch015

Bernardo, M. D. (2017). Smart City Governance: From E-Government to Smart Governance. In L. Carvalho (Ed.), *Handbook of Research on Entrepreneurial Development and Innovation Within Smart Cities* (pp. 290–326). Hershey, PA: IGI Global. doi:10.4018/978-1-5225-1978-2.ch014

Bhatia, P., & Singh, P. (2019). Technological and Gamified Solutions for Pollution Control in Cognitive Cities. In K. Ahuja & A. Khosla (Eds.), *Driving the Development, Management, and Sustainability of Cognitive Cities* (pp. 234–249). Hershey, PA: IGI Global. doi:10.4018/978-1-5225-8085-0.ch010

Bhatt, R. (2017). Zero Tillage for Mitigating Global Warming Consequences and Improving Livelihoods in South Asia. In W. Ganpat & W. Isaac (Eds.), *Environmental Sustainability and Climate Change Adaptation Strategies* (pp. 126–161). Hershey, PA: IGI Global. doi:10.4018/978-1-5225-1607-1.ch005

Bolívar, M. P. (2018). Governance in Smart Cities: A Comparison of Practitioners' Perceptions and Prior Research. *International Journal of E-Planning Research, 7*(2), 1–19. doi:10.4018/IJEPR.2018040101

Bostanci, S. H., & Albayrak, A. N. (2017). The Role of Eco-Municipalities in Climate Change for a Sustainable Future. In W. Ganpat & W. Isaac (Eds.), *Environmental Sustainability and Climate Change Adaptation Strategies* (pp. 213–231). Hershey, PA: IGI Global. doi:10.4018/978-1-5225-1607-1.ch008

Brkljačić, T., Majetić, F., & Tarabić, B. N. (2017). Smart Environment: Cyber Parks (Connecting Nature and Technology). In L. Carvalho (Ed.), *Handbook of Research on Entrepreneurial Development and Innovation Within Smart Cities* (pp. 150–172). Hershey, PA: IGI Global. doi:10.4018/978-1-5225-1978-2.ch008

Cai, Y. Z., Wu, F., Li, J., Wang, J., & Huang, M. (2019). Application of UAV Technology to Planning Study on Chinese Villages in Guanzhong. In J. Vargas-Hernández & J. Zdunek-Wielgołaska (Eds.), *Bioeconomical Solutions and Investments in Sustainable City Development* (pp. 180–195). Hershey, PA: IGI Global. doi:10.4018/978-1-5225-7958-8.ch008

Calisto, M. D., & Gonçalves, A. (2017). Smart Citizens, Wise Decisions: Sustainability-Driven Tourism Entrepreneurs. In L. Carvalho (Ed.), *Handbook of Research on Entrepreneurial Development and Innovation Within Smart Cities* (pp. 20–43). Hershey, PA: IGI Global. doi:10.4018/978-1-5225-1978-2.ch002

Capelo, C. (2017). Exploring the Dynamics of an Energy Service Venture. In L. Carvalho (Ed.), *Handbook of Research on Entrepreneurial Development and Innovation Within Smart Cities* (pp. 269–289). Hershey, PA: IGI Global. doi:10.4018/978-1-5225-1978-2.ch013

Caridi, G. (2018). Calabria 1783: The Orthogonal Grid as a Physical and Ideological Device of Reconstruction. In G. Carlone, N. Martinelli, & F. Rotondo (Eds.), *Designing Grid Cities for Optimized Urban Development and Planning* (pp. 176–187). Hershey, PA: IGI Global. doi:10.4018/978-1-5225-3613-0.ch010

Carlone, G. (2018). Cities and Extension Plans in the Kingdom of the Two Sicilies: Borgo Murattiano of Bari (1812-1859). In G. Carlone, N. Martinelli, & F. Rotondo (Eds.), *Designing Grid Cities for Optimized Urban Development and Planning* (pp. 1–18). Hershey, PA: IGI Global. doi:10.4018/978-1-5225-3613-0.ch001

Carvalho, L. C. (2017). Entrepreneurial Ecosystems: Lisbon as a Smart Start-Up City. In L. Carvalho (Ed.), *Handbook of Research on Entrepreneurial Development and Innovation Within Smart Cities* (pp. 1–19). Hershey, PA: IGI Global. doi:10.4018/978-1-5225-1978-2.ch001

Castagnolo, V. (2018). Analyzing, Classifying, Safeguarding: Drawing for the Borgo Murattiano Neighbourhood of Bari. In G. Carlone, N. Martinelli, & F. Rotondo (Eds.), *Designing Grid Cities for Optimized Urban Development and Planning* (pp. 93–108). Hershey, PA: IGI Global. doi:10.4018/978-1-5225-3613-0.ch006

Castilla-Polo, F., Gallardo-Vázquez, D., Sánchez-Hernández, M. I., & Ruiz-Rodríguez, M. D. (2017). Cooperatives as Responsible and Innovative Entrepreneurial Ecosystems in Smart Territories: The Olive Oil Industry in the South of Spain. In L. Carvalho (Ed.), *Handbook of Research on Entrepreneurial Development and Innovation Within Smart Cities* (pp. 459–490). Hershey, PA: IGI Global. doi:10.4018/978-1-5225-1978-2.ch020

Charalabidis, Y., & Theocharopoulou, C. (2019). A Participative Method for Prioritizing Smart City Interventions in Medium-Sized Municipalities. *International Journal of Public Administration in the Digital Age*, *6*(1), 41–63. doi:10.4018/IJPADA.2019010103

Chatsiwa, J., Mujere, N., & Maiyana, A. B. (2016). Municipal Solid Waste Management. In U. Akkucuk (Ed.), *Handbook of Research on Waste Management Techniques for Sustainability* (pp. 19–43). Hershey, PA: IGI Global. doi:10.4018/978-1-4666-9723-2.ch002

Chawla, R., Singhal, P., & Garg, A. K. (2019). Impact of Dust for Solar PV in Indian Scenario: Experimental Analysis. In K. Ahuja & A. Khosla (Eds.), *Driving the Development, Management, and Sustainability of Cognitive Cities* (pp. 111–138). Hershey, PA: IGI Global. doi:10.4018/978-1-5225-8085-0.ch005

Cılız, N., Yıldırım, H., & Temizel, Ş. (2016). Structure Development for Effective Medical Waste and Hazardous Waste Management System. In U. Akkucuk (Ed.), *Handbook of Research on Waste Management Techniques for Sustainability* (pp. 303–327). Hershey, PA: IGI Global. doi:10.4018/978-1-4666-9723-2.ch016

Collignon de Alba, C., Haberleithner, J., & López, M. M. (2017). Creative Industries in the Smart City: Overview of a Liability in Emerging Economies. In L. Carvalho (Ed.), *Handbook of Research on Entrepreneurial Development and Innovation Within Smart Cities* (pp. 107–126). Hershey, PA: IGI Global. doi:10.4018/978-1-5225-1978-2.ch006

Corum, A. (2016). Remanufacturing, an Added Value Product Recovery Strategy. In U. Akkucuk (Ed.), *Handbook of Research on Waste Management Techniques for Sustainability* (pp. 347–367). Hershey, PA: IGI Global. doi:10.4018/978-1-4666-9723-2.ch018

Crosas, C. (2018). Latin American Cities: Modern Grids From 1850s. In G. Carlone, N. Martinelli, & F. Rotondo (Eds.), *Designing Grid Cities for Optimized Urban Development and Planning* (pp. 39–51). Hershey, PA: IGI Global. doi:10.4018/978-1-5225-3613-0.ch003

D'Onofrio, S., Habenstein, A., & Portmann, E. (2019). Ontological Design for Cognitive Cities: The New Principle for Future Urban Management. In K. Ahuja & A. Khosla (Eds.), *Driving the Development, Management, and Sustainability of Cognitive Cities* (pp. 183–211). Hershey, PA: IGI Global. doi:10.4018/978-1-5225-8085-0.ch008

Dalzero, S. (2018). The Time of the Finished World Has Begun: A New Map of the World – National Borders Partially or Fully Fenced-Off. In G. Carlone, N. Martinelli, & F. Rotondo (Eds.), *Designing Grid Cities for Optimized Urban Development and Planning* (pp. 254–275). Hershey, PA: IGI Global. doi:10.4018/978-1-5225-3613-0.ch014

Damurski, L. (2016). Recent Progress in Online Communication Tools for Urban Planning: A Comparative Study of Polish and German Municipalities. *International Journal of E-Planning Research*, *5*(1), 39–54. doi:10.4018/IJEPR.2016010103

Damurski, L. (2016). Smart City, Integrated Planning, and Multilevel Governance: A Conceptual Framework for e-Planning in Europe. *International Journal of E-Planning Research*, *5*(4), 41–53. doi:10.4018/IJEPR.2016100103

Das, K. K., & Sharma, N. K. (2016). Post Disaster Housing Management for Sustainable Urban Development: A Review. *International Journal of Geotechnical Earthquake Engineering*, *7*(1), 1–18. doi:10.4018/IJGEE.2016010101

Das, S., & Nayyar, A. (2019). Innovative Ideas to Manage Urban Traffic Congestion in Cognitive Cities. In K. Ahuja & A. Khosla (Eds.), *Driving the Development, Management, and Sustainability of Cognitive Cities* (pp. 139–162). Hershey, PA: IGI Global. doi:10.4018/978-1-5225-8085-0.ch006

Deakin, M., & Reid, A. (2017). The Embedded Intelligence of Smart Cities: Urban Life, Citizenship, and Community. *International Journal of Public Administration in the Digital Age*, *4*(4), 62–74. doi:10.4018/IJPADA.2017100105

Djukic, A., Stupar, A., & Antonic, B. M. (2018). The Orthogonal Urban Matrix of the Towns in Vojvodina, Northern Serbia: Genesis and Transformation. In G. Carlone, N. Martinelli, & F. Rotondo (Eds.), *Designing Grid Cities for Optimized Urban Development and Planning* (pp. 128–156). Hershey, PA: IGI Global. doi:10.4018/978-1-5225-3613-0.ch008

Dolunay, O. (2016). A Paradigm Shift: Empowering Farmers to Eliminate the Waste in the Form of Fresh Water and Energy through the Implementation of 4R+T. In U. Akkucuk (Ed.), *Handbook of Research on Waste Management Techniques for Sustainability* (pp. 368–379). Hershey, PA: IGI Global. doi:10.4018/978-1-4666-9723-2.ch019

Dowling, C. M., Walsh, S. D., Purcell, S. M., Hynes, W. M., & Rhodes, M. L. (2017). Operationalising Sustainability within Smart Cities: Towards an Online Sustainability Indicator Tool. *International Journal of E-Planning Research, 6*(4), 1–17. doi:10.4018/IJEPR.2017100101

Ekman, U. (2018). Smart City Planning: Complexity. *International Journal of E-Planning Research, 7*(3), 1–21. doi:10.4018/IJEPR.2018070101

Erdogan, S. (2016). The Effect of Working Capital Management on Firm's Profitability: Evidence from Istanbul Stock Exchange. In U. Akkucuk (Ed.), *Handbook of Research on Waste Management Techniques for Sustainability* (pp. 244–261). Hershey, PA: IGI Global. doi:10.4018/978-1-4666-9723-2.ch013

Eudoxie, G., & Roopnarine, R. (2017). Climate Change Adaptation and Disaster Risk Management in the Caribbean. In W. Ganpat & W. Isaac (Eds.), *Environmental Sustainability and Climate Change Adaptation Strategies* (pp. 97–125). Hershey, PA: IGI Global. doi:10.4018/978-1-5225-1607-1.ch004

Fasolino, I. (2018). Rules for a New Town After a Disaster: The Gridded Schemes in the Plans. In G. Carlone, N. Martinelli, & F. Rotondo (Eds.), *Designing Grid Cities for Optimized Urban Development and Planning* (pp. 157–175). Hershey, PA: IGI Global. doi:10.4018/978-1-5225-3613-0.ch009

Fernandes, C. M., & Dias de Sousa, I. (2017). Digital Swarms: Social Interaction and Emergent Phenomena in Personal Communications Networks. In L. Carvalho (Ed.), *Handbook of Research on Entrepreneurial Development and Innovation Within Smart Cities* (pp. 44–59). Hershey, PA: IGI Global. doi:10.4018/978-1-5225-1978-2.ch003

Fernandes, V., Moreira, A., & Daniel, A. I. (2017). A Qualitative Analysis of Social Entrepreneurship Involving Social Innovation and Intervention. In L. Carvalho (Ed.), *Handbook of Research on Entrepreneurial Development and Innovation Within Smart Cities* (pp. 417–438). Hershey, PA: IGI Global. doi:10.4018/978-1-5225-1978-2.ch018

Gabriel, B. F., Valente, R. A., Dias-de-Oliveira, J., Neto, V. F., & Andrade-Campos, A. (2017). Methodologies for Engineering Learning and Teaching (MELT): An Overview of Engineering Education in Europe and a Novel Concept for Young Students. In L. Carvalho (Ed.), *Handbook of Research on Entrepreneurial Development and Innovation Within Smart Cities* (pp. 363–391). Hershey, PA: IGI Global. doi:10.4018/978-1-5225-1978-2.ch016

Garrick, T. A., & Liburd, O. E. (2017). Impact of Climate Change on a Key Agricultural Pest: Thrips. In W. Ganpat & W. Isaac (Eds.), *Environmental Sustainability and Climate Change Adaptation Strategies* (pp. 232–254). Hershey, PA: IGI Global. doi:10.4018/978-1-5225-1607-1.ch009

Gencer, Y. G. (2016). Mystery of Recycling: Glass and Aluminum Examples. In U. Akkucuk (Ed.), *Handbook of Research on Waste Management Techniques for Sustainability* (pp. 172–191). Hershey, PA: IGI Global. doi:10.4018/978-1-4666-9723-2.ch009

Gencer, Y. G., & Akkucuk, U. (2016). Reverse Logistics: Automobile Recalls and Other Conditions. In U. Akkucuk (Ed.), *Handbook of Research on Waste Management Techniques for Sustainability* (pp. 125–154). Hershey, PA: IGI Global. doi:10.4018/978-1-4666-9723-2.ch007

Gonçalves, J. M., Martins, T. G., & Vilhena da Cunha, I. B. (2017). Local Creative Ecosystems as a Strategy for the Development of Low-Density Urban Spaces. In L. Carvalho (Ed.), *Handbook of Research on Entrepreneurial Development and Innovation Within Smart Cities* (pp. 127–149). Hershey, PA: IGI Global. doi:10.4018/978-1-5225-1978-2.ch007

Goundar, S., & Appana, S. (2017). Mainstreaming Development Policies for Climate Change in Fiji: A Policy Gap Analysis and the Role of ICTs. In W. Ganpat & W. Isaac (Eds.), *Environmental Sustainability and Climate Change Adaptation Strategies* (pp. 1–31). Hershey, PA: IGI Global. doi:10.4018/978-1-5225-1607-1.ch001

Grochulska-Salak, M. (2019). Urban Farming in Sustainable City Development. In J. Vargas-Hernández & J. Zdunek-Wielgołaska (Eds.), *Bioeconomical Solutions and Investments in Sustainable City Development* (pp. 43–64). Hershey, PA: IGI Global. doi:10.4018/978-1-5225-7958-8.ch003

Gruszecka, K. (2019). Ecological Centre of Warsaw as a Development Path. In J. Vargas-Hernández & J. Zdunek-Wielgołaska (Eds.), *Bioeconomical Solutions and Investments in Sustainable City Development* (pp. 117–150). Hershey, PA: IGI Global. doi:10.4018/978-1-5225-7958-8.ch006

Hesapci-Sanaktekin, O., & Aslanbay, Y. (2016). The Networked Self: Collectivism Redefined in Civic Engagements through Social Media Causes. In U. Akkucuk (Ed.), *Handbook of Research on Waste Management Techniques for Sustainability* (pp. 262–276). Hershey, PA: IGI Global. doi:10.4018/978-1-4666-9723-2.ch014

Ikeda, M. (2019). Developing a Sustainable Eco-City in Pre-Olympic Tokyo: Potential of New Methods and Their Limits in an Urban Era. In J. Vargas-Hernández & J. Zdunek-Wielgołaska (Eds.), *Bioeconomical Solutions and Investments in Sustainable City Development* (pp. 196–223). Hershey, PA: IGI Global. doi:10.4018/978-1-5225-7958-8.ch009

İşcan, E. (2019). Strategies of Sustainable Bioeconomy in the Industry 4.0 Framework for Inclusive and Social Prosperity. In J. Vargas-Hernández & J. Zdunek-Wielgołaska (Eds.), *Bioeconomical Solutions and Investments in Sustainable City Development* (pp. 21–42). Hershey, PA: IGI Global. doi:10.4018/978-1-5225-7958-8.ch002

J., J., Samui, P., & Dixon, B. (2016). Determination of Rate of Medical Waste Generation Using RVM, MARS and MPMR. In U. Akkucuk (Ed.), *Handbook of Research on Waste Management Techniques for Sustainability* (pp. 1-18). Hershey, PA: IGI Global. doi:10.4018/978-1-4666-9723-2.ch001

Jiménez, R., Rodríguez, P. L., & Fernández, R. R. (2019). Green Spaces of the Metropolitan Area of Guadalajara. In J. Vargas-Hernández & J. Zdunek-Wielgołaska (Eds.), *Bioeconomical Solutions and Investments in Sustainable City Development* (pp. 151–179). Hershey, PA: IGI Global. doi:10.4018/978-1-5225-7958-8.ch007

Kais, S. M. (2017). Climate Change: Vulnerability and Resilience in Commercial Shrimp Aquaculture in Bangladesh. In W. Ganpat & W. Isaac (Eds.), *Environmental Sustainability and Climate Change Adaptation Strategies* (pp. 162–187). Hershey, PA: IGI Global. doi:10.4018/978-1-5225-1607-1.ch006

Khan, B. (2019). Bio-Economy: Visions, Strategies, and Policies. In J. Vargas-Hernández & J. Zdunek-Wielgołaska (Eds.), *Bioeconomical Solutions and Investments in Sustainable City Development* (pp. 1–20). Hershey, PA: IGI Global. doi:10.4018/978-1-5225-7958-8.ch001

Khanna, B. K. (2017). Indian National Strategy for Climate Change Adaptation and Mitigation. In W. Ganpat & W. Isaac (Eds.), *Environmental Sustainability and Climate Change Adaptation Strategies* (pp. 32–63). Hershey, PA: IGI Global. doi:10.4018/978-1-5225-1607-1.ch002

Khanna, B. K. (2017). Vulnerability of the Lakshadweep Coral Islands in India and Strategies for Mitigating Climate Change Impacts. In W. Ganpat & W. Isaac (Eds.), *Environmental Sustainability and Climate Change Adaptation Strategies* (pp. 64–96). Hershey, PA: IGI Global. doi:10.4018/978-1-5225-1607-1.ch003

Kocasoy, G. (2016). Economic Instruments for Sustainable Environmental Management. In U. Akkucuk (Ed.), *Handbook of Research on Waste Management Techniques for Sustainability* (pp. 192–211). Hershey, PA: IGI Global. doi:10.4018/978-1-4666-9723-2.ch010

Lallo, C. H., Smalling, S., Facey, A., & Hughes, M. (2017). The Impact of Climate Change on Small Ruminant Performance in Caribbean Communities. In W. Ganpat & W. Isaac (Eds.), *Environmental Sustainability and Climate Change Adaptation Strategies* (pp. 296–321). Hershey, PA: IGI Global. doi:10.4018/978-1-5225-1607-1.ch011

Li, Z., Wang, Y., & Chen, Q. (2017). Real-Time Monitoring of Intercity Passenger Flows Based on Big Data: A Decision Support Tool for Urban Sustainability. *International Journal of Strategic Decision Sciences*, *8*(4), 120–128. doi:10.4018/IJSDS.2017100106

Loi, N. K., Huyen, N. T., Tu, L. H., Tram, V. N., Liem, N. D., Dat, N. L., ... Minh, D. N. (2017). Sustainable Land Use and Watershed Management in Response to Climate Change Impacts: Case Study in Srepok Watershed, Central Highland of Vietnam. In W. Ganpat & W. Isaac (Eds.), *Environmental Sustainability and Climate Change Adaptation Strategies* (pp. 255–295). Hershey, PA: IGI Global. doi:10.4018/978-1-5225-1607-1.ch010

López-Arranz, M. A. (2017). The Role Corporate Social Responsibility Has in the Smart City Project in Spain. In L. Carvalho (Ed.), *Handbook of Research on Entrepreneurial Development and Innovation Within Smart Cities* (pp. 439–458). Hershey, PA: IGI Global. doi:10.4018/978-1-5225-1978-2.ch019

Lucas, M. R., Rego, C., Vieira, C., & Vieira, I. (2017). Proximity and Cooperation for Innovative Regional Development: The Case of the Science and Technology Park of Alentejo. In L. Carvalho (Ed.), *Handbook of Research on Entrepreneurial Development and Innovation Within Smart Cities* (pp. 199–228). Hershey, PA: IGI Global. doi:10.4018/978-1-5225-1978-2.ch010

Magrinho, A., Neves, J., & Silva, J. R. (2017). The Triple Helix Model: Evidence in the Internationalization of the Health Industry. In L. Carvalho (Ed.), *Handbook of Research on Entrepreneurial Development and Innovation Within Smart Cities* (pp. 60–79). Hershey, PA: IGI Global. doi:10.4018/978-1-5225-1978-2.ch004

Maiorano, A. C. (2018). Urban Fronts in Murattiano Neighbourhood of Bari: A Selective Survey of the Built Environment. In G. Carlone, N. Martinelli, & F. Rotondo (Eds.), *Designing Grid Cities for Optimized Urban Development and Planning* (pp. 78–92). Hershey, PA: IGI Global. doi:10.4018/978-1-5225-3613-0.ch005

Martinelli, N., & Mangialardi, G. (2018). Cities With Grid Layout: Ubiquitousness and Flexibility of an Urban Model. In G. Carlone, N. Martinelli, & F. Rotondo (Eds.), *Designing Grid Cities for Optimized Urban Development and Planning* (pp. 188–208). Hershey, PA: IGI Global. doi:10.4018/978-1-5225-3613-0.ch011

Mizutani, S., Liao, K., & Sasaki, T. G. (2019). Forest-River-Ocean Nexus-Based Education for Community Development: Aiming at Resilient Sustainable Society. In J. Vargas-Hernández & J. Zdunek-Wielgołaska (Eds.), *Bioeconomical Solutions and Investments in Sustainable City Development* (pp. 224–248). Hershey, PA: IGI Global. doi:10.4018/978-1-5225-7958-8.ch010

Moreira, A., & Ferreira, M. A. (2017). Strategic Challenges of the Portuguese Molds Industry: A Sectoral Innovation Perspective. In L. Carvalho (Ed.), *Handbook of Research on Entrepreneurial Development and Innovation Within Smart Cities* (pp. 534–560). Hershey, PA: IGI Global. doi:10.4018/978-1-5225-1978-2.ch023

Mujere, N., & Moyce, W. (2017). Climate Change Impacts on Surface Water Quality. In W. Ganpat & W. Isaac (Eds.), *Environmental Sustainability and Climate Change Adaptation Strategies* (pp. 322–340). Hershey, PA: IGI Global. doi:10.4018/978-1-5225-1607-1.ch012

Mundula, L., & Auci, S. (2017). Smartness, City Efficiency, and Entrepreneurship Milieu. In L. Carvalho (Ed.), *Handbook of Research on Entrepreneurial Development and Innovation Within Smart Cities* (pp. 173–198). Hershey, PA: IGI Global. doi:10.4018/978-1-5225-1978-2.ch009

Nath, R. (2019). Parametric Evaluation of Beam Deflection on Piezoelectric Material Using Implicit and Explicit Method Simulations: A Study in Energy Engineering. In J. Vargas-Hernández & J. Zdunek-Wielgołaska (Eds.), *Bioeconomical Solutions and Investments in Sustainable City Development* (pp. 65–87). Hershey, PA: IGI Global. doi:10.4018/978-1-5225-7958-8.ch004

Nayyar, A., Jain, R., Mahapatra, B., & Singh, A. (2019). Cyber Security Challenges for Smart Cities. In K. Ahuja & A. Khosla (Eds.), *Driving the Development, Management, and Sustainability of Cognitive Cities* (pp. 27–54). Hershey, PA: IGI Global. doi:10.4018/978-1-5225-8085-0.ch002

Odabasi, A., & Tiryaki, C. S. (2016). An Empirical Review of Long Term Electricity Demand Forecasts for Turkey. In U. Akkucuk (Ed.), *Handbook of Research on Waste Management Techniques for Sustainability* (pp. 227–243). Hershey, PA: IGI Global. doi:10.4018/978-1-4666-9723-2.ch012

Okay, E. (2016). Towards Smart Cities in Turkey?: Transitioning from Waste to Creative, Clean and Cheap Eco-Energy. In U. Akkucuk (Ed.), *Handbook of Research on Waste Management Techniques for Sustainability* (pp. 277–302). Hershey, PA: IGI Global. doi:10.4018/978-1-4666-9723-2.ch015

Pawlikowska-Piechotka, A., Łukasik, N., Ostrowska-Tryzno, A., & Sawicka, K. (2017). A Smart City Initiative: Urban Greens and Evaluation Method of the Sport and Recreation Potentials (SEM). In L. Carvalho (Ed.), *Handbook of Research on Entrepreneurial Development and Innovation Within Smart Cities* (pp. 561–583). Hershey, PA: IGI Global. doi:10.4018/978-1-5225-1978-2.ch024

Raisinghani, M. S., & Idemudia, E. C. (2016). Green Information Systems for Sustainability. In U. Akkucuk (Ed.), *Handbook of Research on Waste Management Techniques for Sustainability* (pp. 212–226). Hershey, PA: IGI Global. doi:10.4018/978-1-4666-9723-2.ch011

Rathee, D. S., Ahuja, K., & Hailu, T. (2019). Role of Electronics Devices for E-Health in Smart Cities. In K. Ahuja & A. Khosla (Eds.), *Driving the Development, Management, and Sustainability of Cognitive Cities* (pp. 212–233). Hershey, PA: IGI Global. doi:10.4018/978-1-5225-8085-0.ch009

Rotondo, F. (2018). The Grid Cities: Between Tradition and Innovation. In G. Carlone, N. Martinelli, & F. Rotondo (Eds.), *Designing Grid Cities for Optimized Urban Development and Planning* (pp. 109–127). Hershey, PA: IGI Global. doi:10.4018/978-1-5225-3613-0.ch007

Sánchez-Fernández, M. D., & Cardona, J. R. (2017). The Perception of the Effect of Tourism on the Local Community before the Ibiza Smart Island Project. In L. Carvalho (Ed.), *Handbook of Research on Entrepreneurial Development and Innovation Within Smart Cities* (pp. 392–416). Hershey, PA: IGI Global. doi:10.4018/978-1-5225-1978-2.ch017

Santos, B. (2017). Improving Urban Planning Information, Transparency and Participation in Public Administrations. *International Journal of E-Planning Research*, *6*(4), 58–75. doi:10.4018/IJEPR.2017100104

Sardà, J. F. (2018). Cerdà/Barcelona/Eixample: 1855-2017 … A Work in Progress. In G. Carlone, N. Martinelli, & F. Rotondo (Eds.), *Designing Grid Cities for Optimized Urban Development and Planning* (pp. 19–38). Hershey, PA: IGI Global. doi:10.4018/978-1-5225-3613-0.ch002

Selvi, M. S. (2016). Physical Distribution Problems of Textile Companies in Turkey. In U. Akkucuk (Ed.), *Handbook of Research on Waste Management Techniques for Sustainability* (pp. 328–346). Hershey, PA: IGI Global. doi:10.4018/978-1-4666-9723-2.ch017

Signorile, N. (2018). Pride and Prejudice: The Murattiano-Modernism. In G. Carlone, N. Martinelli, & F. Rotondo (Eds.), *Designing Grid Cities for Optimized Urban Development and Planning* (pp. 52–77). Hershey, PA: IGI Global. doi:10.4018/978-1-5225-3613-0.ch004

Singh, G., Kapoor, R., & Khosla, A. K. (2019). Intelligent Anomaly Detection Video Surveillance Systems for Smart Cities. In K. Ahuja & A. Khosla (Eds.), *Driving the Development, Management, and Sustainability of Cognitive Cities* (pp. 163–182). Hershey, PA: IGI Global. doi:10.4018/978-1-5225-8085-0.ch007

Singh-Ackbarali, D., & Maharaj, R. (2017). Mini Livestock Ranching: Solution to Reducing the Carbon Footprint and Negative Environmental Impacts of Agriculture. In W. Ganpat & W. Isaac (Eds.), *Environmental Sustainability and Climate Change Adaptation Strategies* (pp. 188–212). Hershey, PA: IGI Global. doi:10.4018/978-1-5225-1607-1.ch007

Sood, N., Saini, I., Awasthi, T., Saini, M. K., Bhoriwal, P., & Kaur, T. (2019). Fog Removal Algorithms for Real-Time Video Footage in Smart Cities for Safe Driving. In K. Ahuja & A. Khosla (Eds.), *Driving the Development, Management, and Sustainability of Cognitive Cities* (pp. 55–86). Hershey, PA: IGI Global. doi:10.4018/978-1-5225-8085-0.ch003

Stone, R. J. (2017). Modelling the Frequency of Tropical Cyclones in the Lower Caribbean Region. In W. Ganpat & W. Isaac (Eds.), *Environmental Sustainability and Climate Change Adaptation Strategies* (pp. 341–349). Hershey, PA: IGI Global. doi:10.4018/978-1-5225-1607-1.ch013

Stratigea, A., Leka, A., & Panagiotopoulou, M. (2017). In Search of Indicators for Assessing Smart and Sustainable Cities and Communities' Performance. *International Journal of E-Planning Research*, *6*(1), 43–73. doi:10.4018/IJEPR.2017010103

Touq, A. B., & Ijeh, A. (2018). Information Security and Ecosystems in Smart Cities: The Case of Dubai. *International Journal of Information Systems and Social Change*, *9*(2), 28–43. doi:10.4018/IJISSC.2018040103

Üç, M., & Elitaş, C. (2016). Life Cycle Costing for Sustainability. In U. Akkucuk (Ed.), *Handbook of Research on Waste Management Techniques for Sustainability* (pp. 96–107). Hershey, PA: IGI Global. doi:10.4018/978-1-4666-9723-2.ch005

Ulker-Demirel, E., & Demirel, E. (2016). Green Marketing and Stakeholder Perceptions. In U. Akkucuk (Ed.), *Handbook of Research on Waste Management Techniques for Sustainability* (pp. 75–95). Hershey, PA: IGI Global. doi:10.4018/978-1-4666-9723-2.ch004

Vargas-Hernández, J. G., López, J. J., & Zdunek-Wielgołaska, J. A. (2019). Entrepreneurial and Institutional Analysis of Biodiesel Companies in Mexico. In J. Vargas-Hernández & J. Zdunek-Wielgołaska (Eds.), *Bioeconomical Solutions and Investments in Sustainable City Development* (pp. 89–115). Hershey, PA: IGI Global. doi:10.4018/978-1-5225-7958-8.ch005

Vázquez, D. G., & Gil, M. T. (2017). Sustainability in Smart Cities: The Case of Vitoria-Gasteiz (Spain) – A Commitment to a New Urban Paradigm. In L. Carvalho (Ed.), *Handbook of Research on Entrepreneurial Development and Innovation Within Smart Cities* (pp. 248–268). Hershey, PA: IGI Global. doi:10.4018/978-1-5225-1978-2.ch012

Velozo, R. A., & Montanha, G. K. (2017). Evaluation of a Mobile Software Development Company. In L. Carvalho (Ed.), *Handbook of Research on Entrepreneurial Development and Innovation Within Smart Cities* (pp. 514–533). Hershey, PA: IGI Global. doi:10.4018/978-1-5225-1978-2.ch022

About the Contributors

António Amaral holds a PhD in Industrial and Systems Engineering from University of Minho and a BSc in Industrial Management and Engineering from the same University. Since 2010, he lectured in several public and private higher education institutions in the fields of Industrial Management and Engineering. Member of CIICESI Center – Center for Innovation and Research in Business Sciences and Information Systems and as a collaborating member of ALGORITMI Research Center, in the IEM line - Industrial Engineering and Management. His main research interests are related with the Sustainable Mobility Design and Management as well as Logistics and Supply Chain Management. Presently, he is an Invited Adjunct Professor at ESTG of Polythecnic Institute of PORTO.

Luís Barreto holds a Ph.D. in Electrical Engineering, a Master's degree in Informatics and a Degree in Computer and Telecommunications Engineering. He is currently an Adjunct Professor and the Director of the direction board of Business Sciences School, of the Polytechnic Institute of Viana do Castelo (Escola Superior de Ciencias Empresariais do Instituto Politécnico de Viana do Castelo). He has been the principal investigator of different R&D projects, nowadays he is the principal investigator of the AltoMinho.SMOB R&D project, in which the main goal is to give a global and integrated response to the problematic of passenger's transportation in the Alto-Minho, ensuring the Passengers Public Transport Service (PPTS) in its different dimensions and sustainably and inclusively. He has published several articles regarding sustainable mobility, mobility as a service, industry 4.0, networking security, wireless mobile health applications and long-life learning, published in peer-reviewed scientific journals and proceedings of international and national conferences.

Sara T. Baltazar holds a degree in Mathematics (2006) and a Ph.D. in Electronics and Computational Engineering (2016). Worked as a university Professor and a consulte until 2018. Nowadays, she is a researcher of the AltoMinho.SMOB R&D project; and collaborates with SYSTEC – Research Center for Systems &Tech-

nologies – from University of Porto, and Institute of Systems and Robotics, from University of Coimbra. Since joining the project group in 2017, her main areas of interest are: sustainable mobility practices, Mobility as a Service (MaaS), with special focus in homologous regions of Alto Minho, and modelling and definitions of the public transport in the region.

João Pedro Cruz da Silva holds a Bachelor, Master and Doctorate in Civil Engineering at the University of Coimbra in 1990, 2001 and 2012 respectively. Started his professional career as a Civil Engineer in a consulting company working mainly in Road Design and Spatial Planning. He participated and was responsible for several projects mostly in the aforementioned areas and also in Structural and Hydraulic engineering. Actually he is an Associate Professor at the Department of Civil Engineering Polytechnic of Leiria, having held the positions of Course Coordinator, Department Coordinator, Member of the Pedagogical Council and Member of the Scientific Council. In parallel with the lecturing activities he has coordinated several practical research studies on Transports and Sustainable Mobility for national and local authorities. He also developed scientific research that resulted in several papers presented/published in Scientific Journals as well as in National and International Congresses.

Luisa M. S. Gonçalves holds a Ph.D. in Civil Engineering in the scientific area of Urban and Regional Planning, and Transports an MSc in Geographic Information Systems (GIS) and a specialization in Mobility Management. She is a senior lecturer at Polytechnic Institute of Leiria where she was responsible for the Laboratory of GIS and Transports and was head of the Civil Engineering Department. Nowadays she is the head of Territorial Planning Division at Town Hall of Leiria. She is also a researcher member of INESCC and a visiting professor of the master´s programme in Science and GIS at the NOVAIMS-New University of Lisbon. She has been involved as a coordinator or team member in research projects, all of which related to the field of urban planning, mobility, remote sensing and GIS, currently she is the project coordinator of the UrbSecurity project- Planning Safer Cities, a European project under the Programme URBACT III. She has authored/co-authored more than 60 publications in peer-reviewed scientific journals and proceedings of national and international conferences.

* * *

Aki Aapaoja has a Ph.D. (2014) in Industrial Engineering and Management. Currently he is a Customer Account Lead and is responsible for sales and customers partnerships for machine and vehicle sectors at VTT Technical Research Centre of Finland Ltd. Prior current position, Aapaoja has gained extensive experience in business ecosystems and living labs, mobility services, construction industry, automated and connected driving.

Adam Cohen is a shared mobility researcher at the Transportation Sustainability Research Center at the UC Berkeley. Since joining the group in 2004, his research has focused on shared mobility, mobility on demand, mobility as a service, urban air mobility, smart cities, and other innovative and emerging transportation technologies. He has co-authored numerous articles and reports on shared mobility in journals and conference proceedings. His academic background is in city and regional planning and international affairs.

Jenni Eckhardt received her M.Sc. (Tech.) degree from the University of Oulu in 2002. She is currently working as Senior Scientist and project manager at VTT Technical Research Centre of Finland Ltd. Her research interests include mobility services, logistics and transport emissions. She coordinated an international MaaS project, MAASiFiE, and leads national MaaS projects.

Emily Farrar joined the Transportation Sustainability Research Center at UC Berkeley after receiving an M.S. in the Energy, Civil Infrastructure and Climate program from UC Berkeley in May 2018. Her research interests include shared mobility, electric vehicles, and climate change mitigation. Prior to joining TSRC, Emily worked as an intern for the Environmental Affairs department of Gulf Power Company. She has previous experience in water quality monitoring and environmental permitting. She received her B.S. in Civil Engineering from Mississippi State University in 2017, and she grew up on the Gulf Coast of Mississippi.

Margareta Friman, Professor and Pro-Vice Chancellor at Karlstad University. Margareta is researching travel behavioral change, accessibility, service quality, satisfaction, attitudes, and wellbeing. She publishes in academic journals in psychology, environmental studies, economics, and transportation. In 2014, she was rewarded the Håkan Frisinger Foundation for Transportation Research Award (given by the Volvo Research and Educational Foundations). She has edited "Handbook in Sustainable Travel" (2014), "Quality of Life and Daily Travel" (2018), and "Children´s Travel and Wellbeing (2019)".

Marian Gogola is affiliate of the Faculty of Operations and Economics of Transport and Communications at the University of Žilina as Associate Professor. His research is focused on the transport planning and modelling, sustainable mobility as well on public transport and non motorised transport, mainly cycling. During his work, he has stayed at/cooperated with the University of Genoa (Italy), Kielce University of Technology (Poland), Kazimierz Pułaski University of Technology and Humanities in Radom (Poland), ETH Zurich (Switzerland), University of Texas at El Paso (USA). Marián took part in several European projects such as Central MeetBike, Benefit, Advance, RECOAUDIT, RUMOBIL, MoTiV.

Harri Haapasalo (Dr Tech) has been Professor in Industrial Engineering and Management over 20 years at the University of Oulu, Finland. He has research interests in business management, product management and also in management of production processes. The list of publications covers more than 300 international scientific publications.

Heikki Liimatainen has strong record of accomplishment in publishing his research in the highest-ranking journals in the field of transport and logistics. He was appointed as assistant professor of transport transformation at Tampere University and docent of future transport at University of Turku in 2016. He has also been the leader of Tampere University's Transport Research Centre Verne since 2016 and has established the biannual International Conference on Mobility as a Service (ICoMaaS).

Timo Liljamo is a doctoral researcher of transportation at Tampere University. His work focuses specifically on the effects of automation and digitalisation on the transport system. He currently works with Mobility as a Service (MaaS) related issues.

Enrica Loria is a PhD student working on Gamification for Smart Cities at the University of Trento (Italy) and the Smart Community Lab at Fondazione Bruno Kessler. Her principal research interest lies in the analysis of gameplay data aimed at the extraction of players in-game behaviors to inform the generation of customized content. Her current application domain is an urban gamified system in the context of persuasive technologies to foster sustainable mobility habits, in which she is involved as both designer and game analyst.

Annapaola Marconi is the lead researcher of the Distributed Adaptive Systems (DAS) unit at the Fondazione Bruno Kessler. Her research interests include distributed adaptive systems, persuasive and motivational systems based on gamification and gameful design techniques. She is actively involved in the implementation and dissemination of initiatives and programs aimed at applying the research results in concrete applications that can create an impact on the local area and improve the quality of life of its citizens.

Lars E. Olsson graduated from Göteborg University with a Ph.D. in Psychology of decision making. After a position as researcher at the Center for Consumer Research at the School of Business, Economics, and Law in Gothenburg, he was in 2009 recruited to Karlstad University. Since 2019 he is Professor of Psychology at Karlstad University. Lars E. Olsson has published research in the areas of sustainability, environmental behavior, consumer experiences, travel behavior and well-being.

Kate Pangbourne works at the Institute for Transport Studies, University of Leeds. Her research and teaching is oriented towards shifting our transport system and individual choices towards greater environmental sustainability, social inclusion and meaningful prosperity. She is particularly interested in the implications of rapid technological change in the transport sector. Current work includes improving the persuasiveness of travel behaviour messages (ADAPT), enhancing the rail passenger experience (SMaRTE) and the societal challenges posed by self-driving vehicles and new concepts such as Mobility as a Service.

Markus Pöllänen works as a lecturer at Tranpsort Research Centre Verne at Tampere University. His broad field of interests in research and education covers the transport system widely and ranges from transport safety to transport transformation and from freight transport and logistics to urban mobility and mobility services.

Susan Shaheen is a pioneer and thought leader in future mobility strategies. She was among the first to observe, research, and write about the changing dynamics in shared mobility and the likely scenarios through which automated vehicles will gain prominence. She is an internationally recognized expert in mobility and the sharing economy and co-directs the Transportation Sustainability Research Center (TSRC) of the Institute of Transportation Studies at the University of California (UC), Berkeley. She is also a professor in Civil and Environmental Engineering at UC Berkeley. She has a Ph.D. in ecology, focusing on the energy and environmental aspects of transportation, from UC Davis and a M.S. in public policy analysis from the University of Rochester. She has authored 67 journal articles, over 120 reports and proceedings articles, 12 book chapters, and co-edited two books. She is working on a third book on Shared Automated Vehicles to be published in 2020.

Dana Sitanyiova, having gained her PhD in Hydrogeology at the Comenius University in Bratislava in 2009, had a brief interlude working as an Environmental manager in a private sector. She joined the Department of Geotechnics at the University of Zilina as an Associate Professor in 1998. She was habilitated in the study program Civil Engineering in 2015. Dana is currently a Head of Department of Geotechnics, with her research interests focusing especially on groundwater modelling, and environmental problems. Dana focus is also on applied research concerning transport planning. Since 2001 Dana has been working as a project manager in many R&D projects concerning urban mobility. She is now in charge of YOUMOBIL project.

Roni Utriainen received the M.Sc. degree in civil engineering from Tampere University of Technology, Finland, in 2016. He is currently a Ph.D. student in transportation systems in Tampere University, Finland. In his Ph.D. research project, he studies potential safety effects of partially and highly automated vehicles in Finland. He also has other research and teaching assignments in Transport Research Centre Verne, Tampere University. Utriainen has published on mobility services (MaaS), partially automated vehicles' safety and serious traffic injuries.

Riku Viri is a researcher and doctoral student at Transport Research Centre Verne in Tampere University. Viri's main focus is to build a calculation model of Finnish car fleet to be used in estimation of energy use of passenger road transport in different mobility scenarios.

Index

B

behavior change 1, 3-6, 13-14, 20, 26-29

C

climate change 1-2, 6-7, 19
Collaborative Economy 17, 30
community building 27
coordination 1, 4, 10-11, 14, 16, 19, 21, 26-27, 30-31

D

demand-responsive transport 11-13, 19, 22, 30, 36

E

edge city 2, 5-6, 15, 17, 19, 30

F

future mobility 2-5, 7, 9, 12-14, 16-19, 21-22, 26

G

game elements 4, 6-7, 11, 13-16, 18, 28, 35
Game mechanics 26, 35
gamification 1, 3, 5-8, 14, 25-27, 35, 51

I

inclusive mobility 19, 26
individualism 1
integrated mobility 1, 3, 5-6, 12, 16, 18, 20, 25-26, 32
integrated mobility services 1, 16, 20, 25
integration 1-21, 23, 30-33, 36, 49
Internet of Mobility (IoM) 1, 3, 26
Internet of Mobility as a Service (IoMaaS) 1, 26
Internet of Things (IoT) 2, 26

M

MaaS 1-23, 25-26, 28-30, 32, 46-49, 51
Mobility as a Service (MaaS) 1-3, 7-8, 10, 12, 26, 30
mobility evolution 3, 18
Mobility on Demand 1-2, 7, 9, 19, 30
Mobility on Demand (MOD) 2, 7, 9, 19, 30
modality 20
motivational systems 1, 3, 5, 25-27

P

peer-to-peer 2, 5-6, 13, 17-18, 20, 22, 30
Persuasive Technology 3, 6, 8, 36
Prosumer 6, 30
public transport 1-25, 29-34, 36, 42, 45-51
Public-Private-People Partnership 1-2, 6, 13, 30

R

ridesharing 5-7, 13, 15, 18, 20, 22, 30-31
ridesourcing 5, 7, 13, 22, 30-31
RUMOBIL project 1, 13-15, 18, 38, 40,

43, 46-47, 49
rural mobility 1, 3, 5, 11-13, 16, 21, 26

S

shared micromobility 7, 16-18, 20, 22-23, 31
shared mobility 1-3, 5-7, 9-10, 12, 14-15, 19-20, 22-24, 30-31, 50
sharing services 2, 4-5, 22, 30
Smart City 2, 6, 8, 36, 49
smart mobility 3, 9, 12, 17, 26, 49
social inclusion 1-3, 5, 9, 11-14, 18-19, 23
Statutory Transportation 12, 16, 30
subsidized transportation 1, 12, 30
suburban 1-10, 12, 15-16, 18-20, 22-24, 30-31, 45
sustainability 1-14, 16-17, 19, 25, 27-29, 46
sustainable mobility 1-6, 9-11, 13, 17, 19-20, 25, 27-29, 36, 42, 46

U

user groups 1-3, 5-6, 8-9, 11, 14-19, 21